Sustainable Urban Housing in China

ALLIANCE FOR GLOBAL SUSTAINABILITY BOOKSERIES
SCIENCE AND TECHNOLOGY: TOOLS FOR SUSTAINABLE DEVELOPMENT

VOLUME 9

Aims and Scope of the Series

The aim of this series is to provide timely accounts by authoritative scholars of the results of cutting edge research into emerging barriers to sustainable development, and methodologies and tools to help governments, industry, and civil society overcome them. The work presented in the series will draw mainly on results of the research being carried out in the Alliance for Global Sustainability (AGS).
The level of presentation is for graduate students in natural, social and engineering sciences as well as policy and decision-makers around the world in government, industry and civil society.

Sustainable Urban Housing in China

Principles and Case Studies for Low-Energy Design

Edited by

Leon Glicksman
Massachusetts Institute of Technology, Cambridge, Massachusetts, U.S.A.

and

Juintow Lin
Fox Lin, Inc., Los Angeles, California, U.S.A.

A C.I.P. Catalogue record for this book is available from the Library of Congress.

ISBN-10 1-4020-5412-2 (PB)
ISBN-13 978-1-4020-5412-9 (PB)
ISBN-10 1-4020-4785-1 (HB)
ISBN-13 978-1-4020-4785-5 (HB)
ISBN-10 1-4020-4786-X (e-book)
ISBN-13 978-1-4020-4786-2 (e-book)

Published by Springer,
P.O. Box 17, 3300 AA Dordrecht, The Netherlands.

www.springer.com

Printed on acid-free paper

An International Partnership

ALLIANCE FOR GLOBAL SUSTAINABILITY

An International Partnership

Alliance for Global Sustainability
International Advisory Board (IAB)

TABLE OF CONTENTS

THE AUTHORS

Leon Glicksman (coeditor) is a professor of Building Technology and Mechanical Engineering and has been the head of MIT's Building Technology Program in the Department of Architecture for the past 17 years. He has worked on research and consulting related to energy-efficient building components and design, indoor airflow, and indoor air quality. He developed the simulation program for heat pumps that forms the basis for one of the most popular heat pump programs available today. He did basic studies to improve thermal insulation for buildings during the period when CFCs were removed from insulation. He coheads a joint study with Cambridge University researching the use of natural ventilation in buildings to improve indoor air quality and reduce energy used for air-conditioning. Glicksman and coworkers are developing a website for advanced envelope systems that can be easily used by architects and developers in the early stages of design. He is the author of over 200 papers in the area of energy and heat transfer. Among his awards are the Melville Medal of American Society of Mechanical Engineers (ASME) and the Robert T. Knapp Award of the Fluids Engineering Division of ASME. He is also an associate editor for the *ASHRAE HVAC&R Research Journal*. He is a Fellow of ASME.

Juintow Lin (coeditor) has worked with the MIT Building Technology Program since 2000 as research fellow, project manager, and project architect in charge of coordinating projects in Beijing, Shanghai, and Shenzhen for the Sustainable Urban Housing in China Project. She also coordinated studio workshops at MIT and local conferences and seminars in China. Lin designed and continues to maintain the project website, http://chinahousing.mit.edu. She is a partner and founding member of the architectural design and consulting firm of Fox Lin, Inc., in Los Angeles, which focuses on implementing innovative building technologies that can concurrently achieve both sustainable and economic viability. Lin has previously worked in the offices of Pei Cobb Freed and Partners in New York, Foster and Partners in London, and Marmol Radziner and Associates in Los Angeles. She was a pioneering member of the Kinetic Design Group at MIT and of the Ocean Design Collaborative (odesco) in Los Angeles, investigating the design and application of interactive and kinetic systems in architecture. Her design work has received numerous awards, has been widely exhibited, and has been featured in various publications. She has also lectured on the topic of sustainability and interactive architecture. Lin received her B.S. in architectural design from MIT in 1995 and her M.A. in architecture from MIT in 2000, where she received the American Institute of Architects (AIA) Gold Medal Award upon graduation.

Leslie Norford specializes in energy studies, controls, and ventilation and is seeking to improve the way buildings use the earth's resources. With Tabors Caramanis and Associates, he consults in the areas of electric utility energy conservation, electricity pricing, and control of thermal storage systems. Before his appointment to the school's faculty in 1988, Professor Norford was a lecturer in the Department of Mechanical Engineering at Princeton University for four years. At that time, he was a research engineer at the Center for Energy and Environmental Studies. From 1974 to 1979, he was a nuclear power engineer with the U.S. Navy and the U.S. Department of Energy. Professor Norford earned his B.S. in engineering science from Cornell University in 1973 and his Ph.D. in mechanical and aerospace engineering from Princeton University in 1984. As a graduate student, he received the Association of Princeton Graduate Alumni Award for outstanding teaching. He is a member of the American Society of Heating, Refrigerating and Air-Conditioning Engineers (ASHRAE).

Andrew Scott has been involved in architectural practice since 1977 and held teaching positions in the United Kingdom and the United States since 1982. He practiced with Foster and Partners in London before establishing his own practice, Denton Scott

Associates, in the United Kingdom, which received several national design awards and competition prizes. Since 1993, he has been an associate professor of architecture at MIT while also maintaining an architectural practice oriented towards design research. His teaching began at the University of Manchester, United Kingdom, and has included teaching assignments at several architectural schools in London as well as Dalhousie University in Halifax, Nova Scotia. His work has also received awards from the Association of Collegiate School of Architecture and the Boston Society of Architects (Unbuilt Design Awards), especially in the field of sustainability and new building technologies. In 1996 he organized "Dimensions of Sustainability," an international conference at MIT; in 1998 he published a book of the same name. At MIT his work has included many research projects including the collaborative research project for Sustainable Urban Housing in China as well as environmental research for British Petroleum in Scotland and Chicago focusing on innovative lab and office design. He is currently writing a monograph on the topic of responsive practice and is on the organizing committee and a jury member for the North American section of the new global Holcim Awards for Sustainable Construction, which began in 2005.

John Fernandez is an architect in private practice and associate professor of design and building technologies in the Department of Architecture at the MIT. His writing and research focus on systems and assemblies of buildings and the resource consumption of the built environment. Professor Fernandez has investigated the performance characteristics and architectural design potential of a variety of emerging materials including high-performance fibers and textiles, natural fibers, new laminated glass assemblies, synthetic and natural composites, and others. This work has culminated in the publication of the book *Material Architecture: Emergent Materials for Innovative Buildings and Ecological Construction* (2005, Architectural Press, Oxford, U.K.). Currently, he is involved in the articulation of the field of industrial ecology as applied to the construction of buildings. Mapping resource flows, documenting building lifetimes, formulating strategies for dematerialization, and offering alternative ways in which design and engineering can contribute to a more efficient use of physical resources is the next radical step for the making of an architecture for the twenty-first century.

Qingyan Chen is a professor of mechanical engineering at Ray W. Herrick Laboratories, at Purdue University, Indiana. He received his B.S. from Tsinghua University, China, and his M.S. and Ph.D. from Delft University of Technology, the Netherlands. He was a research scientist at ETH Zurich, Switzerland, a project manager at the TNO Institute of Applied Physics in the Netherlands, and a former professor at MIT. Professor Chen's current research interests include computations and measurements of airflows in and around buildings, building ventilation systems, indoor air quality, and building energy analysis. He has published over 90 archival journal papers and more than 60 refereed conference papers. Since 1995, he has been either the principal investigator or co-principal investigator of 30 sponsored research projects. He is an author of the book *System Performance Evaluation and Design Guidelines for Displacement Ventilation*. Prizes received include the Best Technical/Symposium Paper Award, Best Poster Presentation Awards, and Distinguished Service Award from ASHRAE; the CAREER Award from the National Science Foundation; the Lecturer Travel Award from the Royal Dutch Academy of Science; and the Best Poster Award at the second CLIMA 2000 World Congress. He has been elected to the International Academy of Indoor Air Sciences. Currently, Professor Chen serves as an associate editor for the *International Journal of HVAC&R Research* and is an editorial board member for the *International Journal of Ventilation* and the *International Journal on Architectural Science*. He was also the vice president of the Indoor Air 2005 conference.

Lara Greden completed her Ph.D. in the Building Technology Program at MIT in June 2005 and is now a consultant with The Weidt Group, an energy and sustainable design consulting firm. Her research focuses on flexible approaches to sustainable building design to address risk using real options theory. Applications include hedging risk in naturally ventilated buildings and implementing other innovative, sustainable technologies. She received dual M.S. degrees from MIT in civil and environmental engineering and in technology and policy, and she holds a B.S. in mechanical engineering from the University of Minnesota. Greden's Master's thesis research aimed to understand the market and policy conditions for sustainable housing in urban China. She is also a team member of the Fab Tree Hab design team (winner Habitat for Humanity/SECCA design competition) and did sustainable design work (renewable energy systems and passive design aspects) for the 1999-2001 Turkey Workshops at MIT. She was a consultant for Arthur D. Little, Inc. (now Navigant Consulting, Inc.) on the topics of renewable energy and energy efficiency for the U.S. Department of Energy. Greden is a National Science Foundation Fellow, MIT Presidential Fellow, and MIT Martin Fellow for Sustainability.

ACKNOWLEDGMENTS

The authors would like to thank the Alliance for Global Sustainability (AGS). The AGS was formed in 1994 by three of the world's leading technical and research universities: the University of Tokyo, the Swiss Federal Institutes of Technology (ETH) and the Massachusetts Institute of Technology (MIT). Since then, Chalmers University has joined the team of engineers that has developed a new approach to strategically addressing problems affecting the "global commons."

This project has also been made possible through the generous contributions from the Kann-Rasmussen Foundation, Kawasaki Heavy Industries (KHI), and, for research reported in chapter 5, the Fundação para a Ciência e Tecnologia.

The Sustainable Urban Housing in China Project is a result of collective efforts from the following individuals and institutions: Klaus Daniels and Alfred Moser from ETH Zurich's Department of Architecture's Air and Climate Group; Jean Bernard Gay from ETH Lausanne's Institute of Solar Energy Research; Shuzo Murakami and Shinsuke Kato from University of Tokyo's Institute of Industrial Science; Jiang Yi, Qin Yuoguo, Yuan Bin, Zhu Yingxin, Wang Peng, Tang Gang, and Wei Qingpeng from Tsinghua University's Department of Thermal Engineering, School of Architecture, and Tsinghua Design Institute; Long Weiding from Tongji University's Department of Facilities Engineering and Management and Architectural Design and Research Institute; and Carl-Eric Hagentoft from Chalmers University.

The designs shown in this book were created and developed within a series of workshops given by MIT's Building Technology Group in the Department of Architecture. Guiding the workshops were professors Leon Glicksman, Qingyan Chen, Leslie Norford, Andrew Scott, and John Fernandez, as well as research fellows Juintow Lin and Zachary Kron.

The success of this project is largely due to student participation. From the first workshop in 1998, students have been actively producing the design and technical investigations that make up the basis for this project. Students and visiting scholars include: Winnie Alamsjah, Ozgur Basak Alkan, Camille Allocca, Meredith Atkinson, Becca Brezeale, Xantha Bruso, Hongyu Cai, Luisa Caldas, Erica Chan, Henry Chang, Catherine Chen, Eva Chiu, Brian Dean, Guilherme Carrilho da Graça, Rocelyn Dee, Shaohua Di, Stephen Duck, Janet Fan, Mingzheng Gao, Lara Greden, Sephir Hamilton, Joy Hu, Elsie Huang, Perry Ip, Yi Jiang, Andy Jonic, Julie Kaufman, Myeoung Kim, Yongjoo Kim, Nobukazu Kobayashi, Sean Kwok, Junjie Liu, Xiaofang Luo, Xiaoyi Ma, Karl Munkelwitz, Eric Olsen, Christoph Ospelt, Sam Potter, Paul Rafiuly, Daniel Steger, Kavita Srinivasan, Carolyn Straub, Pearl Tang, Joli Thomas, Joy Wang, Jesse Williamson, Helen Xing, Jae-ock Yoon, and John Zhai. The editors would also like to thank Jonathan Smith, who developed the Building Energy Calculator program, included on the CD that comes with this book.

Additional contributors to the writing and/or editing of this book include Michael Fox, Zachary Kron, Stephanie Harmon, Guilherme Carrilho da Graça, Sephir Hamilton, Andrew Miller, Ozgur Basak Alkan, Camille Alloca, Luisa Caldas, and Julie Kaufman. Juintow Lin provided the graphic design for the publication. In addition to Lin, Michael Fox, Anshuman Khanna and Yasushi Ishida contributed greatly by providing assistance with figures. The editors would also like to thank Kathleen Ross, Dorrit Schuchter, and Tom Fitzgerald for their assistance throughout the process of putting together this book.

The editors would like to acknowledge the continued encouragement, support, and love of their families in this long endeavor: including Judith, Shayna, Eric, David, Jeff, Tara, Jessica, Carly, Lexi and Julia for Leon Glicksman and Michael, Sha-Li, Lawrence, Buortau, and Shintau for Juintow Lin.

FOREWORD
ENCOUNTERING SUSTAINABILITY IN CHINA

Yung-Ho Chang

What I am writing is by no means a typical foreword. Instead of introducing the work of my colleagues in the Department of Architecture at Massachusetts Institute of Technology (MIT), I will use this opportunity to share my own limited but critical experience with the development of sustainability in China. However, I would like to point out that the authors of this book are among the pioneers for sustainability research and design in China, and they have helped to build the foundation of an ecologically sensitive architecture in China.

A CHANGING PRACTICE

As an architect in China, I have been confronted by the issue of sustainability and have witnessed how it is rapidly changing the practice of architecture. Not only do we now have a more substantial technological content during design, we have also begun to rethink programming and the spatial organization of a building, creating more efficiency for its occupants. Before the era of sustainability, architects used to initiate and carry out schematic design on a project without involvement from engineers until a later stage, such as during the design development. That sequence of the design process is becoming obsolete. One cannot add the concern of sustainability as an afterthought. Today, if an architect intends to develop a building that saves energy, she/he has to collaborate with engineers from the very beginning. Engineers and scientists have truly become our codesigners. This collaboration is well illustrated by the case studies in the following chapters.

SOME EXPERIMENTS

My own practice in China has taken some small steps toward sustainability.

Step I. Split House, Beijing: In 2000, when we designed a house in the mountains outside of Beijing, we enclosed a courtyard with buildings and hillside to create a milder microclimate. We also employed two traditional materials, earth and wood, specifically rammed-earth walls and laminated wood frames, to build a structure that is capable of disintegration and would produce less waste once the house is no longer needed. Although ancient wisdom, rather than advanced technology, was engaged, we were able to address the issue of sustainability for the first time. After all, sustainability is about a way of thinking.

Step II. UFIDA R&D Center, Beijing: For this project for a software company, the sustainable design focused on the reorganization of workspaces. The circulation system was transformed into public areas where communication and socialization is encouraged and located in zones that receive more daylight and better views. Meanwhile, we pushed the flexibility of all the workstations. We learned in this project that if a spatial structure of a building was not established in an efficient and economical manner, technology cannot be added later on to fix it.

Step III. Siemens Shanghai Center competition entry: Recently, we started the design proposal of this office complex with mechanical and energy engineers on our team. How to increase natural light and ventilation, how the building envelope is configured, how energy is produced, and how waste is recycled were some of the questions asked from the very beginning of the project. This is our initial effort to approach sustainability in a more integrated way. However, we already sense that a totally different kind of architecture that is more sensitive to the environment is in the making.

MIXED SIGNALS

Although more positive messages have come from China's Ministry of Construction in recent years regarding the regulation of energy consumption in the building industry, many inefficient high-rise buildings with clear glass curtain walls continue to be built in Beijing and other major cities in China. This indicates that the importance and significance of sustainability has yet to be recognized. This book therefore is not only about specific technical solutions to make buildings or communities more sustainable, but also to advocate an awareness of the pressing nature of the issue and an attitude to take on such challenge.

A NEW LIVING ENVIRONMENT

Sustainability is an urban issue as well as a global issue. Land is one of the most limited resources on earth. Power, water, waste, and the like are all integral parts of the city infrastructure. Pollution, a byproduct of energy consumption, local, regional, as well as global impacts, is an international disaster. For a better tomorrow, sustainability has to be achieved, and it can be achieved with new architecture and new lifestyles as its catalysts and by-products. Hopefully this book will aid in our pursuit of sustainability and will eventually improve the world in which we live.

Yung-Ho Chang
10 January, 2006
Beijing

Head, Department of Architecture
Massachusetts Institute of Technology

Principal, Atelier FCJZ

INTRODUCTION
SUSTAINABLE URBAN HOUSING IN CHINA

Leon Glicksman

OVERVIEW

This book is an outgrowth of a research program carried out over the last five years by MIT, Tsinghua University, Tongji University, and the Alliance for Global Sustainability (AGS), including the Swiss Federal Institute of Technology in Lausanne and Zurich, Chalmers University, and the University of Tokyo. The focus of the Sustainable Urban Housing in China Group has been on sustainable approaches to housing in large urban areas of China. This book summarizes important projects, activities in design and technology, and key findings and recommendations. The goal of the research is to serve as a guideline for future sustainable building projects in China. The case studies and technical findings will also illuminate many principles of sustainability applicable to buildings throughout the world.

Sustainable Development

There has been increased attention to the need for sustainable development. The overarching goal is to preserve and improve the global environment for future generations. This primary goal encompasses, but is not limited to, the reduction in harmful emissions, conservation of nonrenewable resources such as materials and land, and improvements in the quality of life for the present generation. These goals are particularly difficult to accomplish in areas of the world where economies and standards of living are growing rapidly.

China serves as a particularly important research subject because of its accelerated growth rate and improving rank in the world economy, and can therefore articulate the early combined benefits of rapid growth with sustainable design. If properly carried out, sustainable measures can augment the quality of life while aiding overall economic advancement. China's building sector is characterized by rapid growth. There are more than 10 million housing units built each year. The rapid pace of construction and the longevity of the structures mean that today's actions will have a lasting impact on future generations; poorly designed and constructed buildings will be a burden to individual owners and the community for a long time.

Sustainable building development can also have important impacts on China as a whole, on local municipal areas, on building developers and designers, and on individual homebuyers. China can benefit from more efficient use of its resources and capital in future periods of expansion. Investment in energy-efficient building features is often more cost effective than investment in additional energy supplies required for a new building of conventional design. A municipal area can provide better housing for its citizens while retaining livable open spaces and a shared sense of community among its residents.

Opportunity for Leadership

The building sector has undergone radical change with a shift from housing supplied by the employer or government to privatized housing. Chinese building developers have an opportunity to demonstrate leadership in sustainable building development. With a modest increase in cost over conventional developments, developers can establish a unique market position in this field. Their buildings will conform to future as well as present building codes. Buyers will be more satisfied with homes that are more comfortable, have better daylighting, natural ventilation, and more inviting surrounding green space. At the same time, the cost of maintaining comfortable living conditions within the homes will decline substantially. These incentives should increase buyer demand for such developments.

Similarly, designers have a chance to create buildings that will be a benchmark for the future. Buildings that combine innovative technology with the best of both traditional and modern design attributes will be in great demand. Designers who have included thoughtful sustainable design in their projects are being recognized as leaders in their profession.

The goal of this book is to introduce the concept of sustainable buildings for China. Important attributes and advantages will be pointed out, and means to achieve sustainable designs through use of effective technology coupled with site, building, and individual unit design will be illustrated. Technological solutions for particular climatic conditions needed to achieve a comfortable, healthy, and economical building development will be covered, as will examples of specific design projects for various cities in China.

Energy Consumption and Global Warming

Minimizing the energy use needed to achieve desirable goals such as comfort, health, and pleasant surroundings is an important attribute of sustainable design. The consumption of coal, oil, and natural gas leads to the emission of carbon dioxide into the atmosphere that is in turn tied to the danger of global warming.

China's energy use has grown rapidly over the past few decades. Some studies suggest that within the next 30 to 40 years, China will equal or surpass the United States as the world's largest energy-consuming nation (Figure 1). While there may be debate as to the exact rate of this growth, China's energy consumption is undeniably approaching that of the world leader.

The same trend is also true of carbon dioxide emissions in China (Figure 2). Due to the heavy reliance on coal for energy production, the proportion of carbon dioxide emissions is particularly high compared to countries using more oil and natural gas.

Currently, buildings in China consume only about 18 percent of energy in the country. As industries are modernized and the economy develops a service sector, industrial energy use may grow more slowly. At the same time, as people demand larger living spaces and more comfortable living conditions, the use of energy in buildings should rise, and may approach levels in Western countries – approximately one-third of total energy consumption.

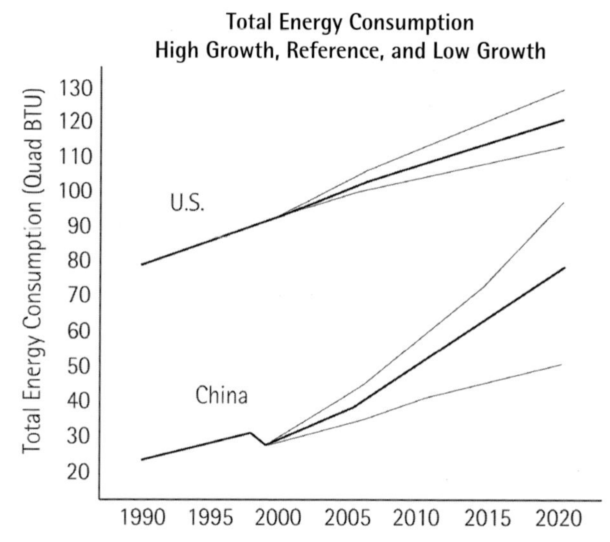

Figure 1 High growth, reference, and low growth projections of future total energy consumption (Source: U.S. EIA 2001)

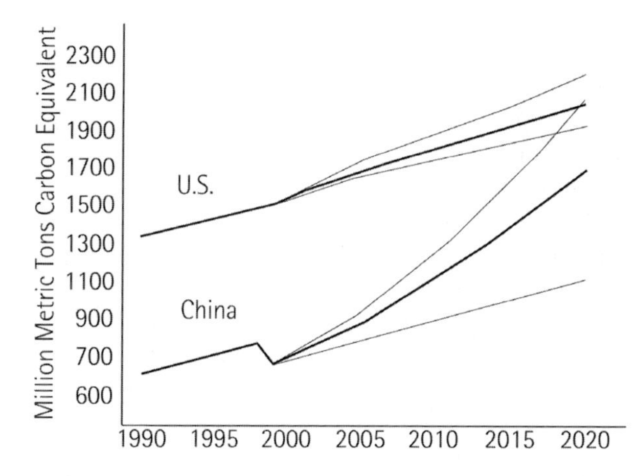

Figure 2 High growth, reference, and low growth projections of future carbon dioxide emissions (Source: U.S. EIA 2001)

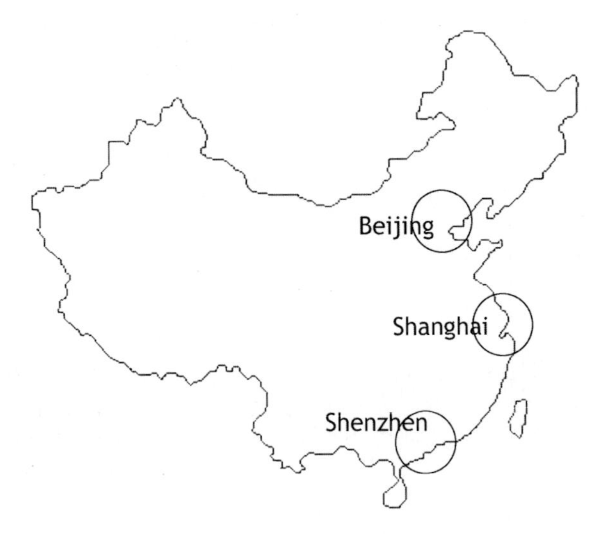

Figure 3 Demonstration cities in China include Beijing, Shanghai, and Shenzhen

The growth of energy consumption in China has caused a critical strain on energy supplies and a substantial increase in fossil fuel imports. Recently, the Chinese government has formulated steps in an attempt to meet the long-range energy needs of the country. The most important element of the recently announced program is increased energy efficiency in all sectors of the economy.

PROJECT BACKGROUND

As environmental concerns become more important at the local, regional, and global level, more awareness must be paid to the development of sustainable buildings. Buildings are accountable for a large portion of resources and energy in addition to producing a substantial amount of environmental pollutants. The rapid growth of the Chinese economy has made the increase of energy efficiency in China vital to the general goal of sustainable world development. The aim of the Sustainable Urban Housing in China Project is to identify design processes that utilize new technologies and new applications of existing technologies that will significantly increase the efficiency of new and renovated Chinese buildings. In addition, strategies for energy efficiency should be appealing to Chinese builders and consumers when they become aware of the advantages in using such strategies.

The research grew from an agreement between MIT faculty, AGS colleagues, and developers and design institutes at Tsinghua University and Tongji University. The goal of this research group is to explore design, technology, and implementation of environmentally responsive urban housing in China. Our focus is on residential buildings in large Chinese cities such as Beijing, Shanghai, and Shenzhen (Figure 3). Some principal goals are to minimize solar gains in the summer, improve air quality and ventilation, and reduce energy consumption of buildings in summer and winter. Investigations are focused upon the careful design of individual building interiors and exteriors, as well as building groupings. Design concepts are tested through accurate modeling, energy simulations, and the use of tools such as computational fluid dynamics (CFD). Designs make use of local materials and building methods as well as local building conditions and lifestyles, while considering innovative technologies as well as incorporating traditional technologies such as shading and natural ventilation. The projects that will be presented are representative of units that a substantial number of urban residents could afford. Emphasis will be on the more cost-effective traditional technologies designed and analyzed using advanced technological tools.

The largest obstacle to the improvement in building energy efficiency is the lack of means to encourage widespread adoption of efficient measures while insuring that sustainable designs are not compromised by poor construction. The focus of this work is on the combined architectural and technical design of sustainable buildings.

Formation of Partnerships

The faculty at MIT and AGS colleagues had numerous exchanges with Chinese colleagues. MIT faculty members and research fellows made numerous visits to China, visiting Beijing, Shanghai, and Shenzhen. We worked with architects at large design institutes and several major developers on specific new projects. The MIT and Tsinghua faculty have held well-attended seminars in Beijing and Shenzhen for interested designers, developers, equipment suppliers, and researchers, during which new sustainable design and technologies were described and discussed. Meetings were held with members of major design institutes at Tsinghua University in Beijing and at Tongji University in Shanghai. These institutes are responsible for the design of many large building projects in China.

Design Methodology

This project focuses on the case studies of design and technology for demonstration buildings that incorporate energy-conservation features. One example is the consideration of natural ventilation to replace or reduce the need for air-conditioning.

Many new residential construction projects in urban areas in China favor high-rise buildings. The widely cited reasons for this preference are code requirements for the separation of buildings and the overall cost of land. It is widely acknowledged however, that lower buildings provide better living conditions for residents and are more easily shaded with vegetation. To test the effectiveness of high-rise buildings relative to low- and mid-rise plans, MIT evaluated a series of alternative layouts on the test sites. These are described in *Part 2, Design Principles* and *Part 4, Case Studies* in this book.

Effective design of natural ventilation requires a detailed knowledge of external aerodynamics. It is not sufficient to consult tables with typical surface pressures. Instead, CFD predictions of the velocity field in the immediate surroundings can provide detailed information about the interaction of wind and building geometry.

For residential buildings in climates such as Beijing, sun-collecting buffer zones, such as terraces enclosed by glass windows or winter gardens, are an effective way to increase usable living space in winter and reduce winter heating loads. When solar irradiation is intense and the elevation of the sun high, the terraces provide shade to the inner zone of the residence. During periods of the year when natural ventilation works well, these spaces should be open to the outdoors. However, in most buildings we observed that the enclosures were permanent, substantially increasing the solar load to the interior. The impact of these "half-open spaces" on the air movement and heat transfer in and around buildings needs careful attention. In the research, AGS colleagues at the Swiss Federal Institute of Technology and the University of Tokyo used CFD analysis to investigate these mechanisms.

Evaluation of Specific Technologies

The research addressed a number of the most promising approaches to sustainable housing in China. The research team considered generic new designs and technologies that made use of solar heating, dehumidification, shading techniques, and other passive measures. Sustainability guidelines developed by the Swiss Federal Institute in Lausanne (EPFL), were used to carry out the evaluation. Based on that evaluation, the team considered promising systems for inclusion in demonstration projects. A number of issues affect whether systems such as natural ventilation will have measurable effects on building performance. The team considered the major implications of each system and developed general rules for their design.

Chinese designers would like to include natural ventilation in the design of their new buildings. This is a complicated issue that involves the design of building interiors and window placement to facilitate airflow. In addition, the proximity of neighboring buildings and the overall shape of the building influences air circulation outside the building and, in turn, natural circulation within the building. Natural ventilation of high-rise buildings is particularly challenging, involving interactions of buoyancy and wind-driven airflows. Several examples will be presented.

Indoor air quality (IAQ) is closely linked to energy consumption and therefore is relevant to building sustainability. There are often trade-offs between improving IAQ, reducing energy consumption, and maintaining thermal comfort. IAQ is one of the important factors that affects the welfare and health of occupants. A predictive model that simultaneously treats these various factors will help produce more sustainable building designs.

Natural ventilation is an example of the interplay between design and technology as it affects the urban setting, individual buildings, and interiors of each residential unit. We will see this interchange play out as a variety of sustainable concepts are described in the case studies.

ORGANIZATION OF THIS PUBLICATION

This project has focused on applications of sustainable design and technologies in three distinct geographical areas of China, Shanghai, Shenzhen, and Beijing.

Part 1 of this book explores sustainability, its role in the building sector, and general cultural and technological conditions in China; it also highlights specific energy concerns and relevant policy issues. Part 2 discusses the specific design methodology that the group has brought to bear on these general conditions. Part 3 defines the outcome of site-specific technical investigations. Part 4 is a summary of several design solutions that the group has proposed for actual residential projects in the demonstration cities. Finally, part 5 summarizes the book and makes recommendations for implementation, policy, and education.

REFERENCES

Energy Information Administration. 2001. *China: Environmental Issues.* Energy Information Administration, U.S. Department of Energy, Washington, D.C. <http://www.eia.doe.gov/emeu/cabs/chinaenv.html>

Energy Information Administration 200a. 2001. *Annual Energy Outlook 2001 with Projections to 2020.* Report No. DOE/EAI-0380 (2001). Energy Information Administration, U.S. Department of Energy, Washington, D.C.

Energy Information Administration. 200b. 2001. *Annual Energy Outlook 2001.* Report No. DOE/EIA-0383 (2001). Energy Information Administration, U.S. Department of Energy, Washington, D.C.

Research Team of China Climate Change Study. 1999. *China Climate Change Country Study.* Tsinghua University Press, Beijing.

World Bank Joint Study Group (National Environmental Protection Agency China, State Planning Commission China, U.N. Development Program). 1994. *China: Issues and Options in Greenhouse Gas Emissions Control*. World Bank, Washington, D.C.

Part One
Background

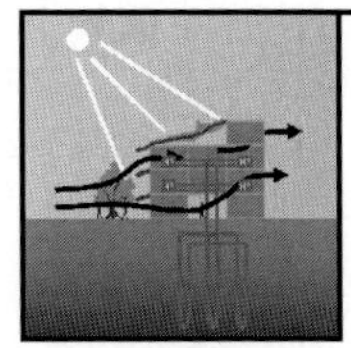

CHAPTER ONE

SUSTAINABILITY AND THE BUILDING SECTOR

Leon Glicksman

INTRODUCTION

Interest in sustainability has continued to grow throughout the world. This encompasses a wide range including reduction and eventual reversal of dangerous environmental impacts caused by emissions to the air, water, and land. Sustainable development also includes the husbanding of nonrenewable resources for future generations. In the broadest sense, this involves intelligent planning of land use as well as concern for materials such as organic feedstocks and scarce metals. To maintain a proper balance of people's needs and available resources requires a simultaneous concern for the demand sector as well as the supply side of the problem.

Nonrenewable Resources

The building sector constitutes an important element in sustainable development from a regional and worldwide point of view. In countries such as the United States, construction is one of the largest industries, and building construction is the largest part of the sector. Buildings consume a large proportion of basic materials such as steel and cement.

In the United States, construction of commercial and residential buildings consumes one-half of the total production of Portland cement (U.S. Portland Cement Association 2004). Construction waste makes up the

L. Glicksman and J. Lin (eds), Sustainable Urban Housing in China, 2-7

largest fraction of landfills. This includes both debris from older buildings that is removed from a site, and waste generated in the construction of their replacements. The amount of construction and demolition waste is 20 to 30 percent by volume and up to 40 to 50 percent by weight for the typical municipal waste stream of developed nations, as well as in China. The consumption of nonrenewable materials devoted to construction (including infrastructure) is between 50 to 70 percent of total materials consumption globally (Wernick and Ausubel 1995).

Land constitutes one of the most precious nonrenewable resources. The expansion of urban areas with the growth of megacities has consumed large portions of the most desirable land that was formerly used for agriculture or recreation. Low-rise developments for single-family homes, shopping centers, and industrial parks are a far less efficient use of our limited land resources than higher-density developments.

Environment

The building sector has an important impact on air pollution, solid waste generation, water consumption, and wastewater production. Much of this is tied to the by-product of energy generation. In the developed world, people spend 90 percent of the time inside buildings. One means to reduce energy consumption in buildings is to reduce ventilation rates from outdoors. Occupants of many new commercial buildings suffer from deteriorating indoor air quality. The challenge is to develop sustainable buildings that reduce waste emissions to the outside while simultaneously improving environmental quality within the building. There are new technologies that will meet both requirements. Examples will be presented in later chapters.

Energy

In the Western world, building operations account for about one-third of the annual energy consumption. Figure 1 shows the energy consumption in the major sectors of the United States. It is surprising to many observers that over the last several decades, the combination of residential and commercial buildings has consumed more energy than the sum of all forms of transportation in the United States. In China, it is estimated that the construction industry accounts for 37 percent of national energy use (*Asia Pulse* 2006).

Buildings also consume well over half of all electricity and are the major source of summertime peak loads and shortages that can cause brownouts. In megacities, the energy consumed ultimately results in heat rejection to the local environment, creating heat islands that exacerbate summertime cooling requirements. The substantial energy consumption not only depletes energy supplies, but the energy consumption and power production produces carbon dioxide, a major source of global warming, as well as other pollutants that cause acid rain.

SOLUTIONS

The solution to increasing energy demands cannot be just corresponding increases in supplies. More efficient use of energy to meet the needs for shelter, comfort, lighting, information technologies, and the like not only reduces the overall demand; in many cases, investment in more efficient consumption technologies is more economical than the corresponding increase in the supply side. For example, older residential buildings in northern China, built circa 1960 and 1970, were constructed of

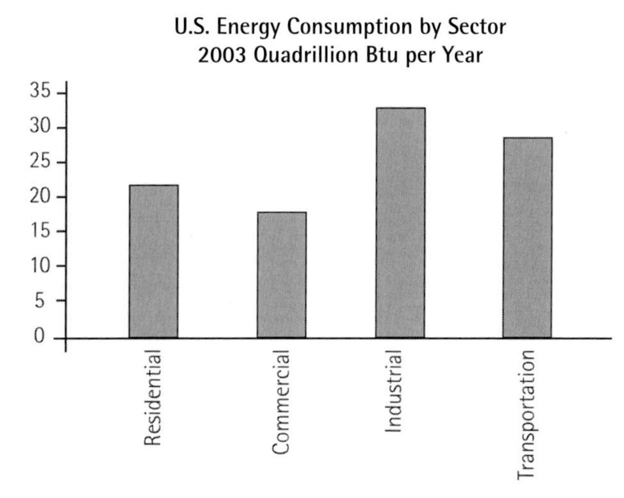

Figure 1 Energy consumption by major sectors in the United States (Source: U.S. DOE Energy Information Administration 2003)

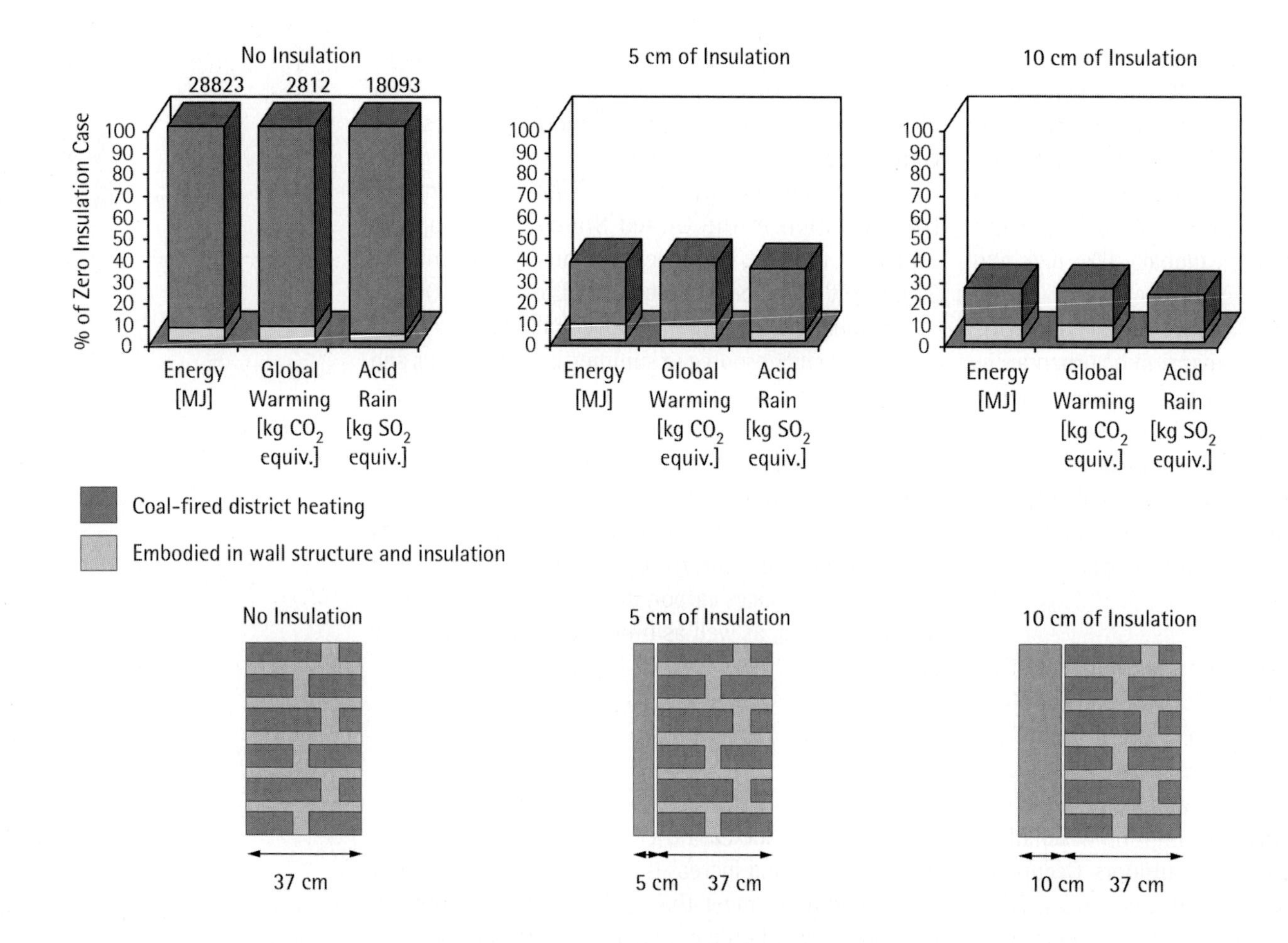

Figure 2 Environmental impacts of 1 m² brick wall over 40 years, for Beijing climate (Source: adapted from Ospelt 1999)

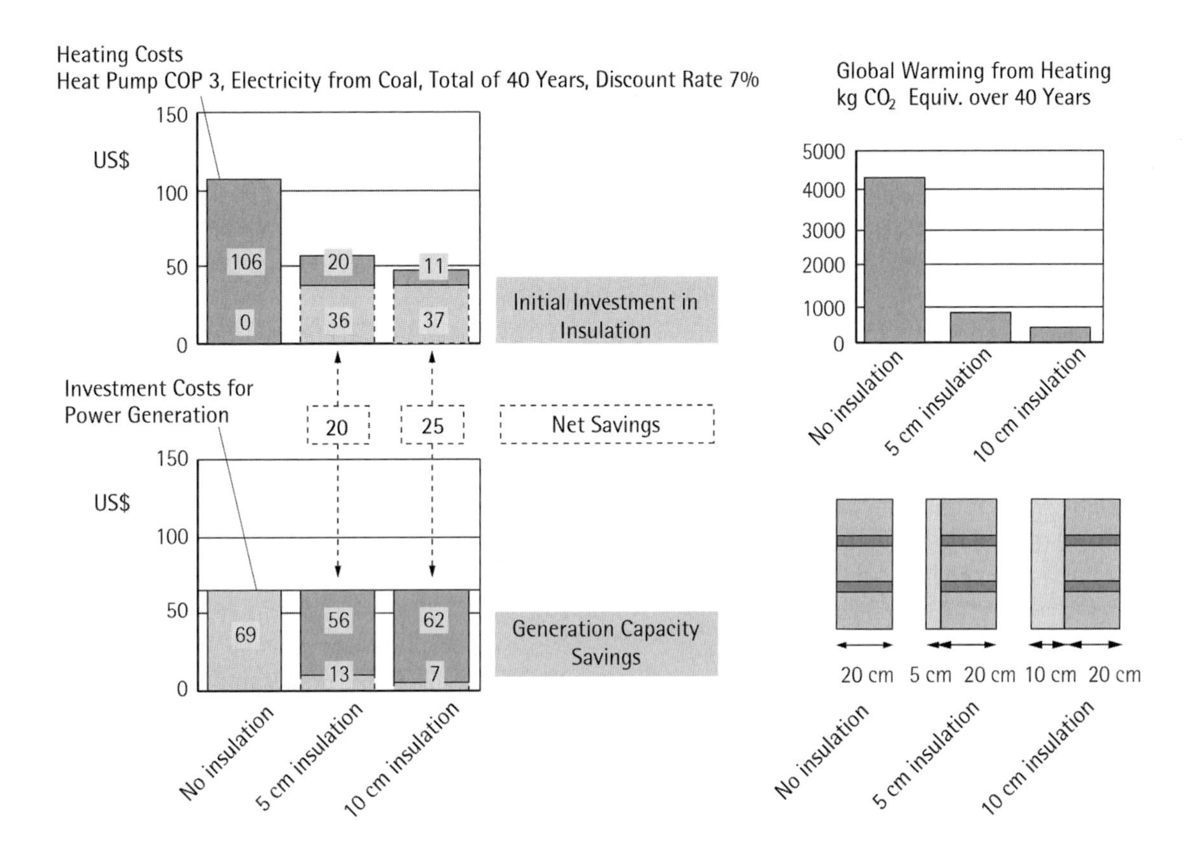

Figure 3 Environmental and economical impacts of 1 m² of block wall over 40 years, for Beijing climate (Source: adapted from Ospelt 1999)

masonry walls without added insulation. Figures 2 and 3 illustrate several scenarios for meeting the heating needs of these buildings. In one scenario, new power plants are constructed that supply heat pumps in the buildings without modifying the wall construction. These are compared to scenarios in which insulation is added to the interior walls. Not only is the energy consumption of the upgraded buildings substantially reduced (Figure 2), the first cost for the insulation is less than the savings in first cost for the new power plants. For example, the additional construction cost of adding five centimeters of insulation to a brick wall is US$36 per square meter of wall area. At the same time, the demand for heating energy is reduced so that the required power plant is smaller, saving the equivalent of US$56 per square meter of wall area. In addition to the savings in first cost for the combined building and power plant construction, the added insulation would save US$86 in operating costs over a 40-year lifetime. The operating cost savings reflect the lower heat loss through the wall assembly for the Beijing climate (Figure 3).

For large building complexes that have their own integrated energy supply systems, increased investments to produce sustainable buildings are generally less expensive, on a first-cost basis, than the increased investments in the supply system to meet the demand of inefficient buildings.

The development of sustainable buildings has been held back by several factors. The building industry is fragmented and has very little vertical integration. The industry is very conservative due to a long life of the product and the concern with liability. Many people associate more energy-efficient buildings with a loss of comfort, although the opposite is usually true: a well-designed sustainable building offers better ventilation, increased comfort in all seasons, greater daylighting, and more individual control. However, there are very few building owners and developers that want to be the first to try designs that are a substantial departure from typical practice. This is true in developed countries as well as the developing world.

The solution for more sustainable buildings requires a proper integration of many factors in building planning, design and technology, and operation. There are no "silver bullets" that will solve the problem. For example, some designers propose to cover the exterior of the building with photovoltaic panels to deal with the energy needs of the building. There is a poor capacity match between building energy demands and available supply from this technology. For example, if all of the roofs of buildings in New York City were covered with photovoltaic panels, it would still fall far short of meeting the electricity needs of the city.

Developing countries such as China are in the midst of a rapid building construction boom. Every year, China is building about 10 to 15 million new residences as well as large tracts of new commercial space. These buildings will probably have a lifetime of 50 years or more. Among the 40 billion square meters of constructed buildings in China, only four percent employs energy-saving strategies (*Asia Pulse* 2006). There is an important opportunity right now to encourage more sustainable building practices in China.

SUMMARY

Sustainability in the building sector must include considerations of nonrenewable resources, starting with land use, as well as construction materials and energy. Buildings in the developed world consume more energy than transportation or the industrial sector. As China develops, the building sector might reach this same level. There isn't a single "silver bullet" in energy – it is clear that energy efficiency is more cost effective in many instances than increase in energy supplies. Well-designed sustainable buildings will not only save energy, they will provide a healthier and more pleasant environment.

REFERENCES

Asia Pulse. 2006. China Sets Energy Efficiency Standards in Urban Construction. Asia Pulse Pte Ltd. Website article published on 26 February 2006 at <http://www.asiapulse.com>.

Craven, D., H. Okraglik and I. Eilenberg. 1994. Construction Waste and a New Design Methodology. Sustainable Construction. *Proceedings of the First International Conference of CIB TG 16*, 6–9 November 1994. Tampa, FL.

Energy Information Administration. <http://www.eia.doe.gov/emeu/aer/consump.html>

Energy Information Administration. *Annual Energy Outlook 2001 with Projections to 2020.* Report No. DOE/EIA-0380 (2001). Energy Information Administration, U.S. Department of Energy, Washington, D.C.

Franklin Associates. 1998. Characterization of Building-Related Construction and Demolition Debris in the United States. Report No. EPA530-R-98-010. U.S. Environmental Protection Agency, Municipal and Solid Waste Division, Washington, D.C.

Ospelt, C. 1999. *Framework for Sustainable Buildings. An Application to China.* Master's thesis. MIT, Cambridge, MA.

U.S. Portland Cement Association. 2004. <http://www.cement.org>

Wernick, I. 2002. Industrial Ecology and the Built Environment, in *Construction Ecology.* C. Kibert,

J. Sendzimir and G. Guy, ed. pp. 177–195. Spon Press, London.

Wernick, I. and J. Ausubel. 1995. National Material Flows and the Environment. *Annual Review of Energy and Environment* 20: 463–492.

CHAPTER TWO

CHINA – ENVIRONMENT AND CULTURE

Leon Glicksman, Leslie Norford, and Lara Greden

ISSUE OF GROWTH

As rapidly developing countries grow, they will consume a larger share of global resources and could produce damaging levels of emissions. In many countries, these problems are compounded by the concentration of the population in large cities. Densely populated areas accelerate the depletion of land and local resources. Figure 1 shows one estimate of the world consumption of energy and the projected consumption one century hence. Currently, a large portion of the world's energy is consumed by the relatively modest population in Western Europe, North America, and the Pacific Rim. If present trends continue, China and India will consume a preponderance of the world's energy by the end of this century. One study projects total energy consumption of the world will increase from approximately 350 EJ (350×10^{18} J) in 1990 to 1300 EJ in 2095. China (China, North Korea, Mongolia, Vietnam, Laos, and Kampuchea) alone is projected to consume approximately 500 EJ by 2095, more than the entire current world energy consumption (Edmonds, Wise and Barns 1995). For this reason, it is vital to ensure that new residential buildings in China's rapidly growing cities are built in a sustainable fashion.

Measures to improve environmental sustainability in China should not and need not impede the improvement of standards of living. Although at present a majority of

L. Glicksman and J. Lin (eds), Sustainable Urban Housing in China, 8-21

energy in developing countries is devoted to industry, a rising standard of living will increase the proportion of energy consumed by buildings, as in the Western world. China's residential and commercial energy consumption is consuming a growing fraction of total energy. As the per capita income and standard of living improves, this fraction should approach a value between the twenty-fifth percentile of Japan and the upper thirtieth percentiles of Western Europe and the United States.

In 1990, only about seven percent of China's energy consumption was in the transportation sector. This is due to the concentration of heavy industry in China. In addition, the work unit, located close to the workplace, has traditionally met the housing needs of much of the population. As new residential projects are developed in urban areas far removed from employment centers, this new urban development will raise the proportion of energy usage due to transportation. Effective planning of sustainable residential development, then, must address the urban scale as well as the individual building.

The development of the Chinese economy encourages the demand for higher living standards in China. At present, air-conditioning is a key symbol of improvement of living standards. In 2001, 14.8 million air-conditioning units were sold, a rise of 41 percent compared to the previous year. The State Statistics Bureau reported that for every 100 urban households, there were 35.7 air-conditioning units in 2001. In eastern China, the average ownership was much higher at 52.5 percent (FriedlNet 2002). Other statistics report that Chinese consumers account for 35 percent of all air conditioners sold worldwide (Smith 2005). In 2006 it is predicted that total domestic market sales would be between 24 and 27 million units (*China Daily* 2005). The air-conditioning industry has increased 13 times since 1985. Energy consumption by air conditioners in the summer has become a main cause of severe electrical power shortage. The electricity consumed by air conditioners in the basin of the Yangtze River is more than one-third of the total power supply during peak summer demand. If the trend continues and all 1.3 billion Chinese approach a living style and consumption level of North Americans, the world's fossil energy reserves will be more rapidly exhausted.

Economic development has fostered a demand for larger living space (Figure 2). In urban residential units constructed in 2003, the average floor space rose to 23.7 m^2 per capita (*China Statistical Yearbook* 2004). The artificial control of the indoor environment using air-conditioning makes it possible to build a building without paying attention to the outdoor climate conditions. As a result, most recent buildings are built without traditional technologies but with mechanical systems. Although many will not accept the comfort levels of the past buildings based on Soviet designs, it is still possible to apply excellent traditional technologies to modern designs. Shading devices, natural ventilation, passive and evaporative cooling, steep roofs, passive heating, and massive walls are useful to improve thermal comfort, reduce the energy demand, and significantly protect the environment.

The trend of massive energy use in China has raised global concerns. Fossil energy consumed by power plants and by building heating plants results in an increase in carbon dioxide and sulfur dioxide emissions. Greenhouse gases such as carbon dioxide lead to global warming. The sulfur dioxide emissions in China bring acid rains to the entire far-east region. Therefore, conservation of energy, protection of the environment, and improvement of indoor air quality and thermal comfort are challenges designers and developers are facing.

Many old Chinese buildings can provide reasonable comfort and indoor air quality without using air-conditioning systems. In the past, in the countryside in

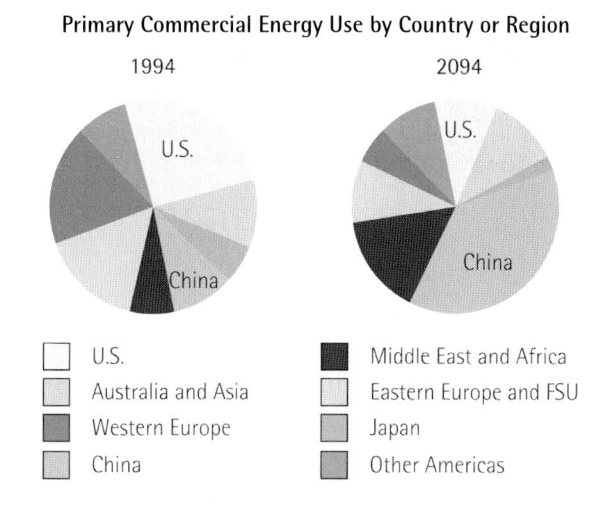

Figure 1 Projections of world energy use in 1994 and 2094 (Source: Edmonds, Wise and Barns 1995)

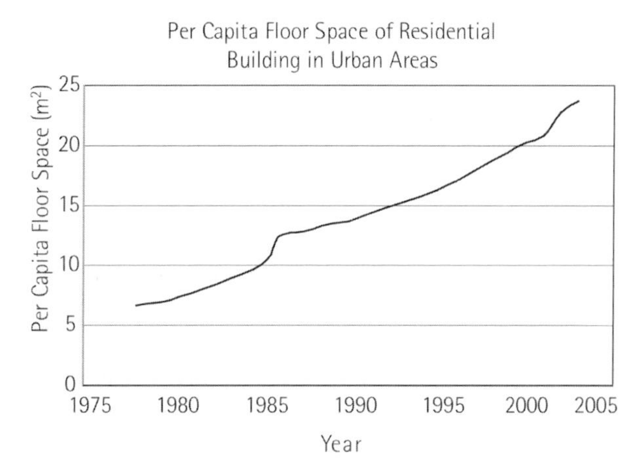

Figure 2 Urban living space per capita has grown rapidly in China over the past three decades (Source: *China Statistical Yearbook* 2004)

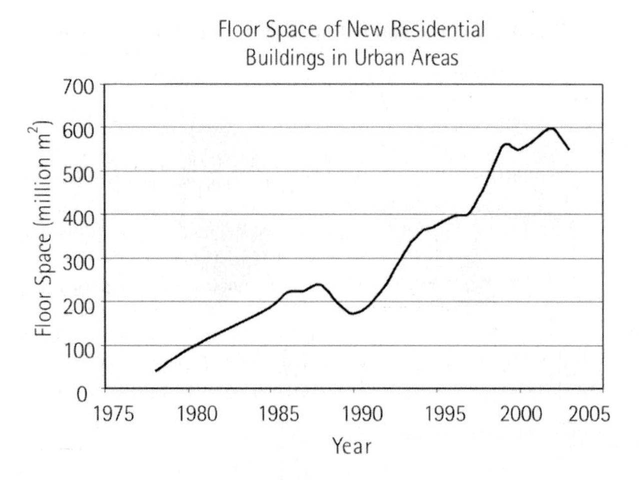

Figure 3 Urban residential floor space built per year has increased rapidly since 1978 (Source: *China Statistical Yearbook* 2004)

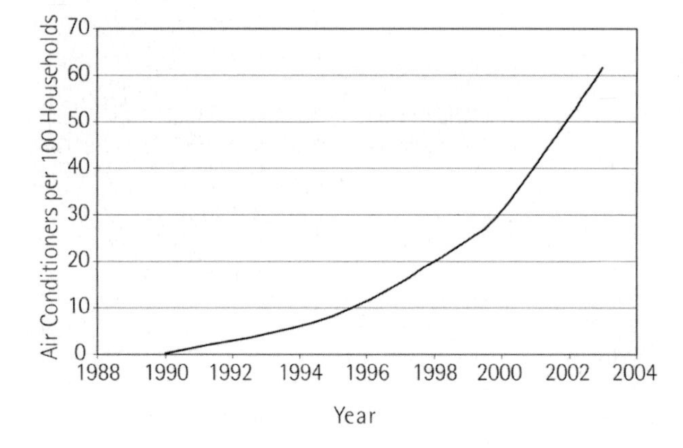

Figure 4 Air conditioner ownership per 100 urban households has increased rapidly since 1990 (Source: *China Statistical Yearbook* 2004)

China, the main objective of Chinese residential buildings was to protect occupants from severe climatic conditions. The designs were simple and in harmony with nature. Natural ventilation and/or thermal mass provided passive cooling and additional heating could be made available when required. The buildings often were built by the owner and occupant of the building. This allowed a continuous improvement of these designs, based on a close feedback between the occupants and builders.

On the other hand, the use of air-conditioning systems may not improve the welfare of the occupants as assumed. More and more occupants are not satisfied with their living environment. The Chinese have realized the problem of sick building syndrome due to poor indoor air quality and thermal environment. *Chapter 6 - Wind in Building Environment Design* describes this further.

Trends in the Chinese Residential Sector

The construction of residential space has increased rapidly in the urban areas of China (Figure 3). In 2003, urban development totaled 550 million square meters in urban areas (*China Statistical Yearbook* 2004). Real estate companies have been formed to develop this housing commercially. In 2003, there were 48,000 construction enterprises in China employing 24 million people. In Beijing alone, there were 560 prospecting and design institutes employing 64,000 people in 2003 (*China Statistical Yearbook* 2004). At a recent press conference in Beijing, the vice-minister of construction stated that annual construction is about 2 billion square meters of floor space for all buildings, or about half of the world's total (*Asia Pulse* 2006). As of 2006, it is estimated that another 20 to 30 billion square kilometers of floor space will be built by 2020 (*Asia Pulse* 2006).

These increases parallel the rise in urban income throughout China from 1,600 RMB per capita in 1990 to over 8,000 RMB per capita in 2003 (*China Statistical Yearbook* 2004). Total residential space built in 2003 exceeded 1.3 billion square meters (*China Statistical Yearbook* 2004). If the average new residence has a floor area approaching 100 square meters, the yearly residential construction exceeded 13 million units. That rate of construction is an order of magnitude greater than the current U.S. rate. Just as in developed countries, the value of that space depends strongly on location. Some of the residents of newly built, subsidized residences outside Shanghai preferred their previous housing near the city center, even though their new units were larger.

The ownership of durable goods has increased rapidly in the urban areas of China. A large proportion of urban households now has fans, refrigerators, washing machines, and color televisions. The surge in ownership of these and other durable appliances has increased energy consumption sharply (Figure 4). For example, China is now the world's largest producer of air conditioners. Guangdong had over 140 air conditioners per 100 households, while Shanghai had 136 in 2003. The demand for energy in the summer of 2005 in Shanghai was projected at a record high of 19 million kilowatts. Air conditioners accounted for a large amount of that demand, 7.5 to 8 million kilowatts, according to the Shanghai Electric Power Company (*People's Daily Online* 2005).

Space heating consumes more than 80 percent of the total building energy use in China by some estimates. In many urban areas, space heating is supplied by district heating systems. Individual units are not metered for space heating use, and in many cases the residents of large residential complexes cannot adjust their own heat supply since turning off the supply to one unit also turns off the supply to units above it. The cost of heating is included in the rent; sometimes heating cost is billed based on the floor area of the unit. This system presents a strong disincentive for conserving energy. Residential buildings in Beijing are estimated to consume 50 to 100 percent more energy for space heating than buildings in the West with comparable climates while providing less comfort (Zhu and Lin 2004). Since 2000, heating price and billing reform has been a primary activity. The Chinese government plans to reform the system so that residents will pay for heating based on consumption (Lang 2004). Newer units sometimes have individual heating units and owners are billed for their consumption levels.

Currently, the energy consumption in a typical Chinese residential unit is an order of magnitude lower than a comparable residential unit in the United States. As the older housing stock is replaced, air conditioners become more common, and preferred winter temperature settings rise, this gap will narrow. For example, in Chongqing, China's most populous municipality, average household energy use increased by about 70 percent between 1990 and 1994. Without incentives to encourage energy-efficient designs, the newer Chinese residential units may soon exceed U.S. energy use per unit.

Recently constructed residential buildings suffer from poor construction quality. Exterior building envelopes have little insulation, loose single-glazed windows, poor siting, and inferior materials. In many ways, traditional buildings in China were more environmentally friendly. They used building mass and wide roof overhangs to shade windows and walls to limit overheating during the summer. Proper site planning of buildings limited exposure to severe winter winds, shading in the summer, and promoted community interaction.

The residents of many new buildings install air conditioners (Figure 5). New buildings typically have few features that would control interior heat gain during the summer, such as shading devices above windows. Most units have outdoor balconies that could serve as sunscreens in the summer, but many residents enclose the balconies with permanent windows to create larger living spaces. In the summer, this creates a greenhouse space that substantially increases air-conditioning requirements (Figure 6).

In Beijing, where the average maximum temperature exceeds 30°C for approximately 40 days out of the year, yearly operating costs of a single-unit air conditioner, if it is operated 8 hours per day, are an estimated 1,280 RMB (US$160) (electricity price of 0.8 RMB/kWh (US$0.10/kWh), consuming 5 kW of energy). This is becoming a modest portion of total income in large cities such as Beijing where the per capita annual disposal income exceeded 13,000 RMB in 2003.

Technical Opportunities

Potential technical innovations must be considered as part of an overall program, one that addresses building design and technology in the context of community needs. For example, tightly built, energy-efficient buildings may have greater problems with poor indoor air quality. Regenerative heat exchangers are necessary to achieve adequate ventilation while maintaining energy efficiency.

In developing countries such as China, Western technology is not always the best answer. These technologies, optimized for countries with high labor costs, a skilled workforce, and mechanized production methods, are often not economically efficient in areas of the world with plentiful unskilled labor and less-developed industries. New designs and solutions must use locally available materials and products. Construction methods must be easy to learn by relatively inexperienced labor. Local tastes and lifestyles have to be considered in the design of individual buildings as well as large new communities.

Figure 5 Many residents install their own air conditioners

Figure 6 Permanently closed-in balconies create a greenhouse space that substantially increases air-conditioning requirements

Simple steps to improve buildings will yield substantial rewards. These should be considered before more capital-intensive advanced technology. Some simple design goals such as natural ventilation and control of solar gains may require substantial background analysis. The resulting design can then be made straightforward to execute.

CLIMATE

Climatic conditions are the starting point for any design process interested in energy efficiency. Climatic conditions vary widely through the year between Beijing, Shanghai, and Shenzhen. Winter temperatures in Beijing are substantially lower than Shanghai, down to -10°C, while Shanghai records only a few dates with temperatures below freezing. Both cities record hot and humid summers, with temperatures somewhat higher in Shanghai. Shenzhen has a more intense cooling season than the other two cities and very modest heating requirements.

Comfort Zones and Building Bioclimatic Charts

The thermal-comfort zone is one where the average occupant will perceive no discomfort due either to heat or cold. There are two distinct and independent sources of heat discomfort: the thermal sensation of heat, and discomfort resulting from skin wetness (sensible perspiration). In Beijing and Shanghai, which have hot and humid climates in the summer, both sources of discomfort are present. Cold discomfort is felt when the skin temperature falls below 32-33°C. Environmental factors that increase cold discomfort include a low ambient air temperature, draft, and low radiant temperatures of adjacent surfaces. Both Beijing and Shanghai have outside winter temperatures below the human comfort zone.

The comfort standards published by ASHRAE were established using data from American and European populations that are located around 40°N latitude and are accustomed to mechanically conditioned buildings. Populations in warmer regions generally prefer higher temperatures, and often show no thermal discomfort at temperatures outside the ASHRAE comfort zone. In a study at Tsinghua University in Beijing, temperature measurements in residential units were recorded and accompanied by a survey of the occupants. None of the units had HVAC systems, but most residents did not complain about thermal discomfort even at indoor temperatures of about 30°C (Jiang 2000).

CODES AND STANDARDS

Building Codes in China

China's building codes categorize its regions in two ways: by zones based on heating needs, and into urban and rural areas. The heating zone is the northern section of the country, including Beijing, where heating of buildings is permitted during the winter. The transition zone is approximately the middle third of the country, including Shanghai, and has significant demand for space heating in the winter and space cooling in the summer. The region south of the transition zone, which includes Guangzhou and Shenzhen, is characterized by a hot climate and significant cooling demands.

The Ministry of Construction (MOC) has authority at the national level over building codes in China, while local governments can develop their own codes as long as they are more stringent than the national code. To date, the emphasis has been on developing energy efficiency standards for residential buildings, although, in the summer of 2002, the MOC approved the development of a national code for commercial buildings.

In the northern zone (heating zone), residential building energy codes have existed since the early 1990s, although they have been enforced only since the late 1990s in Beijing and Tianjin. In the transition zone (formally known as the hot-summer cold-winter region), local residential energy standards were developed in the late 1990s for numerous cities and provinces, including Chongqing, Wuhan, Jiangsu, and Shanghai. This was followed by a national effort that ended in the promulgation of a national energy efficiency standard for the transition zone in October 2001. A residential standard for the southern zone (formally known as the hot-summer warm-winter region) was also initiated in July 2001, with completion planned for the end of 2002 and promulgation in early 2003. Work on the national commercial building energy standard was started in September 2002 (Huang and Tu 2001, Huang 2002).

Green Guidelines for Sustainable Housing in China

In August 2001, China took an important step toward the goal of sustainability in the housing sector with the publication of the voluntary "Green Guidelines for Sustainable Housing in China." Developed within the context of China's stated long-term strategies for sustainable development, the objective of the guidelines is to realize sustainability in the housing sector through the following goals (Nie 2002):

- increased level of functionality and quality in housing;
- promotion of advanced building technology;
- definition and implementation of sustainable housing construction; and
- protection of home ownership rights.

The guidelines are based in part on the technical and policy experience of the U.S. Green Building Council, Canada's Environmental Agency, the U.S. Energy Star Program, the E.U. Ecolabel, Germany's Blue Angel Program, and others. The evaluation methodology consists of five categories, as described in Table 1.

Dissemination of the Green Guidelines will begin with construction of ten developments known as the *Asia Pacific Green Villages*. Vital to the project is an inspection team, led by the China Housing Industry Association, to evaluate and verify building performance after construction. Although these guidelines are not mandatory, MOC officials believe that heavy competition in China's housing market will encourage developers to adopt these guidelines as a way to improve quality and competitive advantage.

Enforcing Building Codes

The seventh five-year plan, from 1986 to 1990, initiated the goal of energy conservation in buildings. The MOC developed a five percent tax incentive for energy-efficient construction. Energy efficiency was not well defined, however, and nearly all building managers found some way to qualify. The end result was a negligible increase in energy efficiency, and the tax incentive was dropped in 1998.

A change to performance-based codes that require builders to achieve an overall performance, while allowing latitude in the individual building systems and designs used, will advance the development of most buildings. However, enlightened building codes will be of minimal use unless enforced. At present, the need for enforcement is much more pressing than the need for development of new codes. Equally important is a consumer education program that emphasizes the key attributes of sustainable buildings. Architects and builders

Evaluation Category	Key Elements
Urban Environmental Planning and Design	Site selection Traffic patterns Facilitation of construction Community greening Air quality Noise pollution Outdoor lighting
Energy Resources and the Environment	Energy-efficient building design Optimization of energy (i.e., HVAC) systems Use of renewable energy sources Environmental impact of energy consumption
Indoor Environmental Quality	Indoor air quality Heating Lighting Sound
Community Water and Wastewater Management	Water supply and wastewater discharge Wastewater treatment and reuse (i.e. gray water usage) Water usage for landscaping Water conserving appliances Sustainable building materials
Materials and Resources	Utilization of local materials Reuse of resources Waste management

Table 1 Key design elements of the 2001 "Green Guidelines for Sustainable Housing in China" (Source: adapted from Nie 2003)

also need more education about ways to achieve energy efficiency, comfort, and good indoor air quality.

One national code requirement states that each residential unit has at least one window that receives one hour of direct sunlight at the winter solstice. This can reduce overcrowding but must be coupled with many other, equally important requirements to achieve the most beneficial results. In some residential developments the winter sunlight code has led to a widely spaced array of tall, isolated towers, a plan not unlike some of the failed low-income housing developed in the United States.

Enforcement of regulations is a primary concern. On-site measurement of buildings designated as "energy saving" shows that 80 percent exceed the energy-consumption limitations of energy codes (Zhu and Lin 2004). Some codes exist, but their correct implementation is left to the design institutes (architectural firms). The design institutes must insure that the designs meet the codes. However, there is little, if any, inspection of the actual construction project or the occupied building. What are the reasons for this lack of enforcement? First, structural tensions in the vertical organization of the Chinese government permit great disparity between how the central government expects firms and local government bodies to behave and how they actually do behave. Local officials theoretically must follow the central government; however, funding and career support, and with it influence, is typically controlled at the local level. Second, local authorities are typically responsible for enforcement of the national and regional regulations; however, local authorities and governments often have financial stakes in companies that might be affected by environmental regulations. When environmental regulations are seen as in conflict with economic development, enforcement and implementation are difficult.

Policy

The building sectors in developing countries share several characteristics important to policy design. Rapid growth or transition to a market economy creates a dynamic economic environment with little history to learn from. The disproportionate distribution of wealth that often accompanies rapid development must be addressed when balancing who will pay for investments. Furthermore, financial markets must be developed to provide the resources needed for such investments. Along with the development of a financial market must come an increased ability and desire to hold debt. Finally, energy prices in developing countries often do not reflect true production costs, thereby creating market inefficiencies.

China's economy is riddled with incentives to ignore or undervalue investments in energy efficiency, or to view such investments as overly risky. Examples include the underdeveloped capital markets and chronic changes in the price of energy. Obsolete and deteriorating equipment remains in operation in many factories and industries, producing environmentally inefficient building products. Other barriers to the widespread adoption of sustainable building practices in China are the limited availability of building products, the lack of information and training in building design and assembly, income disparities, and general market volatility. The key to implementing residential energy efficiency in the long term is to demonstrate to people that energy-efficient investments perform as promised: providing increased comfort while simultaneously reducing costs. People often will not sacrifice their own comfort, income, or perceived quality of life to otherwise make the necessary investment unless its value to them can be demonstrated. Energy-efficient investments must prove themselves, especially since they often require a larger initial capital outlay than the alternative.

Current Policy in China Relevant to Residential Building Energy Efficiency

Policy must play a substantial role in implementing residential energy conservation measures. The most significant achievements to date in residential energy conservation have been achieved through standards and labeling programs for energy-efficient appliances. Currently, the slow adoption of new energy pricing systems, a lack of enforcement of building codes, and the scarcity of educational initiatives are all significant impediments to reform.

The intentions of China's Energy Conservation Law (ECL), approved on 1 November 1997, are to do the following:

- promote energy conservation activities throughout society;
- improve energy efficiency and increase economic benefits thereof;
- protect the environment;
- ensure economic and social development; and
- meet the needs of people's livelihood.

Relevant to building energy efficiency, the law states that the government should develop polices and plans that insure rational energy utilization and strengthen educational activities to increase public awareness of energy conservation. Although the ECL does not give clear guidance on the extent of regulations or the specific goals that are to be achieved, it can be viewed as a sign that energy efficiency is looked on favorably by the national government. At a 2003 meeting on China's energy policy for the next two decades, energy efficiency was cited by the Chinese MOC and colleagues as being of paramount importance. Development of specific measures is left to agencies of the central and local governments. The MOC is the central agency responsible for building construction and codes.

Other provisions of the law mandate the submission of statistics on energy consumption by local authorities, the display of product labels specifying energy consumption, the development of energy conservation plans, and the implementation of energy conservation measures in schools and institutions. The State Bureau of Statistics is responsible for statistical gathering and reporting at the national level.

China's new five-year national development program, covering the years 2006 through 2010, includes a statute for energy efficiency in building construction, which strives to reduce energy use by 50 percent from current levels, and by 65 percent for cities such as Beijing, Shanghai, and several other major cities. In January of 2006, the Designing Standard for Energy Conservation in Civil Buildings was implemented. This standard requires that all civil buildings integrate energy-efficient building materials and adopt energy saving technologies in heating, ventilation, air-conditioning, and lighting systems (*Asia Pulse* 2006).

Reform of Heating Prices

In pre-reform China, prices of all energy products were under unified state control and were allocated according to central plans. Multi-track pricing and partial liberalization was introduced in the 1980s. Energy prices, beginning with coal, were freed to a great extent, in parallel with important new reforms in taxation, finance, and other areas, in the early 1990s. Historically, and often still today, consumers either receive heat as a welfare commodity or are billed for heat based on occupied floor space. They do not pay for the amount of heating energy actually used; therefore incentives do not exist to reduce heat energy consumption. Other disadvantages of the current system include low comfort level due to over-

or under-heating, increased pollution due to unneeded burning of fuel (primarily coal), and low fee-collection rate. Building-area-based billing for heat is a bottleneck for energy savings. However this is beginning to change as residents of new developments pay for their heating bills based on natural gas consumed in their individual unit.

Coal Price Deregulation

One reason for high unit energy consumption is that energy prices do not reflect actual production costs. When fuel prices are lower than production costs, subsidies are needed to maintain production levels. However, subsidization is costly for the government and taxpayers; this leads to inefficiencies and discourages innovation. This scenario describes China's coal-based energy system. Energy is wasted when low energy costs discourage conservation. China has begun to address this problem by reducing subsidies to the coal industry by 50 percent over the past decade. To increase the effectiveness of the market, price increases due to the decreased subsidy must be passed on to the consumer.

Experience with Appliances

Encouraging the use of energy-efficient appliances requires that standards be introduced and that consumers are aware of energy efficiency. Several energy efficiency programs have been introduced in China to date, with mixed success. The China Green Lights program addresses two major barriers to reform, the high initial cost of efficient technologies and poor quality of some of the products manufactured in China. Screw-in compact fluorescent lights have been observed to fail very prematurely, negating the life cycle savings and losing consumer acceptance. The CFC-Free Energy-Efficient Refrigerator Project aims to improve the energy efficiency of Chinese refrigerators through a series of market-oriented measures for manufacturers and consumers. Quality assurance is of primary importance for energy efficiency incentive programs to be successful, an important lesson for designing similar programs for buildings.

Suggested Policies to Drive Adoption of Sustainable Buildings

Policy options are plentiful, and will be most effective if implemented as part of a portfolio of programs. Perhaps the most effective course of action is for demonstration projects to be built, tested by a reputable source, and publicized. This would serve to minimize the perceived risk of new technologies and methods. Demand-side management (DSM) programs, in which utilities offer subsidies for such investments, offer yet another way to leverage incentives for energy-efficient building design. Finally, education and awareness concerning the comfort, health, and cost-saving benefits of sustainable buildings must be spread through public service announcements, workshops, and media coverage.

Stakeholders

Effective policies must be designed according to an assessment of the stakeholder needs. Housing market stakeholders include occupants, potential occupants, developers, design teams, energy suppliers, and policy makers. A critical analysis of each stakeholder's set of needs and frames yields valuable information for policy designers.

Occupants and potential occupants comprise the demand side of the market. In China, two primary ideological barriers hinder the transformation to a market system. First, Chinese households, unlike most Americans, are adverse to debt. The idea of obtaining a mortgage to buy a house opposes traditional Chinese norms. Second, Chinese people are accustomed to thinking of housing and its related services, such as heating, as a welfare privilege. The idea of paying for such essentials is looked upon unfavorably.

Studies are now underway to gauge the Chinese consumer's receptivity to sustainable design features. As the number of privately owned units increases, and as residents become responsible for the operating and maintenance costs, the homebuyer should become more concerned with such design features. The importance of energy efficiency will increase if space-heating costs reflect actual costs to individual units.

Developers and designers comprise the supply side of the residential market. Arguably, the most important factors to energy consumption are decided in the design phases of a project. Investments in design and construction offer many benefits: decreased utility consumption, increased comfort, decreased expenditures for occupants, and lower air-pollutant emissions. Residential building developers in China are receptive to the inclusion of sustainable features in their new projects. However, it is difficult for the developers and their designers to define what this entails. Designers are not well versed in the means of achieving energy efficiency, and access to software tools capable of such analysis is relatively recent. Proper building materials such as insulation are not optimized for traditional wall systems used in China. The design or development firm's perception of the customer's marginal valuation of additional energy-saving features will determine its willingness to supply those features. A real estate firm in eastern China recently estimated the cost increase necessary to achieve energy savings requirements at an additional 100 RMB (US$12.5) per square meter (*Asia Pulse* 2006). A developer is most likely to only make an investment that will be visible to the consumer and warrant asking for a higher price. In this way, developers may be able to use energy efficiency to generate a competitive advantage.

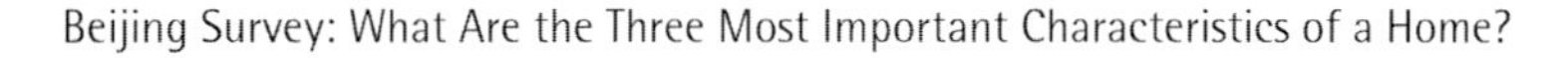

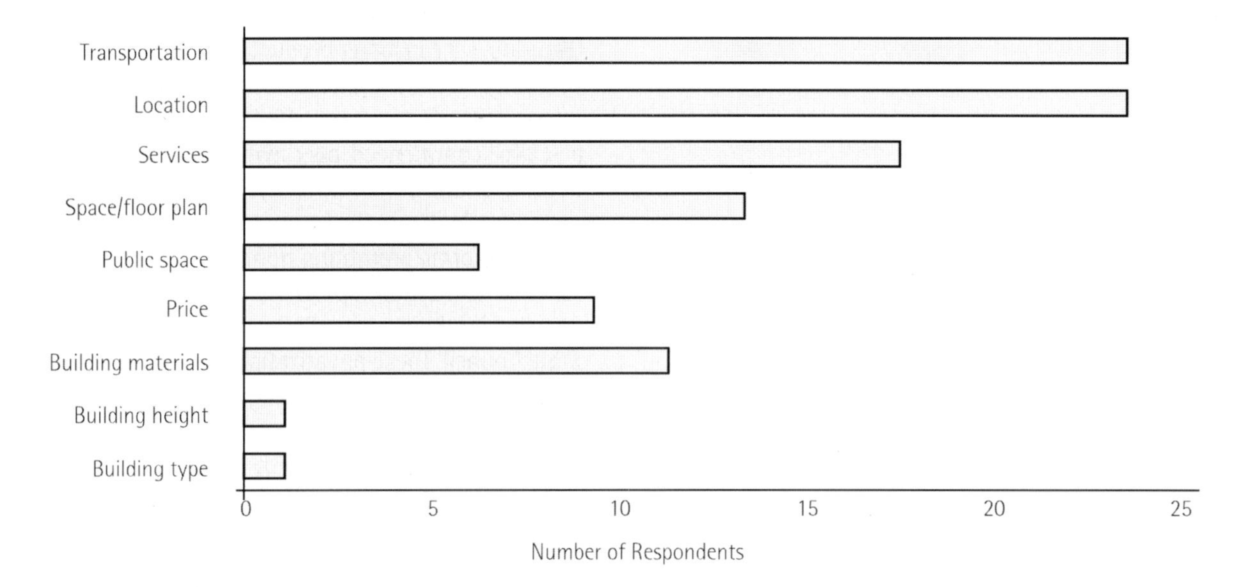

Figure 7 Results of a small-scale survey in Beijing illustrating the most important housing characteristics for homebuyers

A small-scale survey conducted at an affordable housing development in Beijing by Tsinghua University in cooperation with the authors found that homebuyers rate the most important characteristics for a home as location (near workplace), transportation (public), price, space/floor plan, services (safety), and building materials (quality), as shown in Figure 7 (Greden 2001). Willingness to pay more for a "green, sustainable, or energy-efficient" home was limited to less than ten percent of the base house price, and the term "green" generally meant trees and grass to the respondents. There is a desire for more information about residential quality and for advice from a professional consulting service, evidence of the steep learning curve faced by China's first generation of homebuyers. Based on the survey results, during this period of high residential growth, sustainable homes would best be promoted to Chinese home-buyers by marketing the combined benefits of natural lighting and ventilation, better quality, and savings in heating and air-conditioning energy costs. In developments where the heating and cooling energy consumption of each residential unit can be separately metered and energy costs reflect real market costs, the energy costs to the homeowner will be a substantial portion of their income. In this circumstance, it is expected that energy efficiency will merit more importance to the homebuyer.

The Urban Housing Market: History, Progress, and Problems

"Housing is a consumer product, not a welfare product," says Zhao Chen, director of the government's housing reform office. Government efforts to make this statement a reality commenced over two decades ago. Solving housing shortage problems by increasing the quantity of housing built was the focus in the early 1980s. In 1994, several reforms promoting homeownership over public rental and the development of housing markets were unveiled by the Housing Reform Steering Group of the State Council. These privatization reforms have resulted in large changes in the housing sector; China currently has one of the highest homeownership rates in the world (Duda, Zhang and Dong 2005).

Between 1980 and 1985, the average floor area per capita in urban China increased only slightly, from four to five square meters. Today, new, non-luxury residential units are much bigger than the average with sizes ranging from 60 to 140 square meters. Improving the quality of new urban housing by refining design and construction standards was the policy concern of the mid 1980s. Success was minimal, and safety now weighs heavy on the minds of policy makers and consumers alike after major construction flaws were found in many of the buildings born in the early 1990s construction boom. Building collapses are not too infrequent, limiting faith in the safety of buildings. Although price remains a critical factor, the concern for quality is gaining.

A typical affordable housing unit costs 225,000 RMB (US$28,125) whereas a typical lower-middle dual-income family earns only 30,000 RMB (US$3,750) annually (Duda, Zhang and Dong 2005, Kirchhoff 2006). This means that house prices exceed household income by a factor of more than seven for lower-middle-income families, or double the comparable price income ratios of Western markets. Employers are being forced to raise wages to allow employees the means to buy housing, thereby switching the enterprise stake in housing from directly providing the home to providing the monetary means to buy a home. This also means these first-time home buyers must up the security of housing provision in hopes that the new housing market will provide better quality housing and fewer housing shortages.

Compounding the problem for potential homeowners is a lack of available financing options. A government policy in late 1998 offered employees of state-owned enterprises a one-time lump-sum payment for purchase of a new home. Officials are also hopeful that new homeowners will fuel the hefty reported annual economic growth, 9.9 percent in 2005 (Hong Kong Trade Development Council 2006), as they furnish their new homes with goods such as air conditioners and washing machines.

Urban Chinese have four basic housing schemes to choose from (Table 2). Income dictates the choice between these housing types; however, differentiating characteristics among housing developments may influence choice within a type.

Developers are racing to make a profit in a competitive market. The economic reform created a commercial housing development industry. In 2003 there were 48,000 construction enterprises employing 24 million people (*China Statistical Yearbook* 2004). With the magnitude of the expansion of China's urban areas, and the long expected lifespan of the plans and buildings now being made, the need to develop economically competitive models of sustainable development could not be greater.

Housing Type	Family Income Level	Provider	Approximate Price	Description
Public housing (original housing under the old socialist system)	No restriction	State owned or employer provided	- Varies greatly depending on progress in increasing rents and wages - Ranges from 200 RMB to approximately the monthly mortgage payments of an affordable unit (~2,000 RMB/month)	- Subsidized by work units - Leased for a nominal rate - Government plans to sell at discounted price hoping that a secondary housing market will appear as buyers resell homes at market price and upgrade to better-quality homes
Commodity or commercial housing	No restriction	Property developers selling to private purchasers	- Expensive to the average person - Ranges from 5,000-8,000 RMB/m ^{2} and up	- Sold for a profit on the open market - Conflict between government and developers exists regarding marginal return expectations
Economic or affordable housing	Qualifying upper limit	Property developers selling to private purchasers	Ranges from 2,400-4,500 RMB/m^2 and up	- Similar to commodity housing, but exempted from land-transfer premiums - Provides developer returns of 3% - Sells at a discount of 10-12% to commodity housing - Qualifying income raised as incentive to developers
Anju housing	Low income	Government sponsored	Unknown	- Original government-sponsored low-cost housing started in 1995 - Basic accommodation for low- to middle-income families - Construction funded by fixed-rate policy lending

Table 2 Contemporary Chinese housing types and characteristics (Sources: *Business China* 1999, *China Statistical Yearbook* 2000, and author's personal experience)

SUMMARY

The development of the Chinese economy encourages the demand for higher living standards in China. Larger homes and improved comfort standards are leading to much higher energy consumption. Although modern life will not accept the comfort levels of the past, it is still possible to apply traditional technologies to modern designs to reduce or eliminate discomfort during extremes in summer and winter. China needs to encourage energy efficiency in new buildings. New housing codes alone will not meet this need. Demonstration projects of affordable energy-efficient buildings will help. More careful enforcement of housing codes is essential.

REFERENCES

Asia Pulse. 2006. China Sets Energy Efficiency Standards in Urban Construction. Asia Pulse Pte Ltd. Website article published on 26 February 2006. <http://www.asiapulse.com>

Business China. August 1999. China's Housing Schemes, p. 4.

China Daily. 22 September 2005. <http://www.china.org.cn/english/BAT/143042.htm>

China Statistical Yearbook 1997. 1997. China Statistical Publication House, Beijing.

China Statistical Yearbook 2000. 2000. China Statistical Publication House, Beijing.

China Statistical Yearbook 2004. 2004. China Statistical Publication House, Beijing. <http://www.stats.gov.cn/english/statisticaldata/yearlydata/yb2004-e/indexeh.htm>, accessed 21 March 2005.

Choi, 1998. A Housing Market in the Making. *China Business Review*, November–December: 14-19.

Duda, M., X. Zhang and M. Dong. 2005. China's Homeownership-Oriented Housing Policy: An Examination of Two Programs Using Survey Data from Beijing. Paper published by the Joint Center for Housing Studies. Harvard University, Cambridge, MA.

Edmonds, J., M. Wise and D. Barns. 1995. The Cost and Effectiveness of Energy Agreements to Alter the Trajectories of Atmospheric Carbon Dioxide Emissions. *Energy Policy*, vol. 23 (4/5): 309-335.

FriedlNet. 2002. *China Air-Conditioning Industry Research Report*. <http://www.friedlnet.com/product_info.php?cPath=7&products_id=28>, accessed 13 March 2006.

Glicksman, L., L. Norford and L. Greden. 2001. Energy Conservation in Chinese Residential Buildings: Progress and Opportunities in Design and Policy. *Annual Reviews*, Environment and Resources Series, vol. 26 (2001): 83-115.

Greden, L. 2001. The Demand Side of China's Evolving Housing Market and Sustainability: Stakeholder Analysis and Policy Approaches. Master's thesis. MIT, Cambridge, MA.

Hong Kong Trade Development Council. Market Profile on Chinese Mainland. <http://www.tdctrade.com/main/china.htm>, accessed 13 March 2006.

Huang, Y. 2002. Personal communication, 3 October 2002. Berkeley.

Huang, Y. and F. Tu. 2001. *China-U.S. Energy Efficiency Steering Committee Meeting*. Energy Efficiency Buildings Team Report. <http://www.pnl.gov/china/eetbld01.pdf>, no longer available.

Jiang, Y. 2000. Personal communication. Tsinghua University, Beijing.

Jiang, Y. 2002. Current Situation and Important Issues for Building Energy Efficiency in China. In *Proceedings of China Civil Construction Academy* (in Chinese).

Kirchhoff, S. Bubble or Not, High Home Prices Can Hurt. *USA TODAY*. <http://www.usatoday.com/money/perfi/housing/2005-05-10-housing-cover_x.htm>, accessed 13 March 2006.

Lang, S. 2004. Progress in Energy-Efficient Standards for Residential Buildings in China. *Energy and Buildings* 36: 1191–1196.

Long, W. 2000. Construction, Energy, and Environment Situation in Shanghai, China. Presentation to the MIT Department of Architecture. Cambridge, MA.

Nie, M. 2002. China Housing: Sustainable Development. In the *Alliance for Global Sustainability Annual Meeting 2002 Proceedings*, 21–23 March 2002. San Jose, Costa Rica.

Nie, M. 2003. *Sustainable Buildings for Mega Cities*. Presentation at the Alliance for Global Sustainability Annual Meeting 2003. 23-26 March 2003. Tokyo.

People's Daily Online. 9 May 2005. <http://english.peopledaily.com.cn>

Smith, E. 2005. Chinese Snap up Brand-Name U.S. Firms. *USA TODAY*. <http://www.usatoday.com/money/world/2005-06-21-china-usat_x.htm>, updated 22 June 2005, accessed 13 March 2006.

Zhu, Y. and B. Lin. 2004. Ministry of Construction Outline of the 10th Five-Year Plan of Building Energy Efficiency. *Energy and Buildings* 36: 1287–1297 (in Chinese).

Part Two
Design Principles

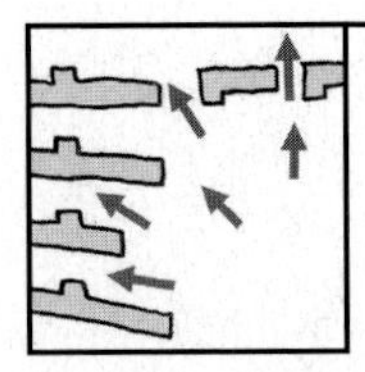

CHAPTER THREE

DESIGN PRINCIPLES FOR SUSTAINABLE URBAN HOUSING IN CHINA

Andrew Scott

THE OBJECTIVES OF SUSTAINABILITY: WHAT ARE WE TRYING TO ACHIEVE?

New urban housing in various areas of urban Chinese cities appears at first review to be relatively unsustainable. Poor performance of the building fabric, a lack of responsiveness in the design to the harshness and forces of the climate, and an absence of a relationship of housing areas to an urban transportation infrastructure stay in one's mind as significant deficiencies in a more comprehensive environmental strategy.

As an architect, one perceives little effort toward designing projects that have clearly articulated environmental objectives, or an awareness that environmental responsibility is on the mind of the typical housing developer. Perhaps a better way of looking at this would be to ask how well substantial new projects that are springing up on the peripheries of cities such as Beijing, Shanghai, and Shenzhen teach us how to build for the future? Which designs might future architects, building owners, and occupants look to as models for a society that needs to be highly responsive to sustainable development and the explosive urban growth in Chinese cities?

Two typologies of buildings dominate the urban landscape of new construction in China; these include five-story walk-up buildings that are arranged as long

L. Glicksman and J. Lin (eds), Sustainable Urban Housing in China, 24-43

building masses, and high-rise towers that have little relationship to the environment and ground plane (Figures 1 and 2).

The Sustainable Urban Housing in China Group has undertaken various projects in order to develop a series of aspiring development guidelines that could point towards a better way of designing with environmental responsibility in mind. We have not sought to impose new "high-tech" solutions or "super-advanced" technologies. Conversely, the attitude that underpins our work has been to rediscover and interpret the traditions of Chinese housing especially in terms of typologies that incorporate social structures and the means by which buildings have traditionally been responsive to working with the climatic forces. An example might be the fortress-like forms of the circular housing in Fujian (Figure 3) or the courtyard housing of the hutongs (Figure 4). For centuries, these forms provided for the needs of the sustainable community, including protecting residents from harsh winter winds.

Of course, contemporary housing has to meet many needs beyond those of the vernacular due to higher densities, better social conditions, stricter building codes and standards, and the increased mobility of the industrialized society – in contrast to those types from the agrarian society of the past. Nevertheless, the design work of our projects has been founded upon designs that reinterpret tradition, that recognize social contexts, and that find building methods that are based upon "appropriate technologies" rather than those that might be perceived as being alien to the prevalent industrial infrastructure.

Therefore, the sustainability of housing is interpreted in multiple modes: social, technological, climatic, and urban. These are both quantitative and qualitative aspects. Some are related to better performance – an efficiency or economy of resources and energy – while

Figure 1 Example of typical low-rise housing block in the Pudong district of Shanghai

Figure 2 Shanghai urban landscape showing building typologies

Figure 3 Circular housing in Fujian; houses were designed to protect inhabitants from the climate and provide security

Figure 4 Courtyard housing in Beijing

others are about quality of life and the ability of people to live together without alienation. At the qualitative building level, the design team at MIT sought design solutions where the housing can be far more energy efficient, principally through the reduction of air-conditioning and improvements in the design of the building fabric, to reduce carbon dioxide emissions through the dependence upon climate-destructive energy sources such as coal-fired power plants, and to maximize the use of natural ventilation and solar energy for daylighting and winter heating. At an urban scale, we sought to develop plans that were more ecologically responsive in relation to water conservation and retention, wind patterns in and around buildings, and to landscape space that built a sense of community. However, rather than developing a design that might contain a large palette of available "sustainability techniques," we sought to develop an understanding of those that would work best, that would be most effective – and therefore would make greatest sense for each project.

Designers should think holistically to ensure all of these aspects are integrated into a project. By inclusively and strategically integrating such techniques into the design at an early stage and developing them throughout the design and construction process, we will yield an end product that will then be a more "sustainable" approach to designing and constructing contemporary urban housing in China.

ASSESSING SUSTAINABILITY: HOW DO WE KNOW IT WORKS?

As sustainability has become a familiar word in the architectural language of practice, many questions remain about how a particular project, rightly or wrongly, claims its environmental credentials. Sometimes the sustainability factor is retrospective because the assessment method that substantiates a project was not integral to the design process. On the other hand, often the assessment method becomes the driver for establishing a set of objectives or goals, and creates for the owner or users a deeper understanding about what might be achieved.

In both Europe and the United States, the past decade has seen the emergence of a number of methodologies for assessing the environmental accountability, usually at the stage of detailed design prior to construction. Typical among these are the BREEAM method (Building Research Establishment Environmental Assessment Method) developed in the United Kingdom in the early 1990s and the more recent LEED (Leadership in Energy and Environmental Design) method in the United States that has gained a significant increase in popularity and is now a common standard for evaluating sustainable design.

Interestingly, the growth of these methods appears to have stemmed more from owner or user demands for an environmental footprint. This may be a result of a company's or institute's underlying value system towards environmental accountability, such as the need for an annual environmental audit, or possibly from a belief that better environmental standards benefit users through a "'healthier building" mentality, and an investment in the "well-being" of occupants or workers. As the theory snowballs, owners or developers see a certified rating attained through one of these assessment methods as a path to a better investment – the higher the rating on a point system, the more sustainable the building is supposed to be.

Most of the assessment methods work on the basis of an extensive list of attributes that projects can achieve through a points-based system. The more points, the better the sustainability index of the project, with performance being broken down into several different categories. The problem that arises is that such ratings may have little to do with the real performance of the project after completion, as the assessment is carried out before construction. By chasing these environmental points, one runs the risk of losing sight of a project's real potential to push the boundaries of sustainable design; "sustainability" can therefore become more about a series of technical, detailed specifications rather than about the concept or form of the building.

Research by Professor Leslie Norford within the Building Technology Group at MIT in which office spaces were monitored shortly after post-construction occupation suggests that the real performance of the building bears scant resemblance to the design criteria, and that a post-occupancy evaluation and adjustment of the building's environmental systems is essential for peak performance.

One of the primary goals of the Sustainable Urban Housing in China Project was to develop building prototypes that could be constructed and monitored for their performance and true sustainability credentials. Only by creating this feedback loop can one truly learn what works best and also understand how buildings and their designers truly learn. In the future, it is hoped that the prototype designs in the work will lead to large scale demonstrations.

THE BARRIERS TO SUSTAINABILITY AND INNOVATION

All buildings should be sustainable and should deploy some form of design strategy or principles appropriate to the design problem. The notion of a "sustainable building" as a unique object different from other forms of architecture is contradictory as it is not a polarized black-and-white problem (sustainable or unsustainable), but rather many shades of gray expressing different degrees of performance or compliance. Some are simply better, some worse, but few are ideal. Perhaps a better way to describe it would be to say that some buildings are simply more unsustainable than others, in other words, designed with no regard to larger environmental concerns in any form.

If high levels of sustainability are achievable with sufficient knowledge, why does it not happen more often? Why aren't all projects pushing the frontiers of innovation to achieve gold or platinum ratings under the assessment methods or, better still, achieving major new goals in integrated design? What are the barriers to achieving better buildings, and are they real or simply perceived?

These questions were central to recent research work by Singh Intrachooto and the author, Andrew Scott, at MIT. Buildings from across the globe that were commonly regarded to have some form of innovation in their production of low-energy or sustainable qualities were studied. The research looked not only at the product of the design, namely, the building and its technology, materials, systems, or method, but also at the design processes and the teams and procedures that produced them. Several factors were found that were incentives as well as barriers to the creation of sustainability. On the positive side, the role of research was often critical to the development of new products, materials, or systems, and the testing of performance through simulations in such projects. Research establishments (universities, institutions, research centers) were able to obtain funding for developing research that is often not possible within the scope and terms of a professional design team's project appointment. Several factors (the barriers) were discovered which inhibited the

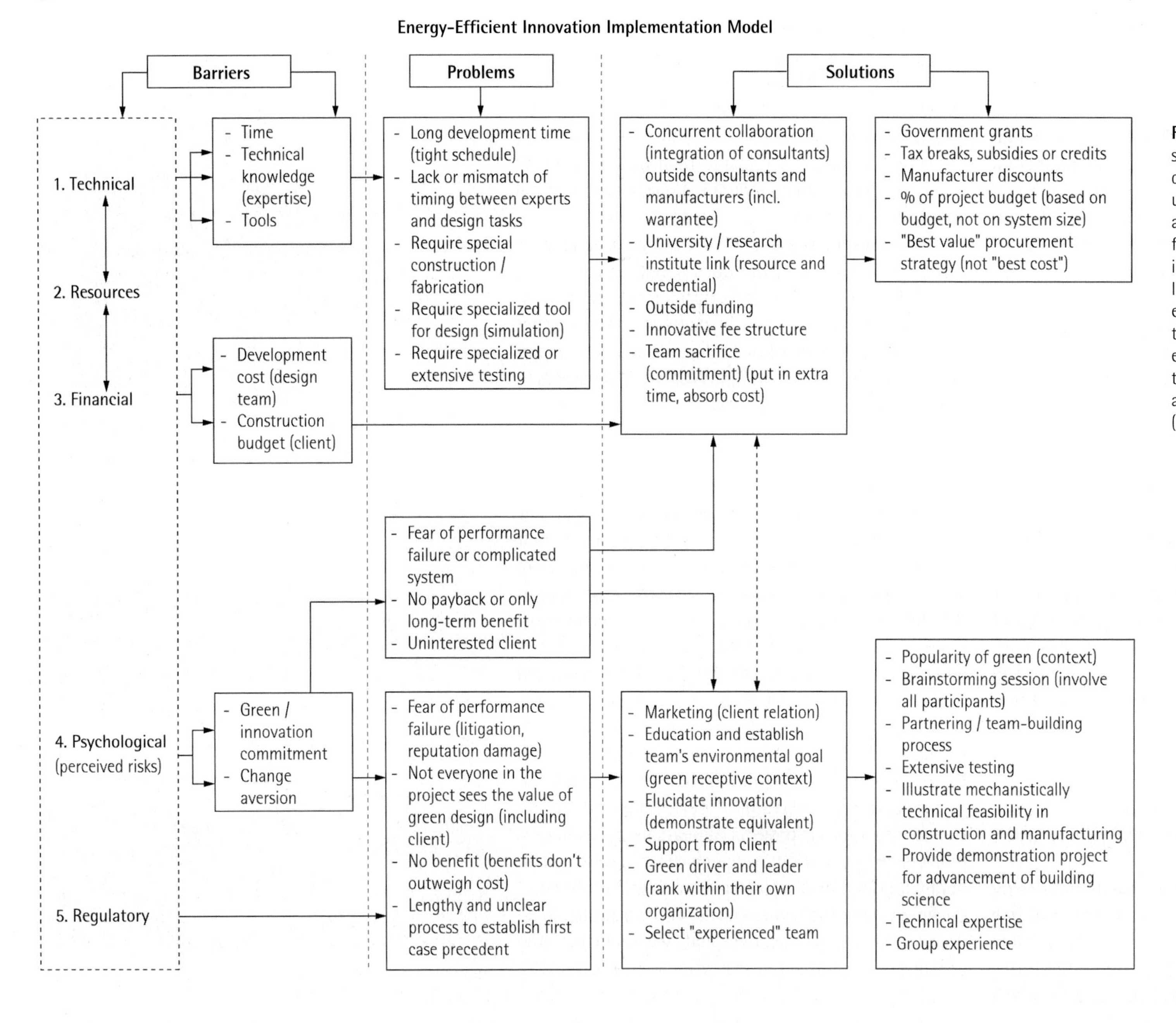

Figure 5 Tracking barriers and solutions to sustainable practice. This diagram is a summary of research undertaken by Intrachooto at MIT that attempted to understand the reasons for sustainability not being fully implemented in architectural practice. It categorizes the perceived barriers or excuses in the generalized sense and then proceeds to define the more explicit problem given. Suggestions for techniques or mechanisms to be used are found under the "Solutions" column. (Source: Intrachooto 2002)

development of sustainable design strategies, and these can be broadly summarized as:

- risk – the perception that departing from the familiar unsustainable project poses different forms of risk for the owner or client and the different members of the design team;
- technical – technical problems stem from a lack of time (to develop new solutions), a lack of knowledge, or a lack of "tools" for testing and simulating solutions;
- financial – a concern for the cost of developing sustainable buildings both in terms of construction cost and the development cost for the design team;
- competence – a lack of knowledge about how to make new ideas work; at the same time, there might also be a lack of agreed-upon objectives between the owner and the design team; and
- regulatory – a lack of appropriate building codes or regulations to enable and reinforce sustainability.

Intrachooto's analysis of the building case studies (Figure 5) suggests that sustainable projects probably need both a different form of collaboration among design team members and, most importantly, a set of agreed-upon objectives among all parties involved in the project. This agreement should be a document that can be updated and monitored throughout the design and construction to provide a comprehensive framework for sustainable design thinking and effective implementation. Without such agreement on the goals, standards, and ambitions, the task will be made considerably more difficult. For the future, one might envision an interactive, web-based project knowledge management guide that navigates all members of the design team through these goals and actions, linking them to cost-effectiveness and relevant research resources, and then enabling a system of monitoring the innovations throughout the design process, construction, and beyond. Some construction firms are now beginning to see such a system as a business opportunity in the service of sustainability.

The role of MIT in the Sustainable Urban Housing in China Project was to attempt to overcome many of these barriers through working with Chinese developers, authorities, universities and construction companies, and also to secure research funding that would enable the design of new projects to move into new territory.

ECOLOGICAL BUILDINGS: DESIGNING FOR DIFFERENT CLIMATES

An essential principle for the Sustainable Urban Housing in China Project is for the designs to respond to, and indeed harness, the local climatic conditions. This strategy could be described as being "bioclimatic" by working with the climatic forces instead of against them and making building forms that are responsive to the forces of wind, sun, and temperature (Figures 6 and 7). But in trying to make each project climatically responsive, one must also look at, and learn from, the different forms of architecture that have developed over the centuries and across the globe as fundamental responses to the need for shelter:

> Around the world people have developed energy efficient building forms that are suited to the climate conditions of their particular location – a form of 'solar vernacular'. They have developed simple solutions to the environmental challenges set before them: heat, cold, rain and wind. These solutions have been developed using a limited range of indigenous building materials, all of them renewable.

Figure 6 Traditional ventilation technology: wind towers

Figure 7 Double layered façade, public building, Manchester, U.K. (architect: Denton Corker Marshall)

> The methods by which we apply technologies in the design and construction of buildings have direct implications for the amount of energy consumed. (Behling 1996)

To illustrate this theme, four "comfort" zones – continental, Mediterranean, subtropical, and maritime – are briefly described below in terms of their climate and the resulting indigenous architectural form.

Continental

Continental climates have significant temperature swings from well below freezing in winter to +30°C in summertime. The building mass provides thermal cooling for intense summer heat and also enables protection for the harshness of the winter cold. Building forms may be protective (recessed courtyards) with respect to strong transcontinental winds and intense sun. Buffer spaces mediate between the inside and outside, enabling inhabitants to move in and out as desired. Inland regions of northern China such as Beijing are typical of this type of climate (Figures 8 and 9).

Figure 8 Mud house, Datong, China (example of *continental* architectural form)

Figure 9 Cave dwellings, Bei Jing Yan Qing, China (example of *continental* architectural form)

Mediterranean

An ideal climate for enjoying outdoor living as it is mild in winter and warm but dry in summer; air-conditioning is rare. Summer breezes enable cooling, especially when combined with evaporation techniques. Building forms are often closely packed, especially in urban settings, in order to provide shade, and shaded courtyards are common. Streets have arcades and verandas to create microclimatic conditions that allow people to be outside, yet protected. Life slows down during the middle of the day to overcome the summer heat. Plant life and landscape flourish with adequate irrigation (Figures 10 and 11).

Figure 10 Plaza de la Corredera, Cordoba, Spain (example of *Mediterranean* architectural form)

Figure 11 Piazza Maggiore, Bologna, Italy (example of *Mediterranean* architectural form)

Figure 12 Porch and balconies as buffer zone to housing, New Orleans, U.S. (example of *subtropical* architectural form)

Figure 13 Inside circular housing, Fujian, China (example of *subtropical* architectural form)

Figure 14 Old Swan House, London, U.K. by Norman Shaw, 1876 (example of *maritime* architectural form)

Figure 15 Oxford University building, Oxford, U.K. (example of *maritime* architectural form)

Subtropical

This climate is warm or hot for most of the year, but with significant humidity that has made air-conditioning popular in the latter part of the twentieth century. Winters can be very pleasant as the warmth is without excess humidity. In the southern United States, buildings have developed the porch as an inhabitable intermediate zone between inside and outside. Traditionally effective cross ventilation is needed to alleviate the humidity. The Gulf region of the United States and coastal regions of China are typical examples (Figures 12 and 13).

Maritime

Maritime climates never get too hot or too cold, but there is significant rainfall due to the effect of land and ocean on weather patterns. Light conditions are not extreme. Buildings usually have large fenestrations for maximizing daylight. There is no need for shading or sun protection but good insulation and thermal mass are essential. Buildings can also be effectively naturally ventilated all year round. Traditional buildings have a fireplace that provides for both winter heat and summer ventilation (Figures 14 and 15).

Figure 16 Windows at Glenn Murcutt's Boyd Education Centre in New South Wales, Australia, showing layering of façade elements, including fins, overhangs, and operable shading devices

Figure 17 Double-skin glass louvered façade for an apartment building in Manchester, U.K. (architect: Ian Simpson Architects)

This cursory look at the relationship of climate to building form and construction reveals a strong sense of regionalism. This regionalism has generated not only particular attributes in building techniques but also the language of the form and the characteristics of the urban form typology. The following words by Roger Stonehouse also describe the attributes of this regional language:

> In buildings designed for passive and inclusive means of environmental control we see many instances of the discovery, reworking, transformation and extension of traditional types and patterns of places and situations that are at the edge, simultaneously inside and outside, or which bring the outside in or create degrees of enclosure: for this is an architecture of well known and half forgotten place, in which we may, in T.S. Eliot's words 'arrive where we started and know the place for the first time.' We see, for instance, developments of window reveals and shutters, window seats, balconies, awnings, bed cupboards, outdoor rooms, lobbies, conservatories, courtyards, atria, pergolas, porches, porticos, verandahs, belvederes, galleria, hanging gardens, deep eaves and glazed and blanketing roofs. (Stonehouse 1998)

The passive nature and environmental control of the contemporary building envelope exemplifies this need for climatically sensitive design based upon an understanding of climatic diversity. The development of complex façade systems - double and / or triple skins - in their many varied forms of depth, control, and ventilation principles, can be seen as an attempt to regionalize the design problem, even if the manufacturers are operating across continents (Figure 16). In Europe, with a more temperate climate, such façade systems have largely taken over the market for commercial buildings where the demand and appreciation for natural ventilation is high, and the financial investment greater. These façades typically allow a degree of passive operation of ventilation and solar control while enabling the enclosure to both trap heat in winter months and provide shade in summer (Figure 17). What remains to be proven in most cases are the long-term economic benefits in terms of energy payback of the façade system with respect to the increased initial cost. Regardless, as an alternative to the early-modernist sealed-skin curtain walling, they are clearly supporting a more climatically oriented form of architecture

In Boston, a project on Thompson Island designed by architects Andrew Scott and Hubert Murray also illustrates how architecture develops from the understanding of local climatic diversity (Figures 18-22). An earlier version of the project shows how the building developed in both plan and sectional form based upon an understanding of local sun angles and orientation, cross ventilation and thermal mass achieved by seating the building in the landscape (Figures 18 and 19). Façade treatments were studied using ray-tracing lighting software to compare summer to winter conditions (Figures 20 and 21). In summer, roof overhangs and shading protect the south-facing façades from overheating, while in winter the low-angle sun was needed to provide passive solar heating.

DESIGN PRINCIPLES

The various Sustainable Urban Housing in China projects were developed by an integrated MIT team comprised of architects, graduate students, environmental engineers, building technologists, landscape architects, and urban planners. In addition, we worked in close collaboration with development companies, multidisciplinary design institutes, and mechanical and

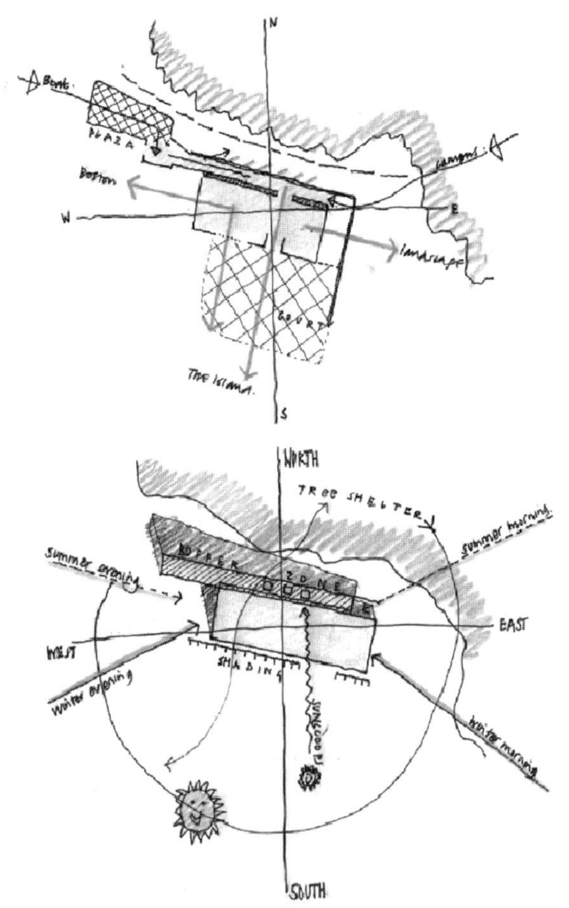

Figures 18a and b Thompson Island project diagrams exploring the relationship of building to site and climatic conditions: light, orientation, wind, sun, and landscape

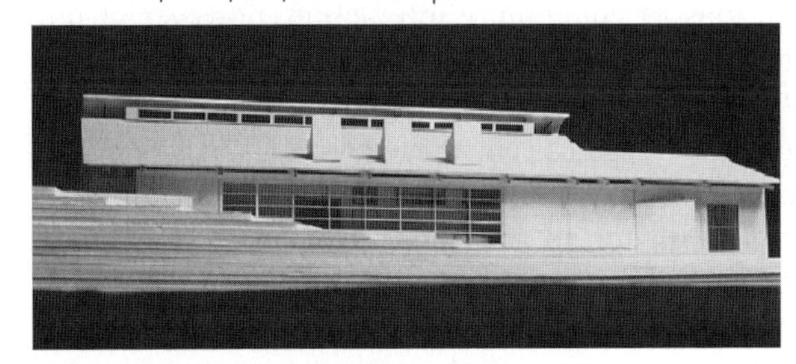

Figure 19 Early Thompson Island project model

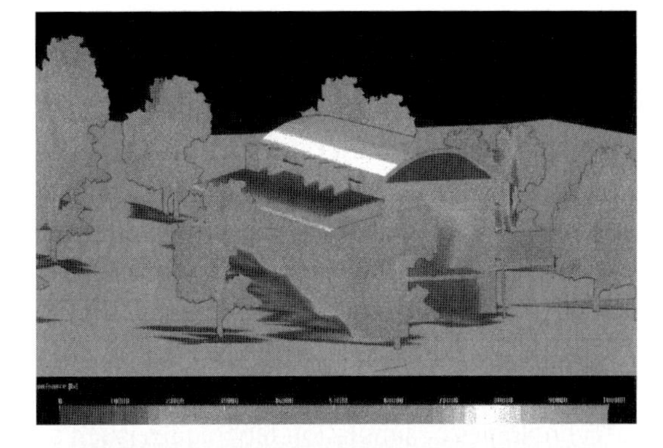

Figures 20a and b Thompson Island project model thermal imaging of radiation on enclosure during mid-summer

Figures 21a and b Thompson Island project model shadow and sun studies of façade during mid-winter

Figures 22a and b Images of completed construction of Thompson Island project

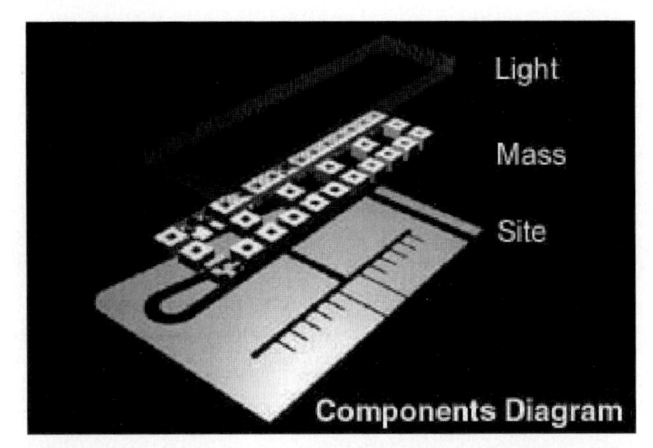

Figure 23 Shanghai low-rise project components diagram showing courtyard scheme on site

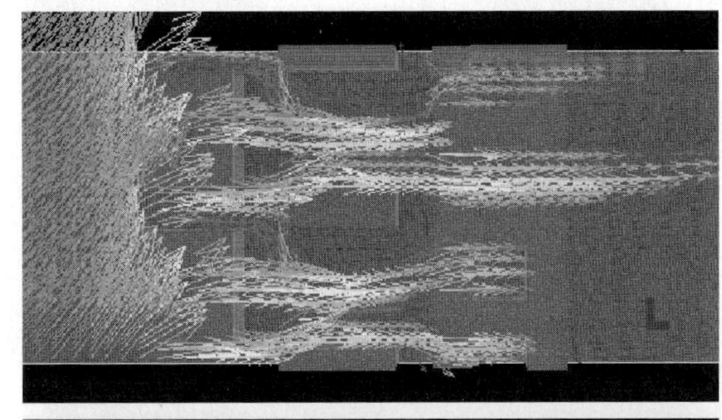

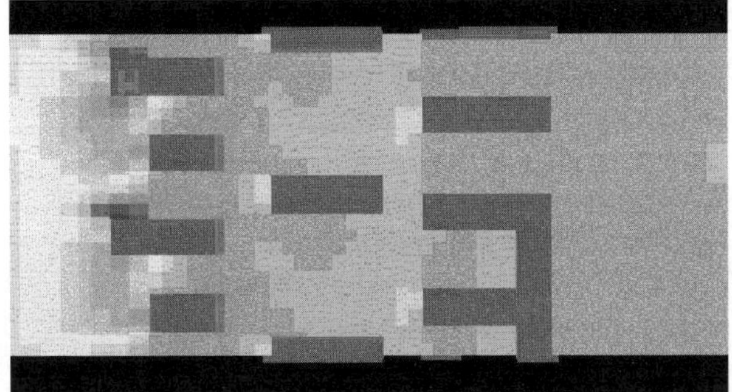

Figures 24a and b Shanghai low-rise project CFD studies of exterior airflow showing (top) velocity and (bottom) pressure

thermal engineers from university research centers in China.

With such an array of talent, it was essential to work in a non-hierarchical manner and to promote an interdisciplinary mode of working. Ideas originated from a mix of disciplines working and innovating together. The design process is not linear but has to be iterative, moving back and forth between macro to micro scales. With the work in China, design ideas were always balanced with technical solutions. Specific physical building technologies or simulation techniques (such as energy spreadsheets, computational fluid dynamics (CFD), or Lightscape) supported more generalized design strategies and suggested the means by which technical solutions could substantiate larger ideas. Similarly, new technical solutions came out of larger design concepts that found their context and meaning in the regional nature of the project. Overall, this approach enabled the technology to be localized, workable, and relevant.

As a principle, and as a part of the design process, members of the design team have to establish and agree upon the environmental or sustainable design objectives for the project. Without this, different disciplines or interests will inevitably be pulling the team in different directions. Such objectives usually emerge over time once a familiarity with the design project or problem emerges. With the projects in China, our focus was to identify the environmental objectives that were both achievable at a local level and that promised to be most effective. This approach often leads to a limited palette of principles and strategies to adopt rather than a large checklist of ideas, some of which may be marginally effective at best.

The design principles are further explained at three integrated scales: site design and planning; building form and typology; and building design and planning. The type and form of building is derived from an approach towards the site planning where each building has a role to play. Similarly, that building or block will suggest a certain plan in which larger ideas, for example regarding natural ventilation, can work successfully.

SITE DESIGN AND PLANNING

Urban Density Distribution and Urban Ventilation

One of our earliest research projects was conducted through a design workshop at MIT in the Department of Architecture. As an introduction to our work, the workshop sought to develop and study design ideas and technologies that might relate to a variety of contexts, for example, high rise/high density, medium rise/medium density, and medium rise/high density. The expectation was that each permutation would create a different set of achievable goals and methods, added to which were different climatic contexts such as Beijing and Shanghai, both of which have different seasonal temperature and humidity characteristics.

A key idea was to relate a study of density with natural ventilation strategies in order to be as effective as possible. Normally, natural ventilation is studied within and around a building form, and so it was particularly challenging to think how patterns of urban ventilation might influence an urban plan. With a medium-rise option, we were also interested in working with the courtyard as a traditional typology that best incorporates notions of community interaction, however, if one develops a tight network of courtyard spaces, there is the likelihood that it will not be particularly effective for using breezes as a cooling method.

A three-dimensional-modeling method was developed for an arrangement of linear blocks with courtyard spaces within (Figure 23). Between the front and rear rows was an intermediate block. Each design iteration was tested for the airflow pattern and movement characteristics using CFD modeling to study how natural ventilation flowed from front to rear (Figure 24). Essentially, what we were searching for was a form that had good "perforation characteristics." What we found was that the orientation and form of each block made a significant difference. For example, orienting the middle building parallel to the ventilation flow enabled air to reach the rear block. Also, cutting away corners of buildings assisted the airflow into courtyards. By moving back and forth between CFD analysis, the plan diagram, and shading studies, we were able to refine the effectiveness of the site planning (Figures 25-27).

The Beijing Star Garden Project was a further exploration of the ideas about density redistribution and urban ventilation. The site had already been designed by an architectural firm in Singapore, and the design was typical of many developments one can see emerging on the urban fringe of the city – namely, a scattering of several high-rise buildings, typically with about 20 to 30 floors, arranged on the site in a regular pattern.

Without much architectural knowledge, one can anticipate several problems stemming from this form of planning, an approach that can be found historically in the West as exemplified by the pioneering work of Le Corbusier's Ville Radieuse project. While the density of the blocks is high, they command a large ground-level footprint in order to avoid problems of overshadowing and to satisfy the winter sunlight code, a building code that requires access to sunlight (a minimum of one hour per day on the first day of winter). The resulting site density is not so large and can be achieved in other forms of land use distribution (i.e., low rise/high density). There are also significant problems associated with wind forces and eddying moving in and around the site and between buildings that can render the ground-level landscape unusable. Finally, at a community scale, the high-rise solution rarely creates a sense of community where

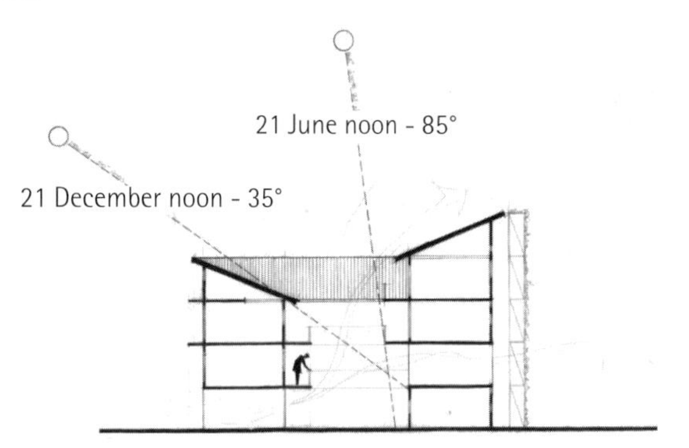

Figure 25 Section showing shading strategies for Shanghai low-rise prototype

Figure 26 Model showing shading strategies for Shanghai low-rise prototype

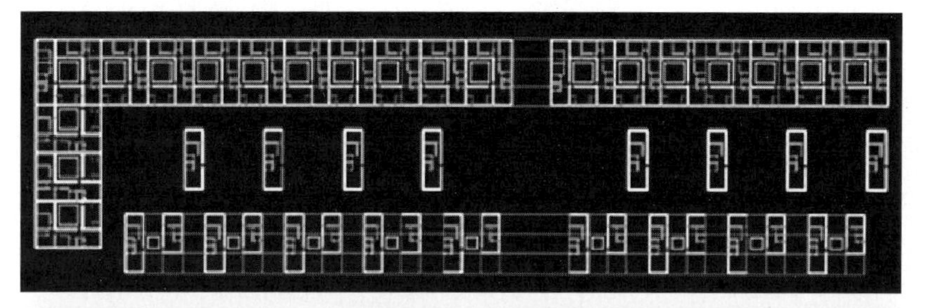

Figure 27a Shanghai low-rise project ground floor unit breakdown, comprised of five types of unit designs: studio, one-bedroom, two-bedroom, three-bedroom, and four-bedroom

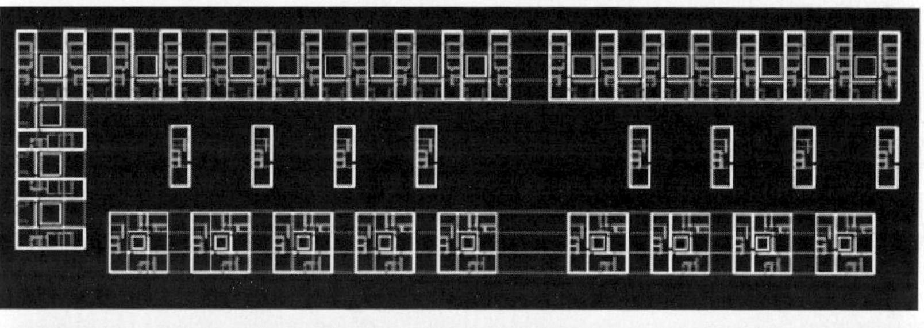

Figure 27b Shanghai low-rise project second floor unit breakdown, comprised of five types of unit designs: studio, one-bedroom, two-bedroom, three-bedroom, and four-bedroom

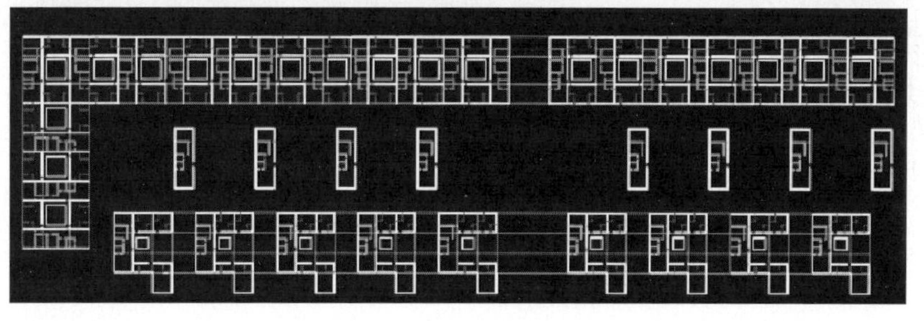

Figure 27c Shanghai low-rise project third floor unit breakdown, comprised of four types of unit designs: studio, one-bedroom, two-bedroom, and three-bedroom

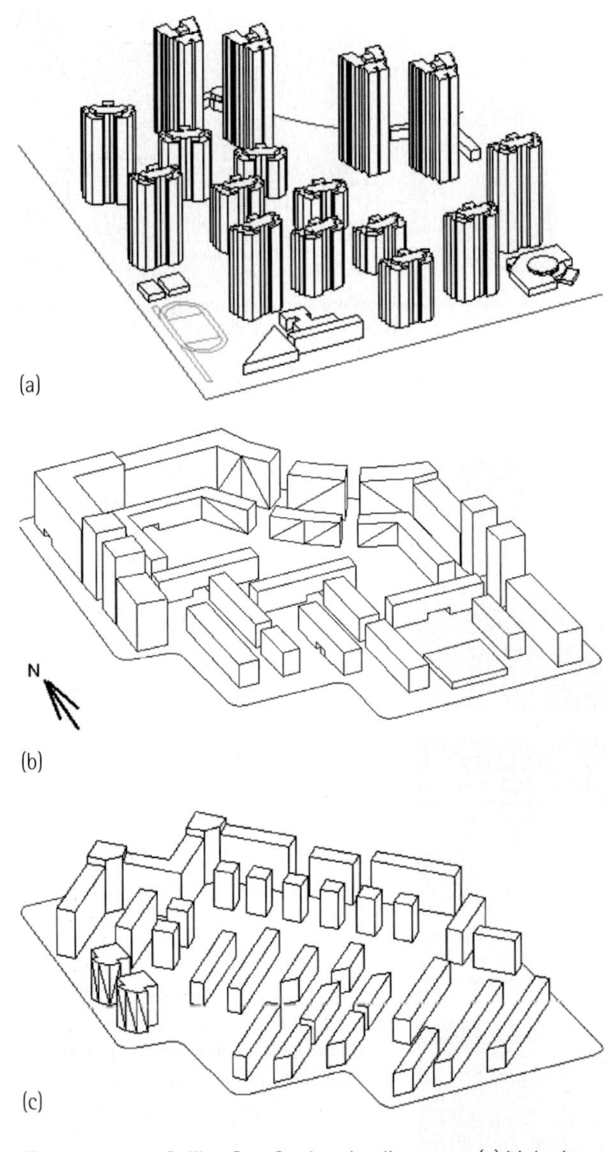

Figures 28a-c Beijing Star Garden site diagrams - (a) high-rise proposal from previous architects; (b) low-rise orthogonal alternative proposal from MIT team; and (c) low-rise angled scheme from MIT team

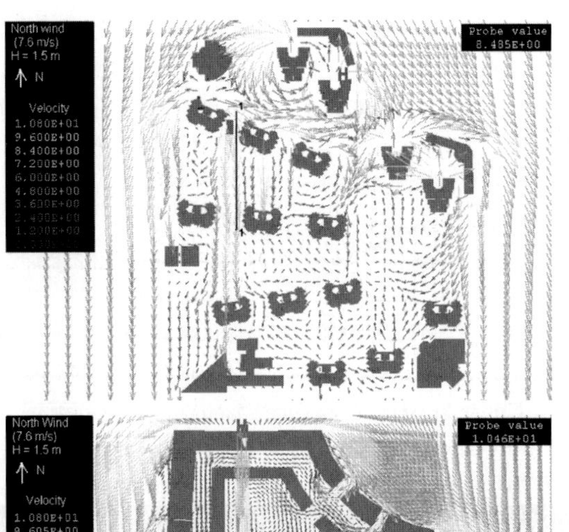

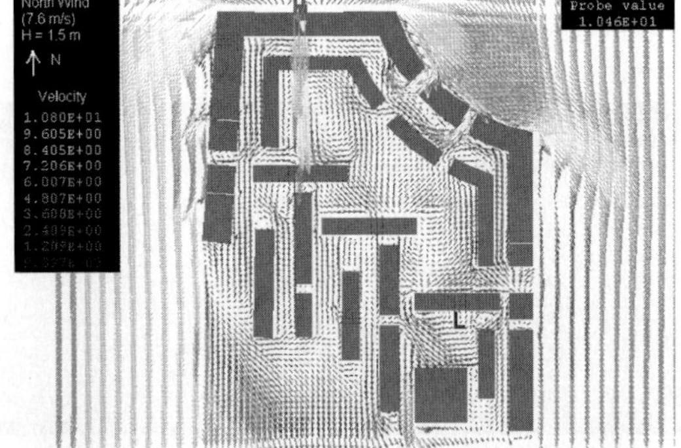

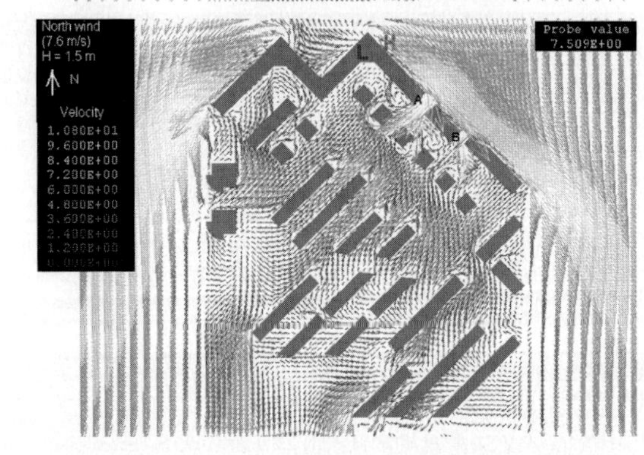

Figure 29 Beijing Star Garden site CFD images by MIT, showing air velocity for the three different site plans: (top) high-rise from previous architects, (middle) low-rise orthogonal scheme from MIT team, and (bottom) low-rise angled scheme from MIT team (see color version in chapter 7, Figures 9-12)

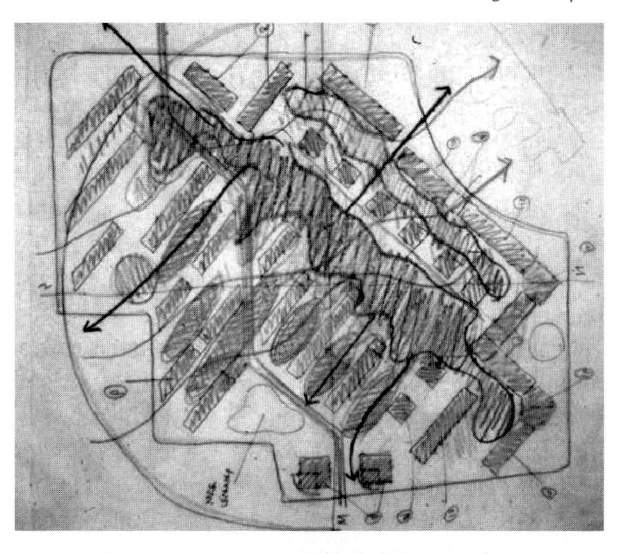

Figure 30 Beijing Star Garden site low-rise angled scheme concept sketch

Figure 31 Beijing Star Garden site low-rise angled scheme daylighting and shadow study

children can play under supervision and where people can interact socially.

Therefore, we started to look at a series of alternative design strategies for the Beijing Star Garden Project that were based on buildings of variable scale, a combination of low-, medium-, and high-rise that were planned to create smaller community spaces between them. The blocks were arranged so that the taller ones created a buffer to the northerly winter winds, while those on the southern edge were smaller in scale and enabled a high degree of penetration by the summer breezes while not enabling the wind to create large eddies in outdoor spaces (Figures 28-30). A process of evolution created a better social and environmental design solution, one that satisfied the needs for density but clearly made more sense as a place to live and grow.

Each design iteration was developed not only for wind effects but also for sun shadows using both physical models and digital modeling (Figure 31). With narrower garden-type spaces, it became imperative to understand the sun and shade patterns and to ensure the compliance with the winter sunlight code. The final solution was designed with buildings oriented southwest-to-northeast, enabling wind penetration while also allowing one façade of each dwelling unit to face south or southeast. See *Chapter 11, Case Study Two – Beijing Star Garden* for more information on how we designed each building to enhance the site planning ideas.

Integrated Urban Ecology Networks

A later project, called Hui Long Guan, was also a satellite community on the urban periphery of Beijing. Our site, called C02 (the name not being related to clean air or carbon dioxide), was one grid of a matrix comprising approximately one square kilometer of development housing sites. When built, this would become a new town of 40,000-50,000 people. As the land was previously an agricultural region, it seemed appropriate for the design to attempt to develop a broader ecological planning strategy for water and landscape use in addition to the climatic design principles described for other projects in Beijing.

The agricultural sites in the area had developed a network of ditches for irrigation because water is scarce in the region, especially during summer months. The problem of water and irrigation is expected to worsen as the area is converted to housing development. Not only would the landscape need more water for sustenance, but water would also be likely to run off into underground culverts and be pulled away from the site.

Therefore, we sought to develop a design and plan where water could be retained and enable the surrounding landscape to be more self-sustaining while at the same time providing an amenity for the community living there. A canal was proposed that would form an inner ring on the site and divide the site between a perimeter zone of smaller residential buildings and an inner core of courtyard blocks, thereby giving a distinction between areas within the project and a difference in scale (Figures 32 and 33). Also, by making the canal a landscape zone, it meant that practically all of the housing was close to water, giving shade and some potential for evaporative cooling in summer. We proposed that the canal could be managed in such a way that it would gather rainwater from the entire site and possibly adjacent highways, and through a process of slow filtration, it could be used to sustain the vegetation and plant growth.

Other parts of the planning strategy, including the pedestrian and road network, building blocks, and courtyards developed from the notion of the water ecology. While the site was perceived as a village in which

Figure 32 Perspective view of proposed site layout for Hui Long Guan C02

Figure 33 Hui Long Guan C02 site plan, with canal in dark grey (see color version in chapter 14, Figure 24)

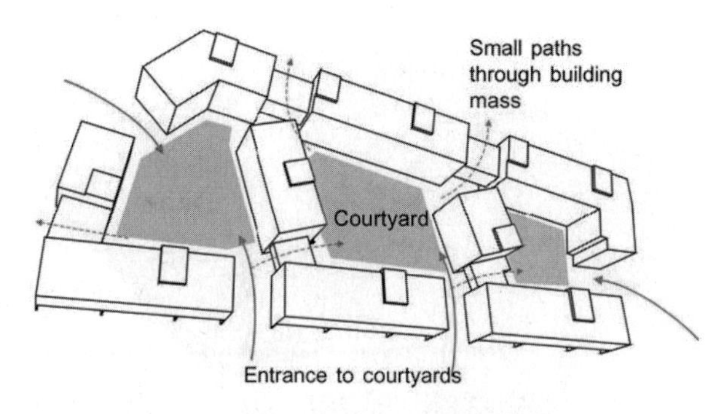

Figure 34 Hui Long Guan C02 courtyard diagram

Figure 35 Hui Long Guan C02 cluster landscape

there would be smaller scale workplaces such as studios, workshops, and offices, it was also broken down into a series of smaller housing clusters and public spaces around which the traditions of community interaction could thrive. Sustainability is an intersection of the environmental, cultural, and social dimensions of design, and we would like to think the site planning at Beijing Hui Long Guan tried to work in this way. See *Chapter 14, Case Study Five – Beijing Hui Long Guan* for more information.

BUILDING FORM AND TYPOLOGY

With most of our design projects in China, there is an intermediate design scale between the urban plan and that of the dwelling unit. It is the scale of the building block or "cluster," which mediates the ideas of the site plan to the operation, mechanics, and technology of the house. This scale of the building cluster or building typology embodies ideas about place, climate, microclimate, and community, and their intersection enriches the previously mentioned notion of sustainability being environmental, social, and cultural.

Urban Housing Clusters as a Response to Climate and Community

With respect to typologies, we have an interest in those building forms and shapes that bring together a series of ideas. The resulting form is the one that enables those ideas or parameters to enrich the project.

An example of this type of approach is the site-planning principles previously explained in the case of the inner clusters in the Hui Long Guan Project in Beijing. Three housing clusters or blocks make up the center of the project, and each is then further divided into three courtyard areas. Roads were maintained at the perimeter so that the inner courtyards formed part of an interconnected, landscaped pedestrian area, creating a community that has a strong sense of identity, climatic enclosure, and visual supervision.

In designing these areas, we wanted to ensure they worked according to the climatic criteria such as wind, ventilation, and solar orientation. The goal was that these courtyard areas be filled with sun in the winter to give solar warmth but also be protected from the harshness of Beijing's northerly winter. Therefore, within each cluster, all units faced south. This resulted in units on the northern edge being accessed from the avenues between the clusters, while those on the southern or western edges are accessed from the courtyards. The entrances to the courtyards would be gated to create gaps that enable breezes summer breezes through the blocks but when closed in winter would protect against cold blasts (Figures 34 and 35).

Diagrams also show how a cluster worked against a multiple of social factors – and how such ideas could be integrated through the building form (Figure 36). These ideas indicate where bikes and cars are parked so as not to be in conflict, where bike paths and walking paths could be constructed and separated from cars, how green elements such as roof gardens and courtyards could be distributed, and how trees are planted so as to shade parts of the elevations in summer. When articulated in this way, the intention of the design was to reach a more sophisticated and integrated understanding of how it can achieve sustainable environmental and social objectives.

A key design decision was to reduce the depth of each unit in order to enhance the effectiveness of natural ventilation in the summer and to allow daylighting to penetrate more of the building interior. This required a rethinking of the master plan to allow adequate space

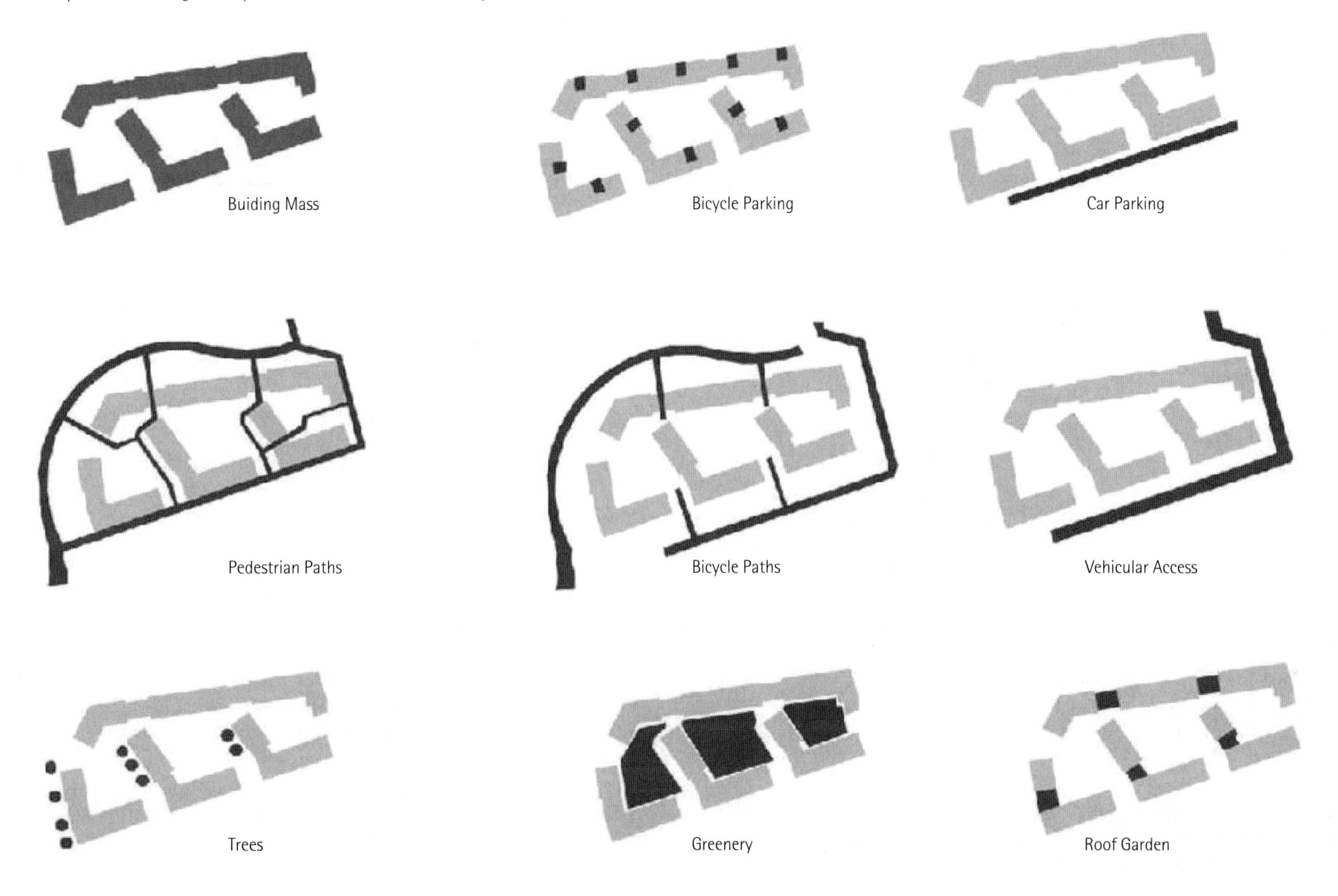

Figure 36 Hui Long Guan diagrams

Figure 37 Shenzhen Wonderland site plan; preliminary site layout for phase IV (designed by MIT) is shown circled

Figure 38 Sketch showing Shenzhen Wonderland shaded balconies

for winter sunlight and summer breezes to penetrate between individual buildings. See *Chapter 14, Case Study Five - Beijing Hui Long Guan* in of this book for more information.

In Shenzhen, the MIT team worked on the design of a housing area for a development named Wonderland. The master plan for the whole development established major roads and an axis that divided the site into two subareas, a northern part and a southern part, with an axial road and a public square in between (Figure 37).

Shenzhen, close to Hong Kong, is significantly further south than Beijing, and this affected our thinking about how the project should respond to the climatic variations through its form. The summer sun is overhead and very strong, while in spring and autumn, there can be significant solar heat gain from the lower sun angle. For this reason, we thought that the shape of buildings could create shade, and the space between buildings when planted with dense trees, would also create valuable shaded conditions away from the harshness of the sun (Figure 38). The prevailing winds were from the east, and therefore the project was conceived as being more porous on the east side and closed to the west in order to harness the passive effects of cooling, especially at night, through the open spaces and into the dwelling units.

On the eastern edge of the project, we designed a series of multi-family "villas" that would flank the road and create an edge. However, instead of being like row houses, they were aligned like stones, separated by gardens that extended into the project. This again enabled the breezes to filter into the courtyards and eventually into the housing units instead of being blocked by a solid line of housing. The effectiveness of such design ideas were tested through the use of CFD airflow simulations.

The two major courtyard blocks were primarily orientated north-south in order to reduce the effects of solar heat gain, and in turn, the necessity for air-conditioning. If the air-conditioning can be limited to the mid-summer usage, then significant gains can be made in respect to energy usage. On the south side of the courtyard blocks, the design team designed several alternate solutions for shading in both vertical and horizontal forms. Shading devices were perceived by the developers as being expensive and possibly difficult to manufacture, and therefore we looked to ways that they could be designed as an integral part of the façade construction. See *Chapter 13, Case Study Four – Shenzhen Wonderland* in this book for more information.

BUILDING DESIGN AND PLANNING

High-Rise Sustainable Building

The MIT design workshop began to look at the design of a medium-rise building typology as an alternative to the all-too-common high-rise building blocks that scatter the new urban landscape in China where there is little responsiveness to orientation, sustainable construction techniques, and sustainable community ideas or reduced energy use. A proposal came out of the workshop for a tall linear building in Beijing that had four underlying ideas, listed below (see also Figures 39 and 40).

- Make the depth of the building narrow enough to allow for natural ventilation through the south-facing façade. The circulation corridor was placed to the north, so a means had to be found to provide cross ventilation through the unit to the corridor and out through the north façade. This was achieved through a ventilation window at high level in the kitchen area so occupants were given control over the openings in the north façade. The units in this case were a maximum of approximately ten meters deep (Figure 39).
- Design each unit as a two-story "duplex." This had a double-height space to the south side for the major living areas, which, in turn, enabled deeper daylight penetration throughout the unit.
- Identify where sun was desirable for winter solar gain, and where it needed to be blocked out through external shading; study the building in relation to the changing sun angles and light penetration.
- Design the glazing on the south-facing "sun space" (typical to most Chinese housing units) to be movable, and relate it to the interior use such that it could also be an extension to the living space when needed. This "sun space" mediates between inside and outside and, while these are sometimes used as external laundry spaces, they are more often enclosed in immovable glass, which creates a solar furnace and excessive solar heat, especially in summer months. This results in yet more demand for air-conditioning. As a part of the workshop studies, we looked at how this space could be more environmentally adaptable and useful (Figure 40).

A later project for the design of a high-rise building scheme, the Beijing Star Garden Project, reexplored some of these themes. The agenda this time was to develop ideas about the social nature of a sustainable residential tower such that there would be places for social and cultural interaction as well as strategies for significantly lower energy use.

The plan developed as two clusters connected by a communal core. The core space would be three stories high (with mezzanines) and would house an elevator and various communal functions that a three-story section of the building would need, such as a lounge, laundry,

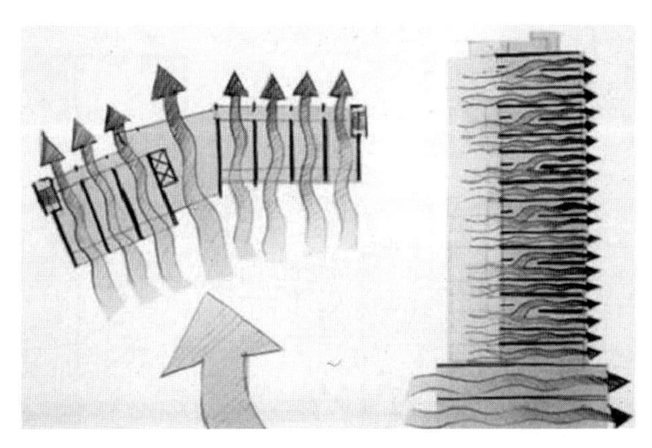

Figure 39 Beijing high-rise sketch of cross ventilation

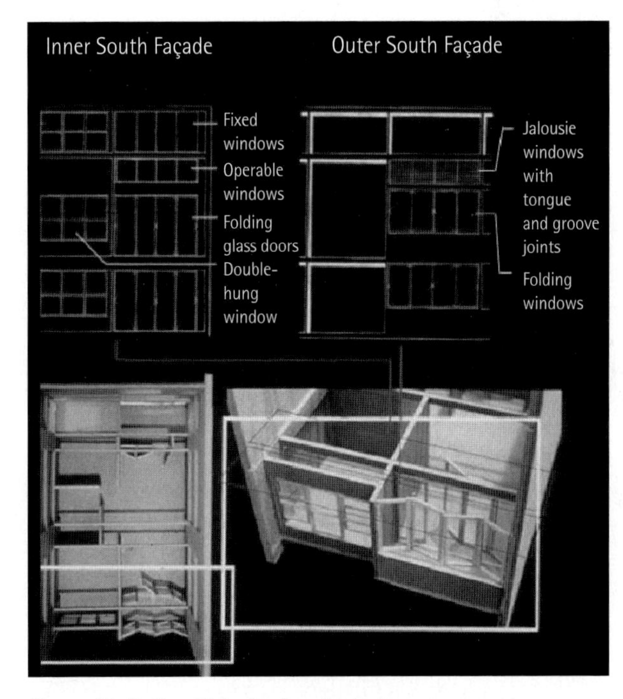

Figure 40 Beijing high-rise façade studies

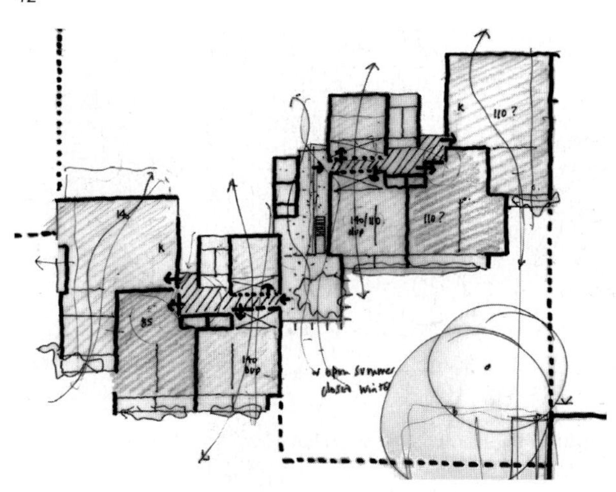

Figure 41 Beijing Star Garden organizational plan diagram (see color version in chapter 6, Figure 10)

Figure 42 Beijing Star Garden elevation study

workspace, child care, and shops (Figure 41). This strategy enabled the elevator to be a "skip-stop" type and stop at every third floor, thereby saving energy. From each core, one would access a common hallway to each of the three units to either side of the core. The shape of the plan then was "perforated," in that it was shaped to allow controlled breezes to blow through these common areas. The units themselves, which were one or two stories, were again designed for natural ventilation.

An added advantage of the plan shape was that it also tended to eliminate the high wind eddying, especially at the base, which was a major problem with the more "stump-like" tower form. The mixture of one-, two-, and three-story spaces also gave it a far more interesting urban scale and detail when viewed in elevation (Figure 42). See *Chapter 11, Case Study Two – Beijing Star Garden* for more information.

Low-Rise, Low-Energy Design

The 12 x 12 house typified the design principles we developed for low-rise building situated as part of a larger housing neighborhood. The house unit gained its name because it was a 12 meter x 12 meter building block that housed four residential units. As such, we were interested to also look into how such a small community of four families could find something to share, as one typically finds in traditional housing forms. Once again, several ideas intersected in developing the design strategies listed below.

- The courtyard space was envisioned as a communal place, an entry space, a potential garden, a place for shade or water, a play space as well as a device to enable better natural daylighting and ventilation of the housing units. It would be a space that mediates between the street and the inner sanctum of the house (Figure 43).
- The four units (two stories each, and two to each side) share a common access stairway. This stairway was also a ventilation chimney for stack ventilation for the units. Each unit would passively ventilate into the stairway and the stack effect would create a displacement airflow designed to reduce the dependence on air-conditioning (Figures 44 and 45).
- The south-facing street elevation is layered with a shading screen. The screen provides variable shade to the housing units to reduce solar gain, and also creates a frame for the growth of climbing plants in summer. This changes the relationship of the building block to the street and creates a more urban feel as well as a sense of arrival and entrance to the building (Figure 46).

See *Chapter 10, Case Study One – Beijing Prototype Housing* for more information.

SUMMARY

The design principles in this chapter were developed over time through several projects that explored similar sustainable design issues relevant for the design of new urban housing in China. It was through the projects and the medium of design that ideas and principles emerged, as opposed to mapping them out at the start in a predetermined manner. To achieve this, the design team defined the environmental objectives at the start of the project and in turn, worked with those strategies that became most viable and effective.

We also believed strongly in the notion of the design principles supporting local technologies and techniques as opposed to importing alien and possibly ineffective "high-tech" ideas. Sustainability in the context of Chinese housing should remain local, support communities, and educate the construction industry to learn from itself.

Design principles for the site or the local neighborhood have to be meshed and integrated with those at the scale of the housing block and of the house. They work together and create more effective solutions. However, there is a need to be innovative and to seek new ideas for making sustainability integrate cultural traditions, improve social infrastructure, incorporate better and appropriate technology, and emphasize a responsible use of resources.

REFERENCES

Behling, S. and S. Behling. 1996. *Solar Power - The Evolution of Solar Architecture.* Prestel, Munich and New York.

Brand, S. 1995. *How Buildings Learn.* Penguin Books, New York.

Intrachooto, S. 2002. *Technological Innovation in Architecture: Effective Practices for Energy Efficient Implementation.* Ph.D. thesis. MIT, Cambridge, MA.

Li, Y., ed. *Old Houses, Traditional Chinese Dwellings of Fujian 1* (in Chinese). Jiangsu Fine Arts Publishing House, Beijing.

Scott, A., ed. 1998. *Dimensions of Sustainability.* E and F Spon, London.

Stonehouse, R. 1998. Dwelling with the Environment: The Creation of Sustainable Buildings and Sustaining Situations through the Layering of Building Form and Detail, in *Dimensions of Sustainability* A. Scott, ed. pp. 126-131. E and F Spon, London.

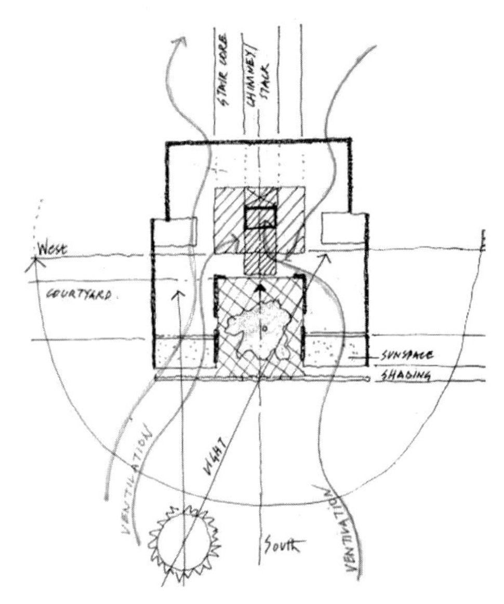

Figure 43 Beijing 12 x 12 house diagram plan showing ventilation and lighting concepts

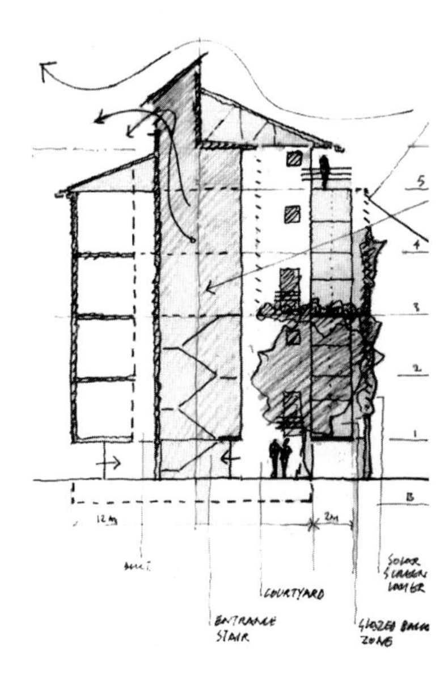

Figure 44 Beijing 12 x 12 house section diagram sketch

Figure 45 Beijing 12 x 12 house section model showing ventilated stair

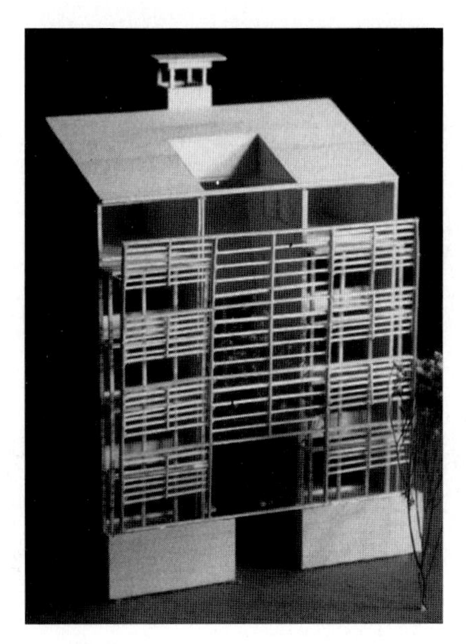

Figure 46 Beijing 12 x 12 house model showing south-facing shading screen

CHAPTER FOUR

MATERIALS AND CONSTRUCTION FOR LOW-ENERGY BUILDINGS IN CHINA

John Fernandez

INTRODUCTION

This chapter addresses strategies for responsible materials use and good construction practices for low-energy residential buildings in China. Achieving well-built, energy-conscious designs will be the result of a myriad of diverse efforts from design and engineering professionals. It will also require the commitment of the local, regional, and national authorities. Therefore, issues that range from materials extraction, land use, and construction materials industry standards to the proper detailing of the exterior envelope must all come together to contribute to the making of energy-efficient buildings.

The intended result of such efforts are that improvements in the quality of design and construction of this type of building, urban housing, will decrease the consumption side of the national energy equation in China. This reduction will aid in reducing the demand on nonrenewable energy sources and lowering carbon dioxide emissions. In light of the extraordinary economic expansion of contemporary China, now is the time to advocate and implement simple and inexpensive improvements in building materials and construction. Better building practices promise to assist China in reducing the dire projections of dramatically increasing energy consumption in the next twenty years. Better construction is a clear and inexpensive way of achieving better buildings. Not only may new energy-efficient

L. Glicksman and J. Lin (eds), Sustainable Urban Housing in China, 44-73

buildings contribute to a substantial lowering of the country's energy consumption, but these buildings may also have a pivotal role in providing a more humane environment for urban dwellers.

To begin outlining the best steps to take toward sustainable construction in China, it is important to describe the current state of affairs. Many building products in China are inexpensive and low-quality materials that underperform. The types of materials and assemblies available - concrete, brick, tile, windows, sealants, insulation, and so on - are of varying qualities. The methods used in the construction of multi-story residential buildings - low-tech using low-skilled labor - are extremely limited, and in some regions quite primitive. Too many residential development projects do not even approach the most common standards in industrialized Western countries. Luxury residential towers are not much better and are constructed with very little regard for practices typical of Western levels of quality, not only in terms of construction integrity and operational performance but also durability, comfort, and appearance. The highest end of the market, detached family homes for the very wealthy, are practically the only kind of residential building that receives a higher level of material quality and construction attention. Unfortunately, this building type is an inappropriate model for urban housing in China.

The reasons behind substandard materials and construction practices in China are complex but not inaccessible to careful scrutiny. The limited availability of a range of appropriate materials and a large, inexpensive and unskilled labor force make construction an unsophisticated activity. In addition, the rapid privatization of the housing market and the continuing migration of thousands from the countryside to the city have resulted in acute housing shortages. These factors have contributed to an extremely rapid and regrettable proliferation of hundreds of undifferentiated and low-performing multi-story residential buildings throughout urban China (World Bank 2001). The construction boom continues today and, even if mildly dampened, by 2015 a full half of the building stock, both residential and commercial, will be comprised of buildings built after the year 2000 (World Bank 2001). Without improvement to the construction quality and the overall building performance, these highly inefficient buildings will produce growing energy consumption demands of the country, with only limited and expensive options for refurbishment and upgrade potential in the future. In addition, the construction industry will continue to be an excessive draw on nonrenewable materials. The use of brick has already been restricted in and around Beijing as a result of soil degradation that accompanies the extraction of the already-limited supply of necessary clay.

In addition, energy-efficient buildings contribute to a reduction of emissions of carbon dioxide and other greenhouse gases and particulates. Therefore, the building industries hold the potential to reduce energy consumption and carbon dioxide emissions and increase the efficiency with which nonrenewable resources are consumed. The strategies that are most important are those that can be implemented sooner rather than later, with greater capacity to affect the growing residential building stock.

It is important to address the lifetime energy consumption of buildings from construction through demolition. It is generally agreed that the energy in operation of most buildings far exceeds the energy embodied within the physical elements and the energy required for construction of the building itself. Most estimates for Western buildings attribute 12–16 percent of the total life-cycle energy expenditure of the building to embodied energy in construction and 84–88 percent to the normal operation during its lifetime (Figure 1).

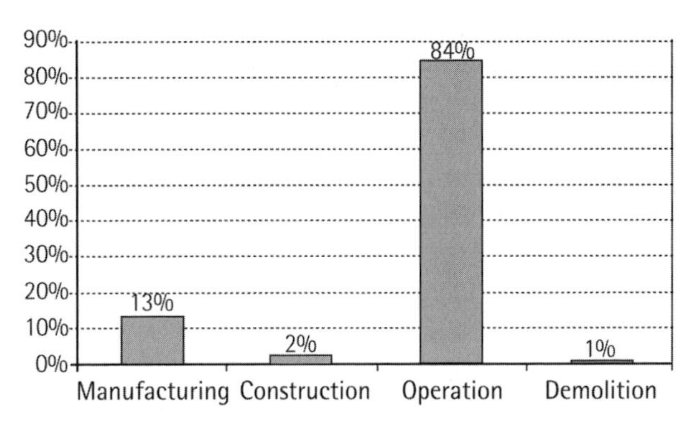

Figure 1 Energy consumption of various stages of a conventional building's life (Source: Thormark 2002)

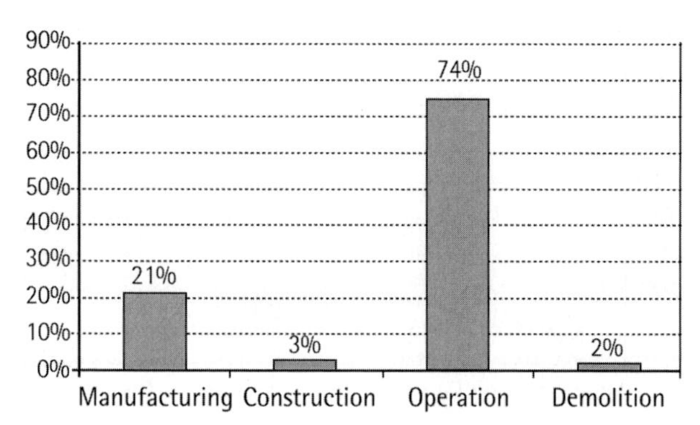

Figure 2 Energy consumption of various stages of an energy-efficient building (Sources: adapted from Keoleian, Blanchard and Pepe 2001 and Thormark 2002)

In many cases, building operating energy exceeds energy in construction after only about four to ten years (Cole and Rousseau 1991). Clearly, lowering lifetime energy consumption may be most easily achieved from improvements in the building's efficiency during operation. This can be accomplished partly through improved construction practices and higher-quality materials. Of course, these improvements will often require an increase in the initial construction cost of the building. However because of the substandard quality of so much construction in China, simple changes and improvements in construction quality will significantly improve energy efficiency during the lifetime of residential buildings. Estimates predict that a mere 5–10 percent added cost premium in construction would easily yield energy savings, in operation, of 50 percent or more. The addition of insulation, better windows, control of air infiltration, the widespread use of caulking and sealants, natural ventilation, and thermal-mass cooling can contribute to achieving this savings. A 50 percent reduction in operating energy would not change the fact that a majority of energy consumed during the lifetime of the building occurs during the operation of the building. The energy in operation would be reduced from 84 to 74 percent of the total (Figure 2). However, the absolute amount of energy would be reduced by 43 percent over the lifetime of the building: certainly a worthwhile goal!

These improvements, improved construction practices, and higher-quality materials will also increase the durability of these buildings by reducing the amount of unwanted moisture and air infiltration that invades the exterior envelope and attacks the materials of the wall and the structure of the building. Improving the durability of the building enclosure is the best way to ensure the fulfillment of the service life of many building components, including barrier materials, insulation, structure, window assemblies, plumbing, and electrical conduits. Simple steps taken today will result in a more durable building stock for many years to come. The investment in housing for increasing urban populations should take into account the continued need for these buildings for the future.

Therefore, a primary strategy for reducing energy consumption and carbon dioxide emissions, while delivering more durable buildings, is designing and constructing higher quality exterior envelopes. This strategy complements a suite of other inexpensive strategies for reducing consumption such as natural ventilation, thermal mass for cooling, optimal solar gain and shading balance, proper building orientation and others mentioned in *Part 3, Technical Findings* of this book. All of these strategies are inexpensive steps that can be taken today.

In addition to the conservation of energy, other important priorities have emerged as part of the dilemma of providing enough housing for the population, while doing so in an environmentally responsible way. Land use, pollution of air and water, minimization of waste streams and the efficient use of nonrenewable mineral resources have become critical issues in the development of China while providing the necessary amount of humane housing.

To properly balance the needs of Chinese society with responsible resource management, the Ministry of Land and Resources (MOLAR) has been given the charge of assessing and planning the existing and future material flows most likely to yield a sustainable future for all of China (Ziran 1999). The Ministry is faced with the challenge of redirecting the existing trend of mineral extraction and land use toward a responsible path. China has exploited its mineral resources ferociously while it has maintained an extremely low materials-recovery rate, lower than any Western industrialized nation and lower than many developing nations. Currently, open pit mining and the dumping of industrial wastes and tailings have scarred 15,000-20,000 square kilometers of land. This area is increasing at a rate of 200 square kilometers per year. Mining and other material extraction methods account for ten percent of the total industrial wastewater produced nationwide (Rongkang 1999).

Compounding the pressures from this history of extraction is the need for suitable land area for agriculture, growing industrial capacity, and housing for an increasing and ever-more mobile population. While China is a large country at 9.6 million square kilometers (slightly smaller than the United States), it is also one of the most densely populated regions in the world, with 1.3 billion citizens (more than four times the population of the United States). As a consequence, pressures on the use and quality of land are quite acute. Per capita reserves of cultivated land, mineral resources, and potable water are below global averages. Cultivated land is 43 percent, mineral resources 58 percent, and potable water 25 percent of per capita global averages. Between 1958 and 1986, land suitable for cultivation was reduced by erosion by forty million hectares. This trend has been lessened somewhat but continues at a rate of 130,000-200,000 hectares per year. In addition, 20 percent of cultivated land has been damaged by polluted water, gas, and waste slag from industrial processes and the abuse of pesticides. Currently, 38 percent of all territorial land nationwide is suffering from water loss and erosion (Ziran 1999).

Pressures to develop land and the opportunity costs of devoting urban space to greater areas of block housing have become critical resource issues. Land grab for real estate development, especially due to the privatization of the housing market, has led to worries that China will face difficult land use decisions for many years to come.

Currently, the most common urban housing model is the five- to six-story multi-family residential block set within a well-defined development parcel. As required by building codes, residential buildings with additional floors require elevators. However, high-rise residential buildings have been built in greater numbers around city centers. Beijing now provides for 24 percent of its total residential floor area with buildings that exceed ten stories (World Bank 2001). The traditional street, the *hutong*, a narrow street of shops and residential entries surrounded by low-scale buildings, has given way to wide boulevards to relieve traffic congestion and multi-story buildings to accommodate ever-growing numbers of new urban dwellers.

It is not clear that the current development rate and the predominant building typology for residential buildings is the most responsible path toward a sustainable solution for the new wave of residential construction. Alternative architectural typologies for housing and more progressive design proposals need to be more widely considered. Traditional housing typologies such as courtyard buildings, high-density and low-scale typologies, and others should be considered. Furthermore, regional planning initiatives need to contribute to a continuing study of the best ways to build needed housing.

The following sections describe the current state of affairs for construction and materials and propose a series of recommendations for improving the buildings that are needed to accommodate growing urban populations.

MATERIALS FOR IMPROVED BUILDINGS

For the most part, the materials used in Chinese construction are similar to primary building materials

Material	Energy Content, GJ/tonne
Aluminum	280
Plastics	85-180
Copper	140, rising to 300
Zinc	68
Steel	55
Glass	20
Cement	3-9.4
Brick	4
Timber	2.5-7
Gravel	0.2

Table 1 Approximate energy content of materials in the West (Sources: Ashby and Jones 2001 and Brocklesby 2000)

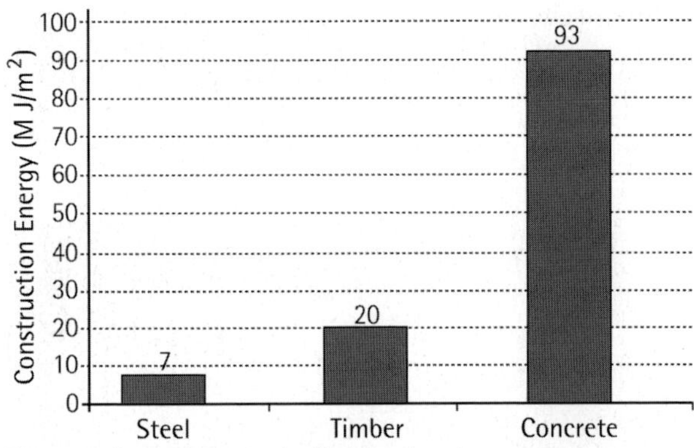

Figure 3 Approximate embodied construction energy for structural systems per unit of floor area (Source: Cole 1999)

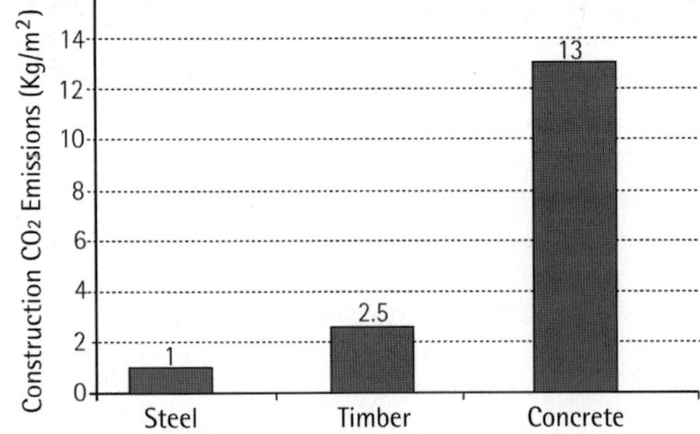

Figure 4 Approximate construction CO_2 emissions of steel, timber, and concrete per unit of floor area (Source: Cole 1999)

found around the world. The primary difference between China and the Western world is the availability in the latter of a large and more sophisticated range of materials for the design and construction of high performing assemblies and building systems. China currently lacks cost-effective material options for better privately developed multi-story residential buildings. With the recent transformation from state-owned housing to a private consumer market, competitive bidding for construction contracts is now becoming a typical process for realizing building designs. This has spawned a slowly emerging market for a greater variety of building materials, especially as cost-effective alternatives to the usual methods are made available. The link between better materials and higher-quality construction and lower energy consumption, while not difficult to understand, requires a sustained level of attention from the government. Currently, there is a general lack of incentives for the private developer to construct higher-performing and more durable buildings.

In addition, while regulations exist for the post-construction inspection of buildings, enforcement of these regulations needs to improve. Tenant alterations are common and often compromise the integrity of the construction, especially the exterior envelope. Regular inspections of housing blocks can vastly improve the lifetime performance of these buildings.

ASSEMBLIES FOR IMPROVED BUILDINGS

Assemblies in China suffer from primitive construction methods and a lack of premanufactured and automated assembly, enforcement of building standards and suitable regulated supervision, and quality control on the construction site. While there is a great deal of interest in improving methods and the results of construction, the typical Chinese residential building is rife with ineffective insulation (if insulation is even used!), air barrier and sealant discontinuities, wide-ranging dimensional tolerances, uncoordinated work from subcontractors, and substandard craftsmanship and material quality. The building systems that should be addressed as soon as possible are the superstructure and the exterior envelope.

The typical multi-story residential buildings are constructed of site-cast structural concrete frames and exterior walls, brick and concrete block exterior walls and partitions, prefabricated floor slabs and roof substrates along with low-quality, single-glazed or insulated windows and doors. Windows are now often insulated glass assemblies in steel (and sometimes vinyl) frames of extremely low quality. Insulation in the exterior wall is rarely used. Caulking and sealing around doors and windows is also often of very low quality, if included at all. Many buildings simply depend on a thick concrete structural wall as the primary thermal and mass resistance material. The quality of concrete varies widely and the inclusion of adequate reinforcing steel in reinforced concrete is an ever-present concern. This lower quality of concrete will undoubtedly bring durability issues to these buildings in the near future.

THE BUILDING STRUCTURE

The use of cement, the primary binding component in concrete and a material of relatively low embodied energy, would seem to be a good choice for responsible construction. Table 1 shows approximate energy contents of a number of commonly used construction materials. Cement possesses vastly lower energy content when compared to steel.

However, when the total energy in construction for various structural systems is tabulated and compared, concrete as a structural system contains a great deal more embodied energy and emits a great deal more carbon dioxide than timber or steel (Figures 3 and 4) (Cole 1999). Reasons for this increase can be found in the following section on concrete.

Figures 3 and 4 show that transportation of constituent materials, extraction energy, on-site equipment, timber formwork, and transportation of a large workforce necessary for casting concrete on site amount to a significantly higher embodied energy total in construction. However, in much of China, timber and good-quality steel are still scarce and therefore require unusual transportation and production energies.

Therein lies the first problem in assessing and reducing embodied energy consumption of typical construction practices. At the moment there is a critical lack of data regarding the embodied energy of basic construction materials in China. For the purpose of this chapter, figures for embodied energy are derived from a number of Western sources, unless otherwise noted.

As a result of the fragmented nature of the Chinese steel industry and lack of good-quality ore, there continues to be industrial limitations in providing enough steel for construction. Therefore, examining ways in which to improve the environmental qualities of reinforced concrete through responsible practices is an important component of sustainable buildings in China, at least for the foreseeable future.

Concrete

Reinforced concrete is the primary structural material in China, as it is in much of the world today (Figures 5 and 6). It is also a primary material in vast expanses of exterior wall construction. Its choice is dependent on a large and inexpensive labor force, readily available quality constituent materials, and a lack of reasonable alternative materials and product diversity for construction. It is also a material that does not require exacting dimensional tolerances. However, the performance results are rarely satisfactory, and the buildings themselves do not contribute to an enhanced sense of well-being or identity for the urban dweller seeking to make a humane home.

Several significant challenges must be addressed in the use of concrete. First, reductions in carbon dioxide emissions from the production of cement should be reduced. For every ton of cement produced, approximately 2,400 pounds of carbon dioxide are released. Carbon dioxide emissions from the production of concrete stem from two primary sources: the energy used in the production of cement and the chemical process of calcining limestone into cement ($CaCO_3 \rightarrow CaO + CO_2$). Each is a major contributor (Table 2).

These two processes contribute to making concrete a significant source of global carbon dioxide emissions. The levels of carbon dioxide emissions from concrete production have been investigated in several studies and, while conclusions do not correlate perfectly, it is generally agreed that a significant portion of global human-generated carbon dioxide emissions may be attributed to the production of concrete. While the actual energy content of cement is relatively low when compared to steel, the total amount of concrete used for construction amounts to a significant contribution to global carbon dioxide emissions. Figures on the order of seven percent of all total global carbon dioxide emissions have been attributed to concrete production (Glavind and Munch-Peterson 2000, Griffin 1987, Lashof 1995, Malhotra 1998, Mehta 1998, Wilson 1993).

In addition to the issues of energy and carbon dioxide, air pollution due to particulate matter released into the atmosphere during concrete processing should also be addressed. It is estimated that 360 pounds of particulate

Figures 5 and 6 Concrete, cast, and block (concrete masonry units), respectively

CO_2 Emissions from	Lbs CO_2 per Ton of Cement	Percent of Total CO_2
Energy use in construction	1,410	60
Calcining	997	40
Total	2,407	100

Table 2 Sources of CO_2 in the production of cement and concrete (Sources: Griffin 1999 and Wilson 1993)

matter are released for every ton of concrete produced. Many concrete production plants in China are not well located and contribute directly to the air pollution of large segments of the Chinese population. The locations of these plants need to be reconsidered before growth swells and engulfs them into the urban fabric.

Furthermore, the construction of concrete frames should be carefully regulated to ensure the use of adequate reinforcing steel. The right amount of steel will substantially reduce catastrophic failure during earthquakes. The supervision of the concrete mix on site will improve the durability of the concrete (Schiessel 1996). Proper quality control is critically important because much of the concrete produced in China still does not meet international standards (Rousseau and Chen 2001).

Despite these issues, concrete will undoubtedly continue to be the primary structural material in much of China because of the availability of low-cost labor and constituent materials and the lack of high-quality alternative structural materials. Therefore, in order to improve the environmental responsibility of concrete buildings, several strategies should be considered, including a reduction of the amount of cement used by substituting alternative matrix materials.

An improvement in the matrix materials of the concrete mix is accomplished by specifying materials that can contribute to a reduction in the amount of cement used while employing materials that would otherwise be sent to the landfill. The most promising use of alternative materials is the use of fine particulates such as pozzolans in the concrete mix. A pozzolan is a particulate admixture, usually a siliceous or siliceous and aluminous material that contributes to the binding of the concrete by chemically reacting with calcium hydroxide to form compounds possessing cementitious properties (Malhotra and Pehta 1996).

A particularly good candidate for use in sustainable buildings is fly ash, a by-product of coal-fired power plants. Used as a pozzolan in the concrete mixture (Bilodeau and Malhotra 2000), fly ash can easily substitute for 15–35 percent of the cement content of concrete mixes without a reduction in strength. Some formulations may substitute close to 60 percent of the cement content with the use of fly ash. This can amount to a reduction of 50 percent of the embodied energy of the concrete. In addition, it has been shown that fly ash can significantly contribute to a structure's durability by reducing corrosive effects on the reinforcing steel (Bijen 1996). Fly ash is already being used in China, though in rare circumstances, to reduce the cement content of the final concrete mix, thereby simultaneously lowering the embodied energy and minimizing the problem of waste material disposal. However, because of the widespread use of coal as a primary energy source and the use of concrete as the primary structural material for residential buildings in China, a much wider use of fly ash can be exploited.

In addition to fly ash, other industrial and agricultural waste materials may be used as pozzolan or aggregate materials in the concrete mix, including recycled concrete, ground granulated blast furnace slag, condensed silica fume, rice hull (or husk) ash, cinders, and other inert materials. However, careful consideration should be taken regarding the binding properties of alternative pozzolans and aggregate and the necessary cement content for the required strength of the concrete (Costabile 2001). The general benefits to the use of alternative matrix materials are maximized when the substitute material lowers the energy-intensive content of cement while utilizing a material that is considered an industrial or agricultural waste product.

In addition, another opportunity for the improvement of concrete as a building material is the use of alternative fuels, including hazardous wastes, for operating the cement kilns (Wilson 1993). Large amounts of energy are consumed during the process of operating the rotary cement kilns necessary for producing cement for concrete. Approximately 90 percent of the energy required to produce a viable concrete is consumed during the process of producing cement in the kiln. As a result of the very high temperatures used in cement kilns (1480°C, or 2700°F), hazardous materials may be burned relatively safely because the high temperatures result in very nearly complete combustion. Waste incinerators operate at much lower temperatures and cannot fully combust many hazardous substances. Used motor oil, tires, spent solvents, and other materials can supply greater proportions of the fuel needed by cement plants. For example, pound for pound, tires contain a higher fuel content than coal itself (Wilson 1993). An integrated industrial waste stream needs to be organized at a national level to take best advantage of and develop the necessary economy of scale to make this process viable.

In addition, most Chinese cement producers use the more energy-intensive wet-process kiln as opposed to the dry-process kiln now common in Western industrialized countries. An important improvement can be made to reduce the environmental impact of making concrete by using higher efficiency dry-process kilns that use less energy and burn alternative and hazardous materials for fuel. The use of a dry kiln can reduce energy consumption by 50 percent compared to a wet kiln.

These improvements can be implemented immediately and will have the effect of reducing the environmental impact of concrete buildings. However, not all buildings in China will continue to be made of concrete. The Chinese steel industry is poised to consolidate into a manufacturer of high-quality steel products for the building industry. Considering the impacts that this may have on the environment, supporting this change should be a priority for the government and building professionals alike.

However, the tradeoffs between steel and concrete should be continuously monitored. In a fast-moving economy like China's, opportunities for reductions in construction energies may appear rapidly, making prior assumptions invalid. Constant assessments of the positive contributions of steel and concrete should be made by authorities in a position to affect the construction industry.

Steel

Since 1996, China has been the leading producer of crude steel in the world. In 2004, China produced 272.5 million metric tons of steel – an increase of 23.2 percent over 2003 – and more than double that of Japan, the second-largest producer at 112.7 million metric tons (Fewtrell 2005). In 2004, global steel production rose by 4.5 percent excluding China, and 8.8 percent including China. Since 1980, gross output in the Chinese steel industry has been growing at an average rate of 11.8 percent per annum (Wu 2000). However, per capita consumption of steel is still significantly lower than other major steel producers, and the use of steel in buildings, especially residential, is minimal.

As Table 3 shows, China's potential for expansion into its own market is substantial, compared with the intensity of use of steel in many parts of the world. However, obstacles to this trend include the fact that China's domestic iron ore reserves are generally of poor quality. The steel industry requires consistent shipments of large volumes of iron ore from beyond its borders. Feeding the growing housing market with efficient and lightweight steel superstructures will require major upgrades to the capacity to produce high-quality structural steel (Figures 7 and 8). Also, standards for steel in China are still based on the outmoded Russian model.

Country	Apparent Consumption of Crude Steel Per Capita
China	229.7
Brazil	113.3
U.K.	252.3
Australia	444.0
Germany	469.3
U.S.	418.8
South Korea	985.1
Japan	629.4

Table 3 Per capita consumption (kg/person) of crude steel in selected countries in 2004 (Source: IISI 2005)

Figures 7 and 8 Structural steel construction

Since its accession to the World Trade Organization (WTO), China has been struggling to meet the requirements of the Technical Barriers to Trade Agreement. Not until a nationwide and mandatory set of standards are enforced will the steel industry be able to strengthen its place in the international market and balance its product lines to include value-added high-tech products that are much in demand domestically (Yan 2002).

Referring back to Figures 3 and 4, steel in construction consumes less energy and emits less carbon dioxide than concrete. The products necessary for steel building frames are higher-quality standard sections that the Chinese steel industry is not producing in large quantities at affordable prices. Part of the reason for this market imbalance comes from the fact that the Chinese steel industry is a fragmented, widely dispersed set of enterprises with 1,570 iron and steel plants located throughout China's 31 provinces. Mergers between companies have been encouraged and continue to occur, a sure sign that the industry will become more efficient (Wu 2000).

Once the maturation of the steel industry occurs, not only will building structures improve, but building enclosure systems will also benefit. Steel is an ideal material in which to produce the refined frames necessary for high-performance exterior envelope systems. The opportunities for enhancing the integrity of the exterior envelope can be promoted through higher tolerances and better quality control with premanufactured modular exterior envelope systems that may be attached to a lighter frame of steel. These systems will bring sophisticated and efficient assemblies to the construction of exterior envelopes for residential buildings. In addition, it is easier to construct continuous air and water vapor barrier systems on a steel frame than on the lower-quality surface of site-cast concrete.

Also, recycling scrap metal is a productive way in which to take advantage of the ease of reprocessing aluminum, steel, and other metals. Using high-quality metals, designing assemblies in which these metals can be separated, and creating economic incentives for the recycling of the material will dampen the pressures on extraction and importation of virgin ores (Tilton 1999).

Timber

Abundant still in Europe and the primary single-family home building material in the United States, timber is rarely used in multi-story construction in China. Though timber can be a fire-rated material, light wood-frame construction raises issues of combustion that are more difficult to manage. In addition, there is greater risk in durability due to the deteriorating effects of water. Therefore, timber will most likely not be a primary material in the superstructure of buildings in China.

However, the use of wood in the form of engineered wood products has good potential, especially with respect to the growing modular housing industry (Brown 2001). Engineered wood uses timbermill waste products such as shavings, sawdust, and chips within a binding matrix. In addition, used formwork can be reclaimed and pulverized and used in another generation of composite products. Products include oriented strand board, particleboard, laminated wood, and other composite materials of good dimensional stability and durability (Figures 9 and 10). Plywood is a readily available product in China. However, formaldehyde is still widely used in China. A proven cause of unhealthy indoor air quality, formaldehyde should be carefully avoided in the making of engineered wood products.

THE BUILDING ENCLOSURE

The exterior envelope is the building system responsible for mediating between the controlled interior environment of the building and the fluctuating conditions of the outdoors (Figures 11 and 12). The assembly of the exterior envelope is specifically designed and built to control:

- heat flow;
- air flow;
- water vapor flow;
- the penetration of rain and other liquid water sources;
- light, solar, and other radiation; and
- interior acoustic environment.

The exterior envelope also needs to be economical, durable, and aesthetically appropriate. While these issues may seem obvious, they point to the need for architects to develop designs that seek to use technology not as a fix to problems, but as the medium through which humane built environments are conceived. Also, the issues of durability and economy remind one of the need to be responsible regarding the resources used in the realization of any building. The durability of an exterior envelope will be a primary determinant of whether the building serves society well or becomes a burden on future resources, both energy and material.

The exterior envelope should also be fire resistant, resist and transfer wind loads and other structural forces, be durable within an appropriate lifetime, and be aesthetically pleasing. Of course, all of these requirements need to be fulfilled within the economic constraints of the building.

It is estimated that new Chinese residential buildings consume between 50 and 100 percent more energy for space heating per unit area than buildings in similar

Figures 9 and 10 Engineered lumber

Figures 11 and 12 The exterior envelope

Location	Regulating Code	Roof	Exterior Envelope	Window
Beijing, China	Conventional design code	1.26	1.70	6.40
	1986 Energy-efficient code	0.91	1.28	6.40
	1995 Energy-efficient code	0.80 or 0.60 [1]	1.16 or 0.82 [2]	4.00
Russian region with comparable heating degree days		0.57	0.77	2.75
U.S. region with comparable heating degree days		0.19	0.45 or 0.32 [3]	2.04
Canadian region with comparable heating degree days		0.40 [4] or 0.23 [5]	0.38	2.86

1. The larger value refers to buildings with shape coefficient smaller than 0.3, while the smaller value refers to buildings with shape coefficient larger than 0.3. Shape coefficient is defined as the ratio between the outer surface area (four exterior walls plus the roof) of a building and the volume of the building
2. External insulation
3. Internal insulation
4. Refers to non-flammable roofs
5. Flammable materials

Table 4 Comparison of building envelope thermal standards (W/m²K) (Source: World Bank 2001)

climates in the United States and Europe. The primary cause of the difference between the consumption of energy by Chinese buildings and those in the West is the quality of the exterior envelope assembly (Table 4). Heat loss through air infiltration and improper or nonexistent insulation of the exterior envelope is the primary cause of poor performance of these buildings. Loss through the wall is three to five times higher in China than in Canadian and other northern-climate countries (World Bank 2001).

Inadequate thermal resistance materials and inconsistent details designed to limit air infiltration, along with substandard construction and low-quality doors and windows, are the primary issues facing the making of good exterior envelopes. Many of these issues can be addressed through simple improvements in the construction practices for exterior envelopes.

Wall Construction

Along with properly specifying thermal resistance and mass barrier materials, the proper construction and inspection of exterior walls is the key link in the process of producing energy-efficient buildings. Currently, China does not properly inspect new construction. In recent years, the Ministry of Construction has been struggling to contend with the rush of residential building development by speculators interested in capitalizing on the young and growing market. Specifications relating to proper construction and best practice, both prescriptive and performance based, have been written and adopted. However, enforcement of these standards is still very much a challenge that threatens to frustrate the aspirations for the construction of low-energy buildings. The reasons for this difficulty reside in the political, economic, and cultural status quo and pose what may be the Achilles heel of delivering low-energy buildings within the next few years (Glicksman, Norford and Greden 2001)

And while the challenges of enforcement and inspection are difficult problems, the issues of proper wall construction are not. The most elemental construction practices for improved buildings are well known and simply need to be promoted and confirmed in new construction. They may be most concisely summarized as the principle of the continuity of barrier materials for all areas of the exterior envelope. Whether it is thermal resistance materials, the air barrier system, the water management system, the superstructural frame, or other assembly, the designer needs to confirm that the system does not contain any systemic discontinuities that would contribute to unwanted mass and thermal transfer. For example, insulation materials should be continuous from the foundation up through the entire surface of the exterior wall, following all geometric features (corners, soffits, curvilinear forms, etc.), up to and continuous with window and door frames, passing from the vertical wall into the sloped and horizontal roof insulation material. It should be sealed and caulked around all openings in any part of the above-grade wall, foundation wall, slab, or roof. This simple principle should be applied to the air and vapor barrier materials as well.

In addition to the principle of continuity, there are several specific and simple rules that every designer of an exterior envelope would do well to remember and apply (Brand 1991). They can be summarized as follows:

- enclose the building in a continuous air barrier;
- provide continuous support for the air barrier against wind loads;
- ensure that the air barrier is flexible at joints where movement may occur;

- provide continuous insulation to keep the air barrier warm and to conserve energy in the building;
- keep the insulation tight to the air barrier;
- protect the insulation with a rainscreen and/or sunscreen supported out from the structure in a way that does not penetrate the insulation with excessive thermal bridges;
- provide enough open space for drainage and construction clearances between the rainscreen and the insulation; and
- drain the wall cavity to the outside and away from the building.

Locations of most common discontinuities are shown in Figure 13. The diagram shows infiltration: (1) through the window seal at the glass edge and the joints between vertical and horizontal window frame members; (2) at the intersection between the window unit frame and the rough opening of the wall; (3) under and around the sill, head, and jams; and (4) through the primary exterior envelope wall material.

Heat gain from roof materials can be easily minimized through the use of light-colored systems. An important and often-overlooked detail is the connection between barrier systems (thermal resistance and mass) in the roof and the corresponding materials in the walls. Discontinuities between these systems can amount to a potentially critical systemic breakdown leading to much higher energy consumption, primarily through air leakage. This kind of discontinuity also leads to durability issues in the exterior envelope.

Typical discontinuities in Chinese residential buildings also include drilled holes through the concrete exterior wall to accommodate the retrofitting of air-conditioning piping and other services (Figure 14). Also, major discontinuities often exist between the window and door frames, and the exterior wall material and insulation.

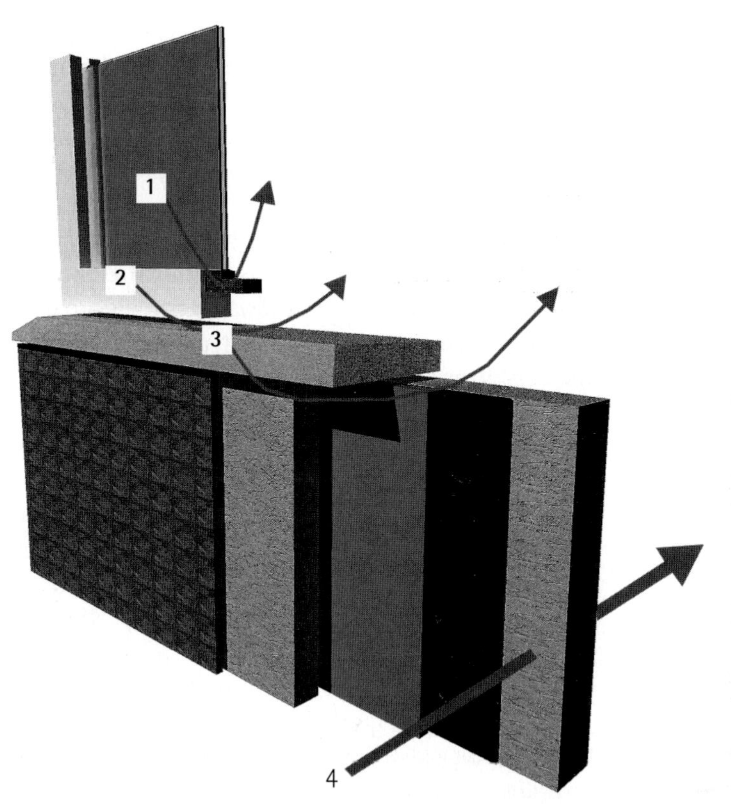

Figure 13 Air infiltration problem areas - the diagram shows infiltration: (1) through the window seal at the glass edge and the joints between vertical and horizontal window frame members; (2) at the intersection between the window unit frame and the rough opening of the wall; (3) under and around the sill, head, and jams; and (4) through the primary exterior envelope wall material

Figure 14 Image of residential building with unsealed through-wall piping from air-conditioning units

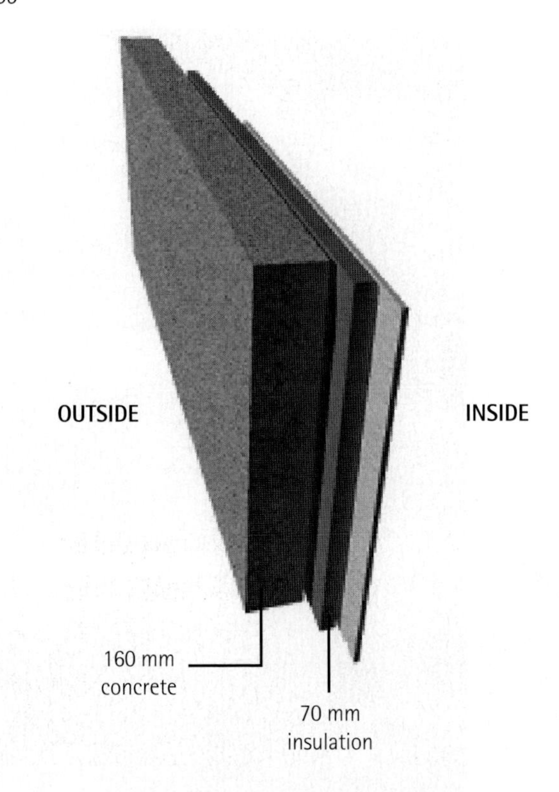

Figure 15 An example of a typical insulated wall assembly in China

Inconsistencies in the quality of construction also contribute systemic discontinuities. Partially grouted mortar beds for exterior block walls allow air infiltration and trap unwanted moisture. Substandard window frames contain thermal bridges and are improperly sealed and caulked at the factory.

Control of the interior environment is substantially assisted by the barrier systems within the exterior envelope. Insulation and air and vapor barriers are used to retard the movement of heat, air, and water through the outside wall.

Insulation Materials

The primary material system for resisting heat flow through the exterior envelope is insulation. All materials in the exterior envelope contribute to the thermal resistance of the wall. However, in walls with good insulation, the great majority of its overall thermal resistance (75–90 percent) is due to this insulating material. For example, the thermal resistance of a good insulator, such as a glass-fiber batt blanket, is 40 times that of a typical concrete. In China, the first step in providing adequate thermal resistance still needs to be taken; that is, insulation needs to be used as a standard material for enhancing the thermal resistance of the exterior envelope.

Therefore, the purpose of insulation is providing an appropriate level of thermal resistance. Different materials provide various levels of thermal resistance. Insulation should never be used as a vapor barrier or air barrier. In cases in which it is used for these purposes, improper detailing will lead to condensation and insufficient ventilation of the wall cavity. As a result, the thermal resistance of the insulation will be critically compromised. Insulation products most appropriate to construction in China are rigid insulated boards that act as the mechanical substrate for building papers and interior finish materials. These boards are also easily used in exterior applications, but the interior application is a more common approach in Chinese concrete construction. Currently, a typical simple exterior envelope assembly in China is composed of 160 millimeters of reinforced concrete exterior wall and 70 millimeters of rigid insulation board on the interior face (Figure 15). This configuration is especially useful as a system that the individual unit owner may purchase to augment the thermal resistance and air barrier properties of the residential unit, and these boards have good potential in China. Various products exist in the United States and European markets for use in large commercial and residential buildings.

Discontinuities are often found at geometric and system boundaries: at a corner, between the roof and walls, between the foundation and the wall, around windows, at soffits, and many other material or assembly junctions. At these points, the barrier material must often be overlapped and fastened such that the performance is not compromised. A variety of materials and methods exist for doing this.

In addition, thermal bridges may also contribute to the overall heat loss through the exterior envelope. These bridges are of many types, including continuous structural elements penetrating through the exterior envelope assembly, localized fasteners, shelf angles, and improperly designed window frames. While heat loss is an important consequence of thermal bridging, often the more critical issue is condensation on the bridges themselves due to leakage of humid air through the exterior envelope. This accumulation of moisture on thermal bridging elements contributes to durability problems inside the exterior wall assembly. Proper design eliminates thermal bridging as much as possible, with particular emphasis on any bridging that may lead to the accumulation of moisture within the exterior envelope. Window frames should always be properly designed with a continuous thermal break around the entire frame. Designers should confirm the design and manufacture of proper thermal breaks in windows they specify. They should also confirm these breaks as windows are delivered to the site and installed.

Air Barriers

Air leakage is the most important problem to address in under-performing exterior envelopes. Air leakage is a major contributor to both heat loss and water vapor infiltration, thus increasing energy consumption and decreasing building durability. Consider the fact that concrete block, one of the most commonly used exterior wall materials in China, leaks air at the rate of 3.4 liters/s/m^2 under a pressure differential of 25 Pa. In a room measuring 3 m x 3 m x 3 m, the air would be changed in a mere 15 minutes (Brand 1991)!

Two conditions need to exist for air leakage to occur. First, there must be a pressure differential; second, there must be a path from a volume of high-pressure air to a volume of low pressure. The primary mechanisms that produce pressure differentials and drive air leakage are wind, the stack effect, and mechanical equipment (HVAC) pressures. These three account for the majority of conditions that lead to a pressure differential between the outside and the inside. In many situations, the differential is a complex combination of two or more of these mechanisms. This complexity leads to variable pressures during the day and night in a variety of locations around the building (Lyberg 1997).

The existence of a path from high pressure to low pressure can be the result of the porosity of the material, construction gaps, lack of sealants, cracks from material contraction or the movement of the building and its components, expansion and structural settlement, and damage to seals and closure mechanisms over time.

Minimizing air infiltration is accomplished by focusing on those areas in a building enclosure that contain primary material or system joints. Figure 16 shows the three primary locations for air infiltration: (1) at any major building joint, for example, between the structural slab and the exterior wall, at corners, and at the roof/ exterior wall interface; (2) through the primary wall material itself; (3) at any exterior envelope discontinuity, especially at doors and windows; and (4) through the window or door assembly itself.

Achieving air tightness is accomplished through simple methods. For concrete masonry, parging is the easiest and best way to insure a well-sealed wall. Parging is the application of a layer of cement-based plaster to the entire surface of the masonry wall. Parging will reduce air infiltration through a block wall by a factor of 100. For other wall types, modified asphalt membranes are the most effective air barrier materials. Two types, self-adhering and torch applied, are commonly used in the West. Asphalt membranes are useful products not only for their air barrier properties, but also their flexibility and ability to travel over construction joints and around difficult geometric and construction intersections. On concrete frame buildings using block infill, these membranes not only serve as the primary air barrier but also accomplish the task of providing an effective vapor-resistant layer.

In addition to these systems, generally applied close to the exterior layer of the wall construction, a simple application of paint on the interior surfaces of the unit will do much to close any remaining paths that allow air leakage.

And yet, even when the exterior envelope is adequately detailed and constructed, the current arrangement of air-conditioning equipment for individual residential units often requires that pipes be run from the exterior, where the individual compressors

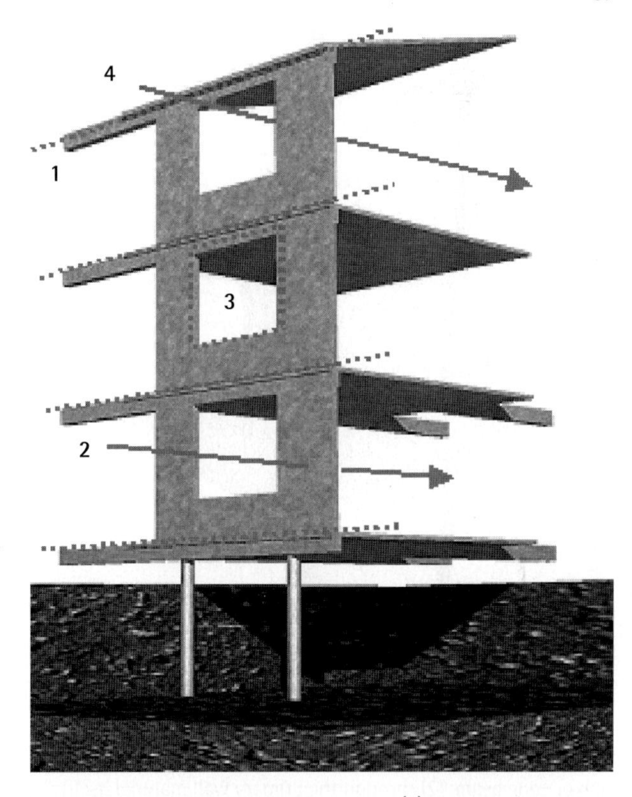

Figure 16 Air infiltration problem areas: (1) between the structural slab and the exterior wall, at corners, and at the roof/ exterior wall interface; (2) through the primary wall material itself; (3) at any exterior envelope discontinuity, especially at doors and windows; and (4) through the window or door assembly itself

Figure 17 Moisture infiltration problem areas: (1) the joint between major systems and materials, especially at the intersection between the top of concrete masonry walls and the slab or edge beam; (2) through the primary wall material itself; and (3) at the corners of windows and doors where the intersection between the vertical and horizontal frame members is never perfect and invites the transfer of air and water between the inside and outside

Figure 18 A proper seal at a material joint

are attached to the building or set on a balcony, to the interior. As mentioned earlier, often these conduits have not been insulated and the space around them through the wall has not been properly caulked and sealed. These retrofits lead to significant continuous paths for heat and mass transfer. Any retrofits need to follow the same principles of eliminating continuous paths for air and water.

For older buildings, air infiltration has also proven to be the single most important challenge in achieving significant energy efficiency. Applying barrier materials to these buildings is a much more difficult challenge. The problems of adding a barrier material into an older exterior envelope usually consist of durability issues, detail inconsistencies, and many other unanticipated material and construction problems. In these buildings, the retrofit of insulating and air sealing interior systems is a more reasonable alternative. Applying an insulation board and vapor and air barrier materials to the interior surface of the exterior envelope is often a relatively easy process. Convincing tenants to give up the few centimeters of floor area to add such a system is the real challenge. A compromise can be reached by simply painting the interior surface of the exterior wall. If nothing else, a painted surface will do well to improve the air and vapor barrier properties of the exterior wall but will not improve the thermal resistance of the assembly.

Moisture Barriers

Moisture control is targeted at two very different forms of water intrusion. First, there is the control of moisture that enters in the form of water vapor. Second, there is the need to control the entry of liquid water from driving rain or other sources. Controlling the entry of water and managing any that has entered is the primary way in which to diminish the failure risk of elements within the exterior envelope assembly and to add durability and lifetime to building products.

As with air infiltration, water infiltration occurs when there is a continuous path between the exterior and interior bounded by a pressure differential. Air passage through the exterior envelope brings in enormous amounts of moisture. The leakage rate through a crack of 3 millimeters x 300 millimeters at a pressure differential of 75 Pa is 10 liters/second. If the relative humidity were 50 percent, this air infiltration would bring 3.8 liters of water into the building in a 12-hour period.

Figure 17 shows the problem areas to focus on in minimizing moisture infiltration. These include: (1) the joint between major systems and materials, especially at the intersection between the top of concrete masonry walls and the slab or edge beam; (2) through the primary wall material itself; and (3) at the corners of windows and doors where the intersection between the vertical and horizontal frame members is never perfect and invites the transfer of air and water between the inside and outside.

The seal between materials should always follow best practice for an effective and durable seal over the designated lifetime of the seal. Too often, building moisture problems can be traced directly back to improper application of sealant materials or the improper design of the most basic weather seal. Most seals on an exterior envelope should be composed of two elements: (1) a sealant, the material that directly resists the exterior environment, and (2) the backer rod, a material that acts as the surface of adhesion, along with the adjacent two edges of the exterior envelope material (Figure 18).

Once the exterior seal is established, water vapor infiltration may be properly addressed. Water vapor infiltration is controlled through two distinct methods. First, porous materials allow for vapor diffusion. It is important to use materials that have a low porosity and therefore a high resistance to the passage of water vapor from one side to the other.

Second, as mentioned earlier, air infiltration brings in moisture as well. Clearly, the minimization of air infiltration significantly reduces the amount of moisture that enters the building and the exterior envelope assembly. Air infiltration is by far the more problematic transport mechanism of the two. Air leakage can typically move up to 100 times more moisture into an envelope assembly than would otherwise occur by diffusion alone (CIBSE 2002). See the previous section on air barriers for suggestions for controlling this type of air movement.

Vapor barriers placed on buildings in cold climates should be on the warm side of the insulation layer, generally toward the interior. In hot climates in which air-conditioning may be used, the vapor barrier should be placed on the outside of the insulation, closer to the exterior surface of the building.

In addition to the minimization of the passage of water vapor, it is also important to design a strategy that provides a barrier to liquid water. Resistance of liquid water from rain and other sources is accomplished through a two-stage water management design strategy. Clearly, water vapor resistance becomes inconsequential if the exterior envelope is not able to resist liquid water leaking into the building from a driving rain or other outside source. The forces that move liquid water into the exterior envelope are:

- gravity;
- capillary action;
- kinetic energy; and
- air pressure differentials.

A two-stage barrier solution addresses all of these forces. The first stage is a material barrier that acts as a screen on the building. This screen is the first line of defense, absorbing the kinetic energy of driving rain and directing a majority of it down and away from the building. Gravity washes liquid water down the face of this screen and proper detailing keeps it from entering further into the exterior envelope. The screen can be made of any one of many materials, including brick, concrete, and metal panels. The screen is not watertight and, as a result, some air and water may travel through the joints into the air cavity behind the panels. This is an acceptable phenomenon and one that is addressed with the second line of defense.

The air cavity contained between the rain screen and the next layer of the envelope is the pressure equalization chamber. This cavity serves to dissipate localized pressure on the building façade from the wind and allow any liquid water to drip down and out the rain screen.

The second stage of the water barrier begins at the interior surface of this air cavity. This layer is responsible for defending against further water penetration into the exterior envelope. The layer is protected from buffeting by wind gusts and the deteriorating effects of ultraviolet radiation. Since this layer can never be a perfectly waterproofed surface, it acts in concert with an air barrier and vapor barrier within subsequent layers of the exterior envelope. It is important to remember that this kind of construction does not waterproof the building but manages the movement of water into and out of the exterior envelope. Attempting to waterproof a single constructed surface will inevitably lead to failure. This two-stage design for water management is a proven method for preventing liquid water from entering into the exterior envelope and the building itself.

Windows

In concert with the construction of the continuous, opaque sections of the exterior envelope, windows are the most important assembly for mitigating the exterior

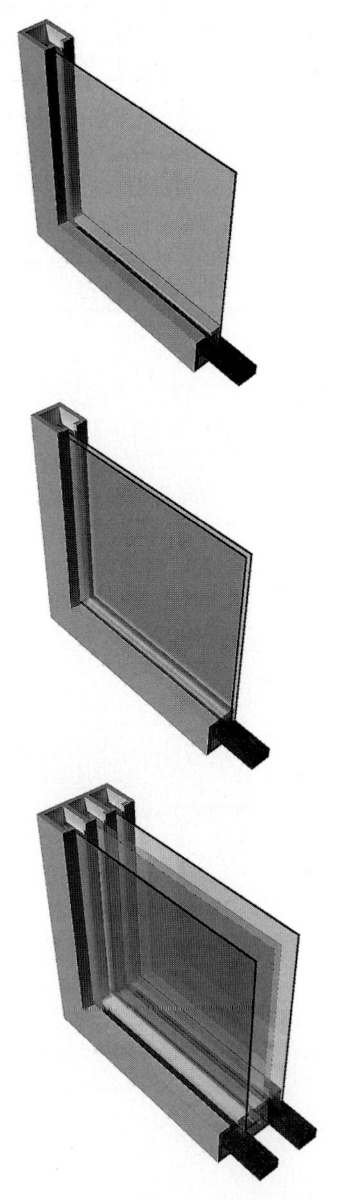

Figure 19 Window types: (top) single-glazing, (middle) insulating double-glazing, and (bottom) triple-glazing (film interior layer)

environment, allowing daylight and fresh air into the building and creating humane and pleasing spaces for the residents. In the exterior envelope, the windows are the best means to interact with the exterior environment in a variety of productive ways. At the same time, windows are the weakest link in the control of thermal and mass transport through the exterior envelope. The quality of the window frames themselves, the seals between glazing and frame, the durability of the gasket seals on operable units, and the sealant design and installation between the frame and the rough opening are all absolutely critical to the proper performance of the unit. In China, all of these elements are often substandard. As such, windows demand particular attention in design and construction. In addition, now that China has become the largest user of air-conditioning in the world and demand for more units will continue to rise, the demands on the integrity of windows will be a more critical aspect of an energy-efficient future. Furthermore, improved windows are critical to good thermal resistance in northern China, where heating is the primary energy use in buildings.

Again, the steps that need to be taken to produce significant energy savings immediately are simple. Two principal improvements in windows should be made:

- ensure continuous and high-quality sealants in the window frame itself and between the frame and the rough opening in the exterior wall for better protection against air infiltration; and
- ensure the widespread use of double- and triple-glazed insulated window units for better thermal resistance.

The first and most important step in making significant improvements is minimizing air leakage through the exterior wall. The air leakage through an average weather-stripped window is often on the order of 0.4 liters/second per linear meter of frame, with a pressure differential of 25 Pa. In a room with a volume of 27 cubic meters and two windows measuring 1 meter x 1.2 meters, the leakage would change the air in the room in a little over two hours (Brand 1991)!

The second step is the use of insulated glazing (Figure 19). Insulated double-glazing significantly improves the performance of a window unit. Double-glazed windows can save up to three times as much heating energy as single-glazed units and low-emission (low-e) double-glazed windows can save up to four times as much heating energy as single-glazed units. Figures 20 and 21 and Table 5 show the advantages of a double-glazed, low-e assembly over other window types. A lower U-value ultimately results in less energy loss. The use of gas filler between panes also improves performance.

Other higher-performing types of windows are available but require much higher initial investments that may place these systems beyond the reach of many Chinese builders. However, it is important that good-quality insulated windows be used to ensure a robust and durable assembly. Aluminum and PVC window frames perform better than timber and steel frames, however many aluminum frames will continue to be prohibitively expensive for developer residential buildings at least for the foreseeable future (Garvin and Wilson1998). In addition, because of the high thermal conductivity of aluminum, it is imperative that thermal breaks in the frame be included as a standard feature.

Also, inert gas filler between the insulating panes is another improvement to the performance of the window but again, availability and cost may prove insurmountable obstacles, at least in the short term (Figure 21, Table 5).

A system with good potential, both in terms of performance and cost, is the "2+1 window system" originally developed in northern Europe for heating

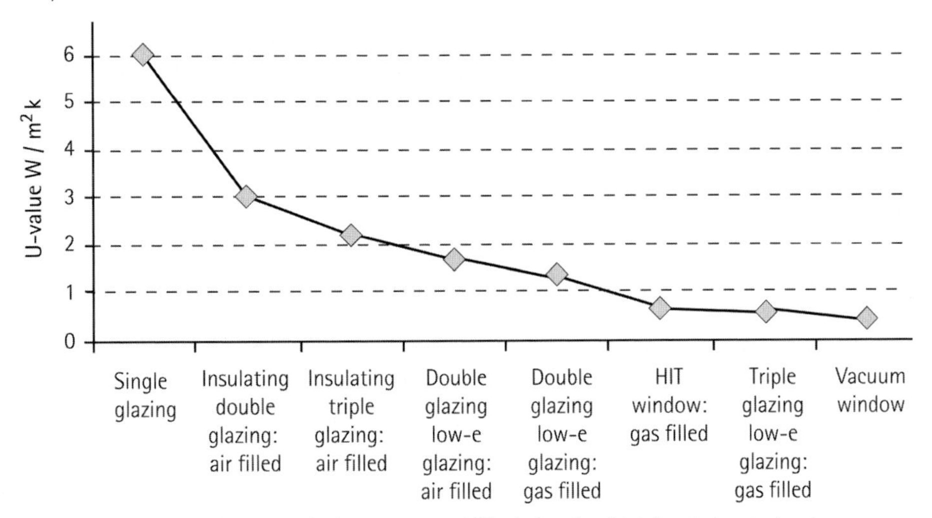

Figure 20 U-values of various glazing systems; a HIT window is a high insulation technology window, which has two layers of a polymer film between two panes of glass (with low-e coatings) (Source: Oesterle et al 2001)

Window Type	U-Value (W/m^2K)	
	Center of Pane	Whole Window
Conventional double-glazed, 16 mm air fill	2.59	2.8
Double-glazed, low-e, 16 mm argon fill	1.45	2.0
Double-glazed, with super low-e, 16 mm argon fill	1.05	1.75
2+1 window with single pane outer, cavity blinds, and double-glazed inner, low-e, 16 mm argon fill	0.95	1.15

Note: U-values based on a 3 m x 2.8 m window unit; whole window U-values will vary according to window size

Table 5 Effect of window type on U-value (Source: CIBSE 2002)

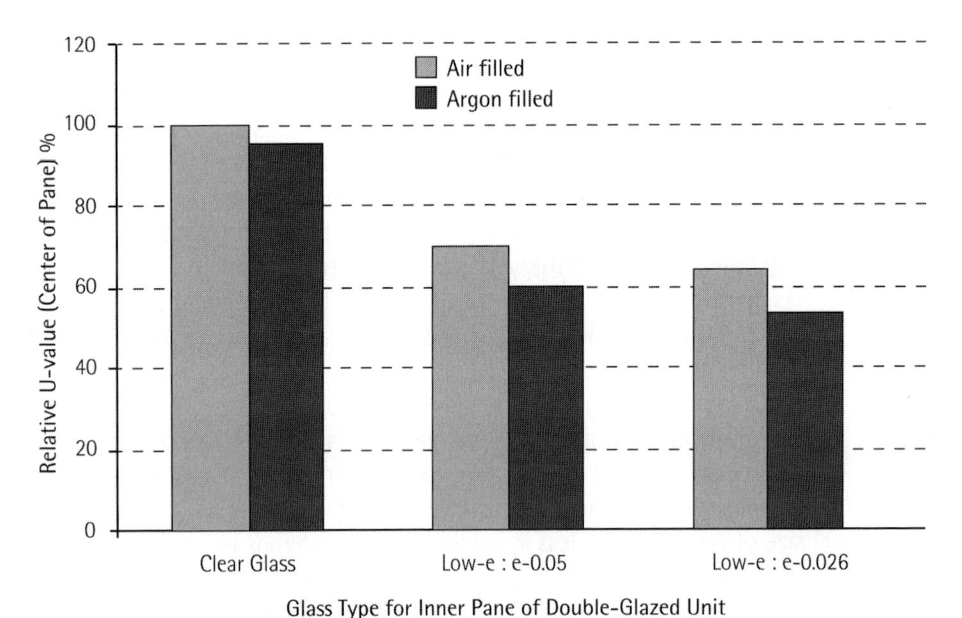

Figure 21 U-values of various double-glazed units (Source: adapted from Kubie, Muneer and Abodahad 2000, Amstock 1997, and Daniels 2003)

Figure 22 Illustration of 2 + 1 window system

climates (Figure 22). The window is simply an enhanced insulated window assembly through the addition of another single pane on the exterior side. The single pane is always operable, allowing it to serve as an additional thermal resistance material that also provides an air cavity during heating and cooling periods. The additional pane comes as part of the window unit, thereby avoiding any additional on-site coordination or special skills during installation.

This kind of assembly holds promise over more sophisticated systems because of the low cost of the unit in comparison to insulated triple-glazed units. Triple-glazed low-e is a sophisticated solution, and undoubtedly brings improvements to the window assembly, but is likely too expensive a solution for widespread use. The 2+1 window system is a good solution for parts of northern China with significant heating seasons. For the rest of the country, double-glazed, low-e windows will bring substantial energy savings.

Shading windows from direct solar gain during the summer while allowing needed gains in the winter is an important strategy for assisting in the control of the interior climate. Simple additions to the building façade may satisfy optimal shading conditions. Shading from balconies can also satisfy this requirement, if the residents do not succumb to the temptation to enclose the landings to increase the interior area of the residential unit.

On south-facing walls, a shaded window size of 50 percent or more of the total wall area increases winter solar gain without raising interior temperatures in summer. On east- and west-facing exposures, a window size of 20 percent or less in combination with blinds is recommended to achieve a reasonable comfort level while minimizing solar heat gain (MIT China Housing Workshop 2001).

Finally, another way to improve the performance of windows is simply through proper detailing in terms of their placement in the exterior envelope. Figure 23 shows the placement of the window frame in relation to the insulating layer and the exterior and interior walls of a typical concrete masonry cavity construction. The insulation is contained within the cavity in each example. In each case, it is critical that sealant be applied in an unbroken line to both the exterior and interior window frame to wall edges (indicated with arrows in Figure 23). In addition, the cavity must be adequately closed to prevent moisture from moving from the cavity into the window's rough opening.

Example A shows a configuration in which the cavity is properly sealed and the window frame can be sealed along its interface with the wall opening. B and C are the best options, bringing more of the window frame toward the interior and thus increasing its lifetime durability. Option D should be never be used because of the lack of overlap between the opening and the frame. The window frame does not overlap any of the wall materials, allowing a continuous gap that would run around the entire frame and allow air and moisture to easily penetrate.

In China today, low-quality exterior wall construction of multi-unit residential buildings often do not contain cavities. In fact, typical construction merely consists of placing window assemblies directly into openings cast into solid concrete walls. The absence of a cavity does not easily allow for a change of plane and the making of a robust air and water seal between inside and outside.

When this kind of construction is necessary because of cost, contractor preference, or local practice, it is critical that a continuous seal be applied around the perimeter of the window assembly (Figure 24). Best practice would include the construction of an additional assembly that overlapped both the wall material and the

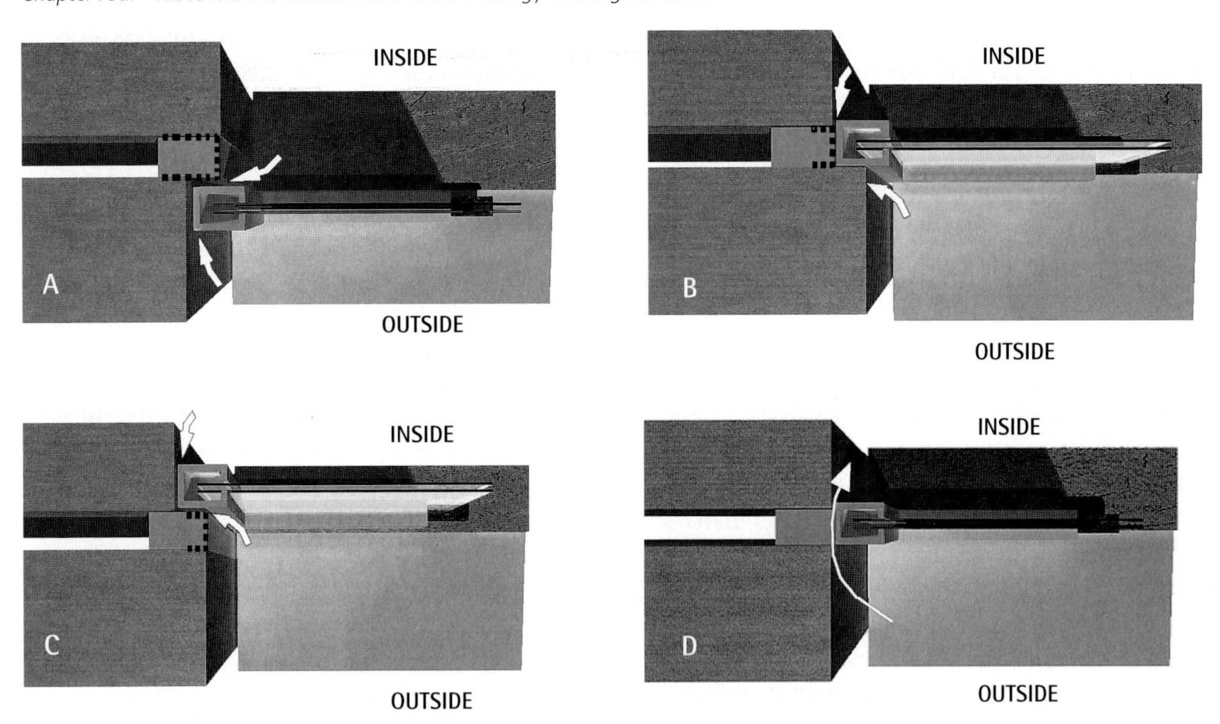

Figure 23 Window placement positions - it is critical that sealant be applied in an unbroken line to both the exterior and interior window frame to wall edges (indicated with arrows).

Figure 24 The simple insertion of a window assembly into a concrete cast in place rough opening requires a continuous air and water seal around its perimeter. Additional trim pieces may be required to provide continuous and redundant protection at the interface between the window assembly and the wall at area "A". The view shows the window from the interior.

window assembly. When this is not possible, a continuous bead of elastomeric sealant (ideally silicone) should be applied, inside and out. Either solution should eliminate all direct paths for air leakage between the interior volume of the residential unit and the outdoors.

POLICY AND REGULATION RECOMMENDATIONS FOR IMPROVED BUILDINGS

Using better materials is limited by the deficiencies in the building materials industries. Availability of high-quality materials and assemblies in China is limited and the demand for more sophisticated products is low. When developers are motivated to build with better materials, either by regulation, persuasive governmental suggestion (as for example, through the Ministry of Construction), or economic incentive, they are often frustrated by a lack of availability and prohibitive costs. At the same time, material producers cannot supply larger quantities of these products until there is clear evidence of expanding and sustained demand for such systems (World Bank 2001). It is possible that this situation is merely an indication of the growing pains of a nascent market for such products. However, unlike automobiles, appliances, or other energy-consuming devices, buildings - once constructed - are part of the energy consumption infrastructure for many decades to come. Every building built in the old style locks another energy sink into the building stock. Future improvements to existing inefficient buildings (and substandard buildings built today) are difficult, expensive, and materials intensive, involving selective demolition, materials removal, and often compromises in efficiencies easily achieved in new, better construction.

Among recommendations put forth by various organizations, the following are most important in establishing both economic incentives for a diversity of building products and a culture of better building practice throughout China (World Bank 2001, Gilboy and Heginbotham 2001):

- certification of building materials;
- bulk procurement of important materials and systems; and
- trade group promotion.

Each of these will contribute to fostering the maturation of the building industry and promote good construction practice. The government needs to be involved not only in the process of drafting and adopting specifications and regulations that call for energy-efficient buildings, but also in the process of facilitating the economic and educational context for making these buildings possible now.

POTENTIAL FOR SUSTAINABLE MATERIALS

The use of "green" materials in China is not generally a priority widely adopted by developers or architects today. As is true in many countries, one of the primary obstacles to the use of sustainable materials is clearly establishing the metrics and the overall criteria for determining a "green" material, and therefore making good choices between alternatives. In addition, reliable information about the embodied energy, off-gassing properties, recycled content, and other criteria of materials and industrial processes used to produce building products are difficult to obtain in many countries, China included. Furthermore, as mentioned earlier, difficulty with the availability of a wide variety of materials in China makes choices very limited; therefore the aspirations to use sustainable products and materials may be stifled early in the design process. This chapter cannot fully address the issue of sustainable materials in China, but can offer some guidelines to the designer in beginning to discriminate between a variety of materials and products that may meet appropriate criteria for Chinese construction.

China has begun the process of organizing research and policy studies to advance the use of sustainable materials in all sectors of the national economy (Zuo, Wang, and Nie 2001). The National Education Committee, the Chinese Ecomaterials Society, and the Chinese Center for Materials Life Cycle Assessment have been established with the mandate to support research of environmentally positive materials. These organizations are particularly interested in working with the mandates contained within ISO 14000.

Compiled by the International Standards Organization, ISO 14000 is a set of generic standards concerned with the environmental management of a variety of enterprises and industries. These standards are formulated to assist industries in improving their environmental performance; ISO 14000 has become the international guideline for many countries that are establishing specific industry and materials standards. Japan, the United Kingdom, Sweden, Spain and Australia are global leaders in implementing various aspects of ISO 14000, especially in the area of eco-labeling and building-ratings systems (Ball 2002). ISO 14000 also specifies a set of life cycle assessment (LCA) standards that have become the leading method in a variety of regions around the world.

However, there remain significant obstacles to a useful integration of a Chinese environmental management system with the mandates of ISO 14000. These obstacles echo the experiences of other countries in their attempts to derive specific standards from the general guidelines of ISO 14000. The most notable difficulties are the development of eco-labeling schemes that have been formulated throughout Europe, the United States and Asia (Hong Kong in particular). The BREEAM system in the United Kingdom was the earliest, followed by the "Blue Angel" and the "Green Dot" in Germany, the "NF-Environment Mark" in France, and the "AENOR Medio Ambiente" system in Spain. These and other regional systems are absolutely necessary in achieving reliable assessments of embodied energies, waste production in manufacturing, and other detailed materials data. In addition, regional systems are often critical to the development of regulations and the binding legal restrictions necessary for effective enforcement. However, over time, a proliferation of contrasting and sometimes redundant criteria brings with it inconsistencies and growing incompatibilities between the systems themselves. As a result, coordination between such systems may become an impossible task of synchronizing criteria independently developed and based in contrasting economic and cultural contexts. The Chinese effort would do well to carefully scrutinize the evolution of these and other ratings systems in a variety of countries globally.

Therefore, China is well positioned to learn from mistakes of the past to formulate inventive and appropriate standards based on the guidelines of ISO 14000. China is large enough that a single national standard may not address specific regional issues well enough. On the other hand, it is important that the industry in China has a consistent set of guidelines that cut across the 31 provinces. The regions in China are distinct enough that data from one region will not satisfy another. Individual sets of data, consistent with the conditions in each region, need to be collected.

One way in which these needs could be achieved would be the development of standards at two very distinct levels. First, one level would be national and would act as the link to the international community through ISO 14000. The next level would be organized in terms of climatic regions of the country. This regional set of criteria would be responsible for the development of useful standards based on reliable local data. In addition, the regional sector would be responsible for enforcement. Because eco-labeling of many kinds is often based on an LCA model, a combination of national and local systems is necessary in bringing together reliable data with internationally accepted evaluation methods. However, before a system such as the one just described can be formulated, the following issues should be understood.

First, it is important to acknowledge that the various criteria used to definitively assess a material's environmental impact have not been completely quantified, as of yet. There is currently no definitive method for determining the life cycle environmental impacts of all materials in all contexts using quantitative metrics that allow for a clear decision-making process. Conclusions are still arrived at with a good deal of subjective judgment. In essence, the determination of sustainable materials in construction is still a combination of art along with the science.

Second, many life cycle environmental impact guides for materials are highly localized and as such have limited utility in other contexts. General principles may be gleaned from sustainable material guides, but specifics are hard to transfer from one region to another, not to mention inadvisable from one continent to another. The United States and Europe currently use a variety of tools for the determination of the sustainability of materials in various contexts. However, these guides are of limited use in China because of the very different economic and industrial context of both Chinese cities and the countryside. Therefore, sustainable materials guides need to be produced for the various regions of China, taking into account available local resources, existing production capacity, local skill sets, and sustainable economic scenarios that can complement, not oppose, the inevitable and necessary development of the country over the next decades. In addition, the development of these guidelines should take into account the continuing research in the theory and practice of LCA for buildings (Adalberth 1997, Bogenstatter 2000, Chevalier 1996, Cole 1996, Thormark 2002).

Only these kinds of locally relevant guides have much of a chance of fostering true reductions in materials consumption and effective and appropriate materials reuse and recycling, without compromising other – potentially more effective – sustainable strategies.

Third, once studies have been undertaken and some measure of certainty acquired regarding the overall assessment of sustainable materials in China, certification of materials becomes the next step in codifying and disseminating this information. Mentioned previously in *Policy and Regulation Recommendations,* this step needs to be catalyzed and supported by the government.

However, for use here, it is possible to establish some general principles that can lead to a set of criteria for sustainable materials in China. MOLAR has been leading various efforts to survey existing resources and materials flows for the purpose of guiding regions through the best strategies for sustainable development. This agency has identified land management and reclamation, mineral and water resource management, and marine environment protection as key elements to a sustainable future (Ziran 1999).

In addressing materials flows in construction, a sustainable future can be approached by giving priority to the five most important aspects of the extraction, production, use and reuse of materials for buildings. These aspects are:

- manufacturing energy: lowering the energy in production, transportation, and assembly on site affects both the local air quality and global warming;
- energy in operation: using materials that lower the amount of energy used in the operation of buildings;
- outdoor air pollution and toxicity: reducing the amount of outdoor pollution and eliminating sources of toxicity created through the full life cycle of the material;
- indoor air pollution and toxicity: reducing the amount of indoor pollution and eliminating sources of toxicity created through the cradle-to-grave life cycle of the material; and
- minimizing waste and maximizing reuse: through recycling, reclamation, adaptive reuse and other techniques, the useful life of materials may be extended and redirected to minimize the amount of materials that end up in the landfill.

These five priorities can be implemented through a variety of methods. It is the formulation of these methods that is most appropriately done through additional research that takes into consideration the specific context of distinct regions of China. However, the following list provides a framework for such a study (Wilson 2001). Sustainable materials may satisfy one or more of the following criteria:

- products and materials can contribute to minimizing resource extraction and the use of raw materials through salvaging items from existing buildings;
- products and materials that use post-consumer recycled content;
- products that contain post-industrial recycled content;
- engineered wood products that use reclaimed waste wood from a variety of wood-using industries;
- rapidly renewable products;
- products that use agricultural waste material;
- minimally processed products;
- products that reduce material use;
- alternatives to ozone-depleting materials;
- products that reduce the impacts of new construction;
- products that reduce the impacts of demolition;
- building components that reduce heating and cooling loads;
- products with exceptional durability or low-maintenance requirements;
- products that prevent pollution or reduce waste;
- products that reduce or eliminate pesticide treatments;
- products that do not release significant pollutants into the building;
- products that remove indoor pollution;
- products that support passive servicing of the interior environment, for example, concrete for high thermal mass; and
- products that improve light quality of the interior.

POTENTIAL FOR ALTERNATIVE CONSTRUCTION: EARTHEN AND STRAW-BALE BUILDINGS

Alternative construction techniques and building materials may offer a wider range of solutions for residential buildings in China. While there are many to consider, the two highlighted here have both proven themselves as successful building techniques for residential buildings in various regions of China. These two methods have shown good potential in bringing together inexpensive and plentiful labor with local materials to provide for appropriate and necessary housing as well as employment opportunities (Melendez 2001).

Earthen buildings are constructed of clay mixtures (loam) set into a variety of forms. Loam in construction is not a new material for large-scale construction and residential buildings in China. In fact, the compression of clay soil into solid walls has a long history in China, reaching back literally to the very foundations of the Great Wall in the third century B.C. A great many vernacular building types have used rammed earth as the superstructure of the building, including multi-story residential buildings, a variety of farmhouses, enclosing compounds and fortifications for villages and cities. The methods – *hangtu*, *zhuangtu*, and *banzhu* (Knapp 2000) – have been used to produce structures as tall as nine meters. These techniques are still commonly used in northern China where bamboo and timber are scarce. Even in the humid south, tamped-earth buildings are common in rural villages. The imposing circular houses of Fujian are large four- and five-story buildings of tamped-earth walls and timber floor structures (Qijun 2000). Sun-dried and kiln-dried loam bricks – *tupi*, *zhuanpi*, or *nipi* – are also commonly used in a variety of buildings in many regions.

The variety of construction techniques and the various forms produced through the use of loam are too numerous to fully describe here. Loam is an attractive material primarily because of its low expense, ease of construction with small-scale teams, the history of use in much of China, and the low energy used in construction – a mere one percent of the energy needed to cast on-site concrete (Minke 2000). In addition, loam construction easily adapts to and fulfills the cultural needs for forms appropriate to residential buildings.

However, loam construction is a labor-intensive method that does require a high skill and knowledge level. Unlike standardized materials such as concrete and steel, loam construction is highly dependent on the soils of the region and the climate. It is absolutely necessary that those responsible for using these construction techniques have substantial experience and that they produce details that are appropriate to the system. For example, it is critical that buildings that use loam also incorporate substantial eaves to reduce the amount of rainwater that reaches the exterior surface of the wall. Standardized guides have been produced in New Zealand and further international guidelines are being written. Efforts to advance earthen materials have focused on the measure of strength and predictions of durability (Heathcote 1995).

Another interesting building system is straw-bale construction. Several straw-bale projects have been successfully built and more projects are being proposed, especially in northern China where the need for warm and inexpensive housing is acute, and agricultural materials and labor are readily available. Straw-bale construction is low in embodied energy and utilizes an agricultural by-product for the making of seismically resistant buildings. A series of projects has shown that the technology is both appropriate to the skills of the population and beneficial to the fuel-conservation needs of homeowners. Four northeastern provinces have assembled building teams to be trained and deployed with the mandate to build 1000 straw-bale houses within two years (Lerner 2000, Lerner 2001).

Earthen materials, straw bale, and other biocomposites are appropriate to both urban and rural conditions. However, these techniques have not been introduced into urban environments and substantial obstacles exist before introduction can proceed. One consideration is the thickness of the walls in both systems. Clearly, these construction methods result in thicker walls – sometimes up to a meter for a multi-story edifice. Developers will have to decide whether this additional construction dimension will be prohibitive – certainly, in tight urban situations, it may be. However, for many contexts in China, there is no good reason why these materials should not be considered as viable systems for multi-story urban residential construction. Furthermore, the need for housing is also acute in rural areas of China, where these methods can expect wider application in the short term (Knapp 2000).

Finally, the widespread use of natural materials in construction will require the formulation of design standards that will guide the design professional through an understanding of these unique materials. It has been suggested that one strategy in approaching the variability of material quality, composition, and structural properties is an increase of the average safety factor for these materials to a range of three to five, as opposed to the range of two to three commonly used for standardized structural materials (Straube 2001).

SUMMARY

Improvements in both construction practices and building materials will yield important advances toward significantly improving the energy efficiency, comfort, and long-term durability of China's growing housing stock. Chinese politics has given rise to a market economy of impressive proportions. Output from private companies now exceeds 40 percent of all industrial production. Private employment of nonagricultural workers is now at 30 percent (Gilboy and Heginbotham 2001). The housing market is rapidly privatizing and a flood of workers is invading the cities. Change is coming fast to China. The potential for important changes is dependent on a series of clear choices that the Chinese government and design professionals must make (Sha, Deng, and Cui 2000). The following conclusions are important to supporting the aspirations of sustainable low-energy buildings through appropriate building materials and systems.

There are important opportunities for turning the tide of the construction of inefficient energy-consuming buildings through the use of higher-performing materials and better construction practices and inspection. However, active measures need to be implemented to facilitate the availability and cost-effectiveness of systems to improve energy efficiency in the operation of residential buildings. Three important measures include:

- bulk procurement of important materials and certification of building materials;
- bulk procurement of important materials and systems; and
- trade group promotion.

There are clear advantages to affecting the present situation through a limited but forceful campaign to implement several simple changes. These are generally intended to improve the integrity of the exterior wall through improvements in thermal resistance and minimization of air in-/exfiltration. Specifications that clearly state the need and establish standards for high-performance windows, continuous insulation and, where appropriate, air and vapor barrier materials, cavity wall, and rainscreen construction and sun shading should all be included.

Of particular importance are improved windows, a technology that is absolutely necessary for significant improvement of the situation. Manufacture and nationwide distribution of insulated windows would greatly improve the energy efficiency of residential buildings in China. Along with effective insulation materials, insulated windows that are easy to install, seal, caulk, and maintain will be crucial to meeting aspirations of low-energy residential buildings.

Specifications already formulated and approved need better enforcement. New buildings have commonly disregarded the requirements established by the Energy Conservation Design Standard (ECDS). Simply enforcing the ECDS will bring great benefits. Crucial to this enforcement is post-construction inspection that has the power to delay and prohibit the sale and occupation of substandard units until standards have been met (Glicksman, Norford and Greden 2001).

Premanufactured systems for construction should be promoted for the advantages that a "cleaner," more materials- and energy-efficient process brings to the construction industry. Various types of "premanufacturing" may bring significant savings including a greater use of pre-cast concrete sections and shapes, modular units of high-quality bathrooms and kitchens, modular exterior envelope panels, and other products. Premanufacturing allows for greater quality control and less materials waste.

Future large-scale and uninterrupted supplies of good-quality steel for construction are still a challenge that Chinese industry has not met. To be realistic, concrete will always occupy an important role in the making of multi-story buildings in China. However, in seismically active zones where a light frame is much preferred and where a higher level of performance of the exterior envelope may be achieved, the option to use steel should be available.

Adoption of standards and regulation of certification for "green" building materials will improve the odds that the development, design, and engineering disciplines will act responsibly in building, operating, and demolishing residential buildings. Working with ISO 14000 and 14001 has demonstrated to be one of the best ways in which to achieve regionally useful standards. These globally recognized systems allow for local information to be implemented in national and regional plans for environmental management systems. However, a persistent challenge is maintaining the overall coherence of international guidelines while developing regionally useful standards. With the increasing complexity and interconnectedness of the global economy, coherence within international standards is a necessity for measuring the progress toward a sustainable architecture in China (Ball 2002). The development of standards for environmentally responsible construction and materials needs to take stock of the various national attempts to implement ISO 14000. A possible model for development in China would be the formulation of distinct national and regional standards.

Further studies need to be conducted in the field of alternative building materials and techniques, especially those techniques that have been in use in China for extended periods of time and those that may take advantage of the reuse of an existing industrial or agricultural waste stream (Zuo, Wang and Nie 2001). Low-embodied energy materials should be investigated further, including earthen construction and biocomposites, including straw-bale buildings. In addition, ecocomposites such as agricultural waste boards (from straw and grass) should be promoted as an environmentally sound interior partition and finish material. Regionally available resources and labor skill sets should be carefully considered in the investigation of appropriate construction solutions. Assessments of these alternative systems need to acknowledge the powerful ability of locally derived and rediscovered construction methods to link the local population with a cultural heritage. The significance of cultural connection is an often-discounted set of issues. However, the ultimate success of sustainable architecture is, to a great extent, determined by a simple resonance with the history of the local culture and the value that contemporary designers and developers place on a legacy of traditional forms and buildings. By working in this way, designers and developers alike will link the need for housing to the desire for humane environments.

REFERENCES

Adalberth, K. 1997. Energy Use during the Life Cycle of Buildings: A Method. *Building and Environment*, vol. 32, no. 4: 317-320, 321-329.

Amstock. J.S. 1997. *Handbook of Glass in Construction.* McGraw-Hill, New York.

Ashby, M. and D. Jones. 2001. *Engineering Materials I.* Butterworth Heinemann, Oxford, U.K.

Ball, J. 2002. Can ISO 14000 and Eco-Labelling Turn the Construction Industry Green? *Building and Environment*, vol. 37, no. 4: 421-428.

Bijen, J.1996. Benefits of Slag and Fly Ash. *Construction and Building Materials*, vol. 10, no. 5: 309-314.

Bilodeau, A. and V. Malhotra. 2000. High-Volume Fly Ash System: Concrete Solution for Sustainable Development. *ACI Materials Journal*, 97-M6, January-February 2000.

Bogenstatter, U. 2000. Prediction and Optimization of Life-Cycle Costs in Early Design. *Building Research & Information*, 28(5/6): 376-386.

Brand, R. 1991. *Architectural Details for Insulating Buildings.* Van Nostrand Reinhold, New York.

Brocklesby, M. and J. Davison. 2000. The Environmental Impacts of Concrete Design, Procurement and On-Site Use in Structures. *Construction and Building Materials*, vol. 14, no. 4: 179-188.

Brown, R. 2001. The China Challenge for Western Home Builders. *Automated Builder*, edition 365, November 2001.

Chevalier, J. and J. Le Teno. 1996. Requirements for an LCA-based Model for the Evaluation of the Environmental Quality of Building Products. *Building and Environment*, vol. 31, no. 5: 487-491.

Ching, F. 1991. *Building Construction Illustrated*, 2nd edition. John Wiley & Sons, New York.

CIBSE. 2002. *HVAC Strategies for Well-Insulated Airtight Buildings.* TM29, Department of Trade and Industry, London.

Cole, R. 1999. Energy and Greenhouse Gas Emissions Associated with the Construction of Alternative Structural Systems. *Buildings and Environment*, vol. 34, no. 4: 335-348.

Cole, R. and P. Kernan. 1996. Life-Cycle Energy Use in Office Buildings. *Building and Environment*, vol. 31, no. 4: 307-317.

Cole, R. and D. Rousseau. 1991. Environmental Auditing for Building Construction: Energy and Air Pollution Indices for Building Materials. *Buildings and Environment*, vol. 27, no. 1: 23-30.

Costabile, S. 2001. Recycled Aggregate Concrete with Fly Ash: A Preliminary Study on the Feasibility of a Sustainable Structural Material. *Proceedings of the First International Conference on Ecological Building Structure*, 5-9 July 2001. San Rafael, CA.

Daniels, K. 2003. *Advanced Building Systems: A Technical Guide for Architects and Engineers.* Birkhauser (Princeton Architectural Press), New York.

Fewtrell, J. 2005. "World Produces 1.05 billion Tonnes of Steel in 2004." Press release, International Iron and Steel Institute, Brussels, Belgium, 19 January 2005.

Garvin, S. and J. Wilson. 1998. Environmental Conditions in Window Frames with Double-Glazing Units, *Construction and Building Materials*, vol. 12, no. 5: 289-302.

Gilboy, G. and E. Heginbotham. 2001. China's Coming Transformation. *Foreign Affairs*, July/August: 26-39.

Glavind, M. and C. Munch-Petersen. 2000. Green Concrete in Denmark. *Structural Concrete*, March, vol. 1, no. 1: 19-25.

Glicksman, L., L. Norford and L. Greden. 2001. Energy Conservation in Chinese Residential Buildings: Progress and Opportunities in Design and Policy. *Annual Review Energy Environment*, vol. 26: 83-115.

Griffin, R. 1987. CO_2 Release from Cement Production 1950-1985. Report published by the Institute for Energy Analysis, Oak Ridge Associated Universities, Oak Ridge, Tennessee.

Heathcote, K. 1995. Durability of Earthwall Buildings. *Construction and Building Materials*, vol. 9, no. 3: 185-189.

IISI. 2005. Steel Statistical Yearbook. *International Iron and Steel Institute*, Brussels.

Keoleian, G., S. Blanchard and P. Reppe. 2001. Life-cycle Energy, Costs, and Strategies for improving a Single-family House. *Journal of Industrial Ecology*, vol. 4, no. 2: 135-156.

Knapp, R. 2000. *China's Old Dwellings.* University of Hawaii Press, Honolulu.

Kubie, J., Muneer, T. and N. Abodahad. 2000. *Windows in Buildings.* Architectural Press, Oxford, UK.

Lashof, D. 1995. Intergovernmental Panel on Climate Change (IPCC) Working Group II. Unpublished report. IPCC Secretariat, World Meteorological Organization, Geneva.

Lerner, K. 2000. Introduction of Straw-Bale Construction to Mongolia and China, in *Alternative Construction: Contemporary Natural Building Methods.* L. Elizabeth and C. Adams, eds. pp. 339-356. John Wiley & Sons, New York.

Lerner, K. 2001. Straw-Bale Development on the Eastern Frontier. *Proceedings of the First International Conference on Ecological Building Structure*, 5-9 July 2001. San Rafael, CA.

Lyberg, M. 1997. Basic Air Infiltration. *Building and Environment*, vol. 32, no. 2: 95-100.

Malhotra, V. 1998. Role of Supplementary Cementing Materials in Reducing Greenhouse Gas Emissions. Paper presented at CANMET/ACI International Symposium on Sustainable Development of the Cement and Concrete Industry, 21-23 October 1998. Ottawa, Canada.

Malhotra, V. and P. Mehta. 1996. *Pozzolanic and Cementitious Materials.* Gordon and Breach Publishers, New York.

Mehta, P. 1998. Role of Pozzolanic and Cementitious By-Products in Sustainable Development of the Concrete Industry. *Proceedings of the 6th CANMET/ACI/JCI Conference: Fly Ash, Silica Fume, Slag and Natural Pozzolans in Concrete*, Ottawa, Canada.

Melendez, M. 2001. Ecomaterials Create Local Jobs. *Proceedings of the First International Conference on Ecological Building Structure*, 5-9 July 2001. San Rafael, CA.

Minke, G. 2000. *Earth Construction Handbook.* WIT Press, Southampton, U.K.

MIT China Housing Workshop. 2001. Results from a workshop given during spring semester 2001 by the Building Technology Group in the Department of Architecture. Available at <http://chinahousing.mit.edu>.

Oesterle, E., R. Lieb, M. Lutz and W. Heusler. 2001. *Double-Skin Façades.* Prestel, Munich.

Qijun, W. 2000. *Ancient Chinese Architecture: Vernacular Dwellings.* Springer-Verlag, Vienna and New York.

Rongkang, Z. 1999. Natural Resources and Sustainable Development. *Resources and Industry*, 6:6: 22-31.

Rousseau, D. and Y. Chen. 2001. Sustainability Options for China's Residential Building Sector. *Building Research & Information*, 29(4): 293-301.

Schiessel, P. 1996. Durability of Reinforced Concrete Structures. *Construction and Materials*, vol. 10, no. 5: 289-292.

Sha, K., X. Deng and C. Cui. 2000. Sustainable Construction in China: Status Quo and Trends. *Building Research & Information*, 28(1): 59-66.

Straube, J. 2001. Alternative Building Materials and Systems - Understanding Technical Risk and Uncertainty. *Proceedings of the First International Conference on Ecological Building Structure*, 5–9 July 2001. San Rafael, CA.

Thormark, C. 2002. A Low Energy Building in a Life Cycle - Its Embodied Energy, Energy Need for Operation and Recycling Potential. *Building and Environment*, vol. 37, no. 4: 429-435.

Tilton, J. 1999, The Future of Recycling. *Resources Policy*, 2: vol. 25, no. 3: 197-204.

Trechsel, H., ed. 2001. *Moisture Analysis and Condensation Control in Building Envelopes*. ASTM: MNL40.

Wilson, A. 1993. Cement and Concrete: Environmental Considerations. *Environmental Building News*, 2(2), at <www.buildinggreen.com/features/cem/cementconc.html>.

Wilson, A. 2001. Building Materials: What Makes a Product Green? *Environmental Building News*, special reprint, (9)1.

World Bank. 2001. China: Opportunities to Improve Energy Efficiency in Buildings. *Asia Alternative Energy Program and Energy and Mining Unit, East Asia and Pacific Region*. World Bank, Washington, D.C.

Wu, Y. 2000. The Chinese Steel Industry: Recent Developments and Prospects. *Resources Policy*, vol. 26, no. 3: 171-178.

Yan, C. 2002. ASTM International and the Chinese Steel Industry, *Standardization*, July: 24-27.

Ziran, Z. 1999. Natural Resources Planning, Management, and Sustainable Use in China. *Resources Policy*, 25: 211-220.

Zuo, T., T Wang and Z. Nie. 2001. Ecomaterials Research in China. *Materials and Design*, vol. 22, no. 2: 107-110.

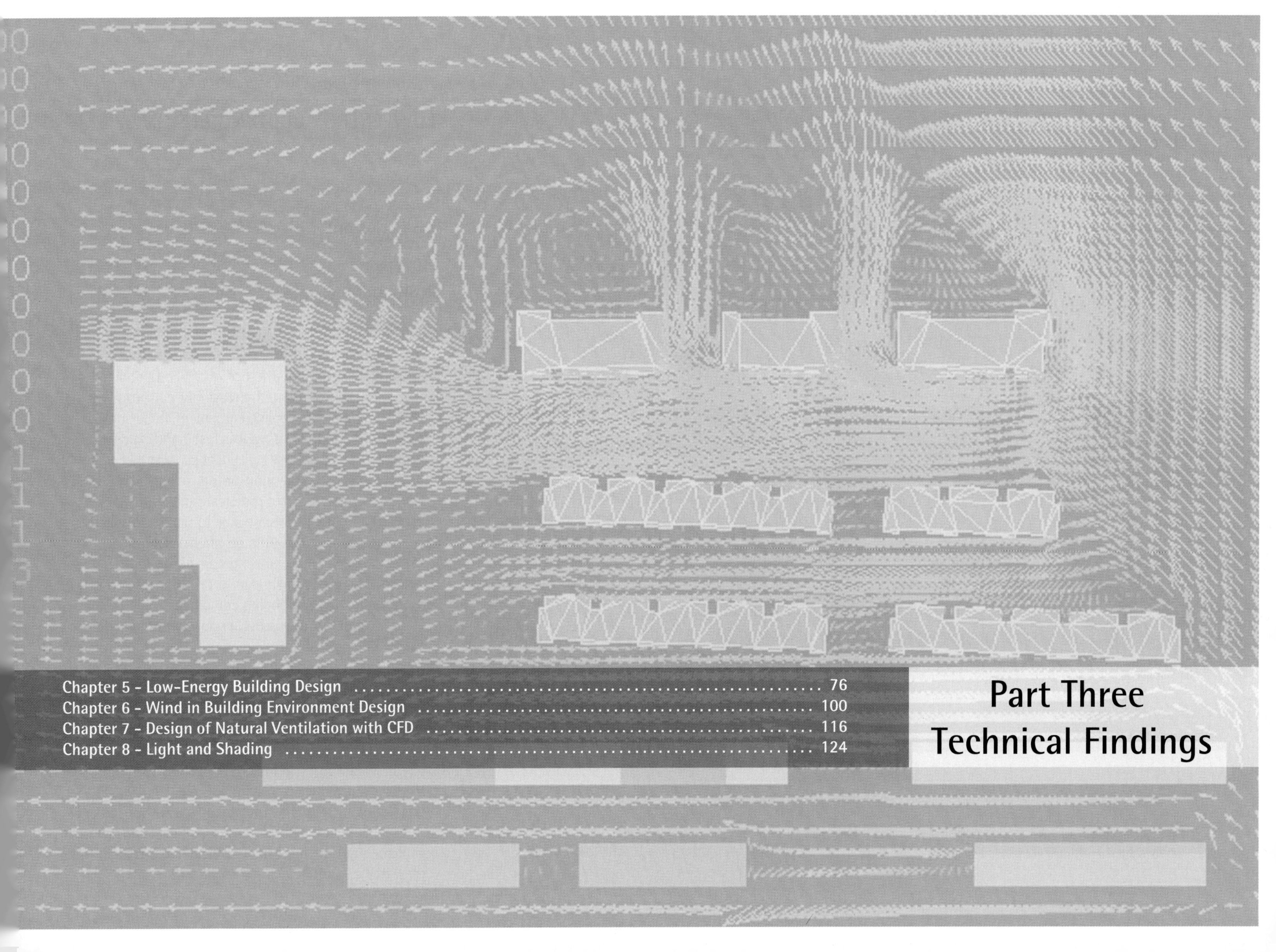

Part Three
Technical Findings

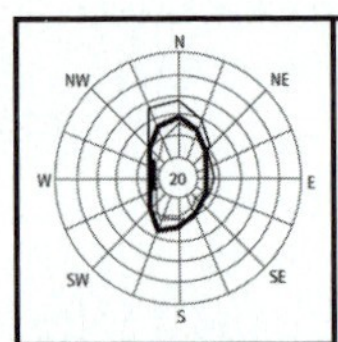

CHAPTER FIVE

LOW-ENERGY BUILDING DESIGN

Leslie Norford

INTRODUCTION

While the reduction of heating energy remains an important goal in China, there is now a pressing need to address cooling energy. The use of air-conditioning is burgeoning as occupants increasingly can afford to purchase and operate cooling equipment. In Beijing, for example, formal surveys as well as first-hand experience of the authors and Chinese colleagues suggest that while many people use air-conditioning, others who are unaccustomed to it or are more careful about shading and ventilating their units are able to remain thermally comfortable without it. This research strengthens an understanding of how housing can be designed and operated to provide a viable (comfortable and affordable) alternative to air-conditioning equipment.

This chapter first discusses thermal comfort and then presents findings from three technical studies for three different cities and climates. The first study (Carrilho da Graça et al 2002) examined wind-driven ventilation in Beijing and Shanghai dwellings without mechanical ventilation or cooling. Using widely applied thermal-comfort standards and appropriate wind speeds, this study showed that the number of uncomfortably warm hours could be reduced substantially in Beijing and, to a lesser extent, in Shanghai. The second study (Norford et al 1999) focused on thermal comfort in Beijing housing, complementing the first study by including mechanical

L. Glicksman and J. Lin (eds), Sustainable Urban Housing in China, 76-99

ventilation to guarantee airflow even when neighboring buildings screened a given dwelling from winds that might otherwise provide cooling. The second study also examined the impact of increasing the allowable maximum temperature, as has been found appropriate by recent work on thermal comfort. It was found that thermally comfortable indoor conditions without air-conditioning could be maintained throughout a typical summer in the limit of eliminating solar gain and internal loads. This limit can be approached via careful shading and choice of building materials and appliances. The third study concerned an energy-efficient housing development in Shenzhen, with the goal of providing guidelines to the developers about the site and the construction of the dwellings. To aid in the initial selection of low-energy buildings, a simple-to-use numerical program is included in the CD at the end of this book. This will allow a first indication of the value of good building design, construction quality, and proper material selection on overall energy use.

Thermal Comfort

Understanding what range of temperatures is considered by building occupants to be comfortable is an important first step in predicting building-energy consumption. This will help identify how such opportunities as using natural ventilation in lieu of vapor-compression cooling affect energy savings and the comfort of the occupants.

The international thermal-comfort standard identifies 26°C as the upper bound of the thermal-comfort region under humid conditions (ASHRAE 2001, ISO 1994). This upper bound increases with airspeed at the skin. For a relative humidity of 50 percent, light clothing, and sedentary building occupants, Fanger's predicted-mean-vote (PMV) method for estimating thermal comfort predicts a thermally neutral temperature of 26°C for an air speed of 0.2 m/s and 28°C when the air speed is 1.5 m/s. Lechner associated an air speed of 1 m/s with an equivalent temperature reduction of 3.3°K (Lechner 2001).

While the thermal-comfort model embodied in the standard accounts for air speed, it is based on laboratory tests, not field measurements in the workplace or home, and has its origins in Western countries. Many have observed that the ISO thermal comfort zone simply does not apply in all regions of the world (Khedari et al 2000). It is not a matter of doing with less cooling and less comfort in response to economic pressures, but rather of establishing conditions that people prefer. A study of adaptive thermal comfort in Thailand showed acceptable indoor temperatures of 28°C for air-conditioned offices and 31°C for naturally ventilated offices (Busch 1992). A separate study in Bangkok established an upper limit to thermal comfort of 31.5°C (Jitkhajornwanich et al 1998), 5.5°K beyond the upper boundary presented by ASHRAE. Motivated by a desire to reduce air-conditioning loads in Thailand without sacrificing comfort, Khedari recorded thermal sensation votes of college students cooled with desktop fans. The thermally neutral temperature was 30.6°C when the air speed was 1 m/s and the relative humidity was between 50 and 60 percent and increased to 33.5°C under more humid conditions (50–80 percent relative humidity) when the air speed was increased to 2 m/s. Kwok (1998), working in Hawaii, found that occupants of naturally ventilated classrooms were comfortable in conditions beyond ASHRAE specifications and noted the potential impact of such results in a state that is highly dependent on imported fuel. Oseland (1998) found that occupants of naturally ventilated offices in England had a wider thermal-comfort range than did those in air-conditioned offices.

These and other studies have led to proposals for adaptive thermal-comfort models and standards in which an indoor comfort temperature is related to the outdoor

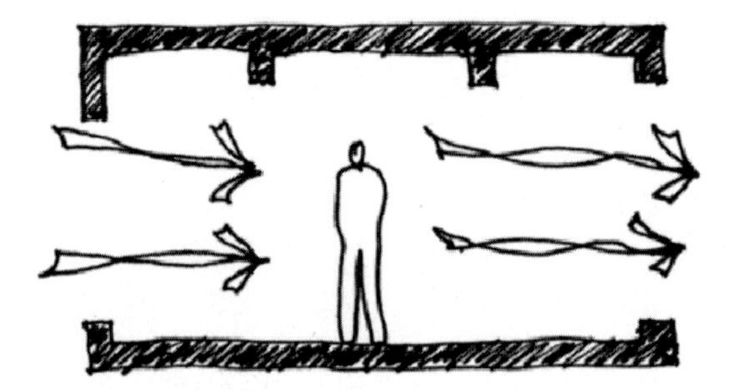

(a) Daytime ventilation - airflow removes heat from the indoor environment

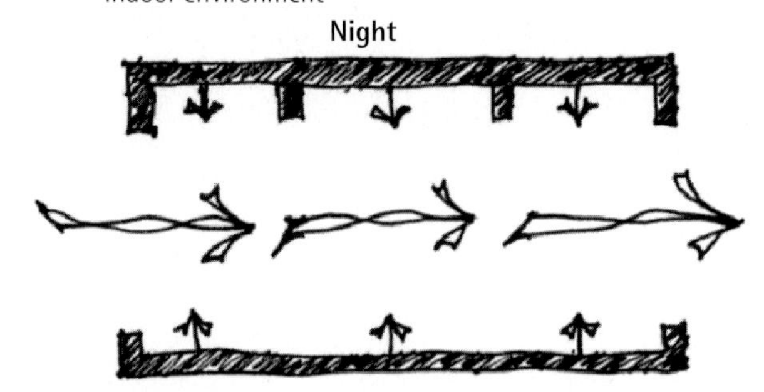

(b) Night cooling - walls release heat, maximum ventilation

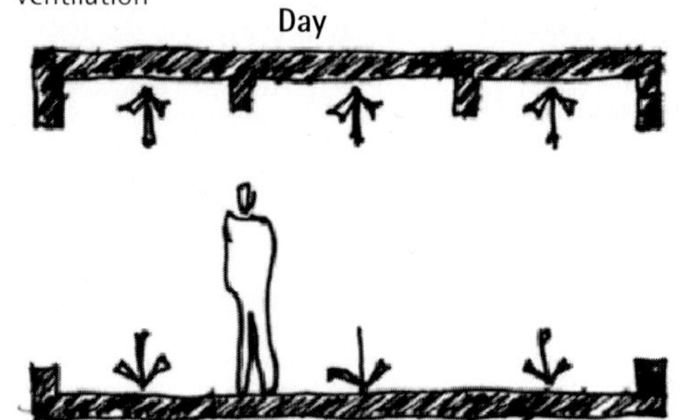

(c) Night cooling - walls absorb heat, minimum ventilation

Figure 1 Principle of daytime ventilation and night cooling: (a) outdoor air removes the heat gains indoors, (b) outdoor air cools the thermal mass during the night, and (c) the thermal mass absorbs heat during the day (Source: adapted from Carrilho da Graça et al 2002)

mean monthly temperature and the availability of heating and cooling equipment (Nicol et al 1995, Humphreys and Nicol 1998, Nicol and Kessler 1998, Malama et al 1998). Based on a review of 21,000 sets of raw data from 160 thermal comfort studies in 160 office buildings in four continents, de Dear and Brager proposed an adaptive thermal-comfort standard for naturally ventilated buildings in which the 90 percent acceptability limits for indoor comfort temperature increased with outdoor temperature and were 23–28°C at a monthly mean outdoor temperature of 25°C and about 26–31°C for mean monthly temperatures above 33°C (de Dear 1998, de Dear and Brager 1998, Brager and de Dear 2000). This adaptive standard has recently been incorporated into the ASHRAE thermal-comfort standard (ASHRAE 2004).

Recent studies in China provide collaborative evidence that temperatures above the current international standard may be preferable. A survey conducted by Tsinghua University, directed toward occupants of 83 Beijing units, found that 80 percent of occupants considered 30°C to be tolerable (Xia and Zhao 1999). It makes sense to use the newer thermal comfort criteria in areas of the world where air-conditioning is not yet well established and codes can be more enlightened.

Passive Cooling Strategies

In order to remove heat gains from buildings, it is necessary to use at least one of the heat sinks available in nature, such as the atmosphere and the ground (including large water bodies such as rivers, ponds, and underground water). Passive cooling systems transfer part of the building's internal heat gains into these heat sinks with little or no use of mechanical systems. Passive-cooling systems can be grouped into five main types (Givoni 1994).

1. Daytime ventilation (comfort ventilation) is the most common passive-cooling system. The system uses outdoor air during daytime to remove the heat gains in the room air in a way shown in Figure 1a. The system also increases the occupant's thermal comfort by increasing convective and evaporative heat transfer between the occupants and the room air. The maximum indoor air velocity is approximately 2 m/s (Givoni 1998). If the outside temperature is high, the indoor air temperature will then be too high to be accepted by the occupants. This system works better in climates with a mild summer.
2. Night cooling uses cold outdoor air during the night to cool the building thermal mass (the building's internal partitions and structure). The thermal mass functions as a heat sink during the day, absorbing the internal heat gains. Figures 1b and 1c show the operation of night cooling. In order to reduce the heat gains due to ventilation during the day, the windows should be kept closed. In this period, infiltration normally provides an acceptable indoor air quality. A ceiling fan may be used to enhance indoor air circulation and to increase heat exchange between the occupants and the thermal mass. The lower the outdoor air temperature during the night, the more effective the night cooling system. This system works better in climates with a minimum temperature below 22°C during the night.
3. Evaporative cooling evaporates water into the air. The evaporation process removes a large amount of sensible energy from the air, providing a cooler indoor environment. This system is effective for a dry climate.
4. Radiative cooling removes heat from the building's external surfaces through long-wave radiation to the sky during the night. The cooled building then absorbs part of the heat gains during day. The system works for regions with clear skies and a building with a large

exterior surface, such as a detached single-family house.

5. Ground-coupled cooling uses the ground as a heat sink. Because the ground temperature is relatively low and stable, the heat gains in the building can be transferred to the ground (normally using a mechanical system). Generally, this system is more expensive than the other passive-cooling systems.

Compared to air-conditioning systems, passive-cooling systems consume little energy, require little maintenance, have low first-costs, and are environmentally friendly. However, the effectiveness of a passive cooling system depends on the climate and building type, as discussed previously. In Beijing and Shanghai, the climate is hot and often humid, and most residential buildings are medium- and high-rise buildings. Therefore, daytime ventilation and night cooling seem to be the best options, and are the subject of the first two sections of this chapter.

Climate Data

To narrow the simulation period for Beijing to warm weather, we examined a typical meteorological year (TMY) hourly weather file for Beijing; temperature data is plotted in Figure 2. The hourly dry-bulb temperatures during warm months were similar to those obtained from the Medpha (Meteorological Data for HVAC Analysis) data available for Chinese cities and typically used by Chinese researchers (DeST 2006). Taking outdoor dry-bulb temperatures of 25°C as being sufficiently warm to assess passive cooling and thermal comfort, we selected the period of May through September as relevant for our study. The same definition of the cooling season was applied to Shanghai. Climate data for both cities are shown in Table 1. The climate in Shenzhen is warmer than in Beijing and Shanghai, and the goal of our study

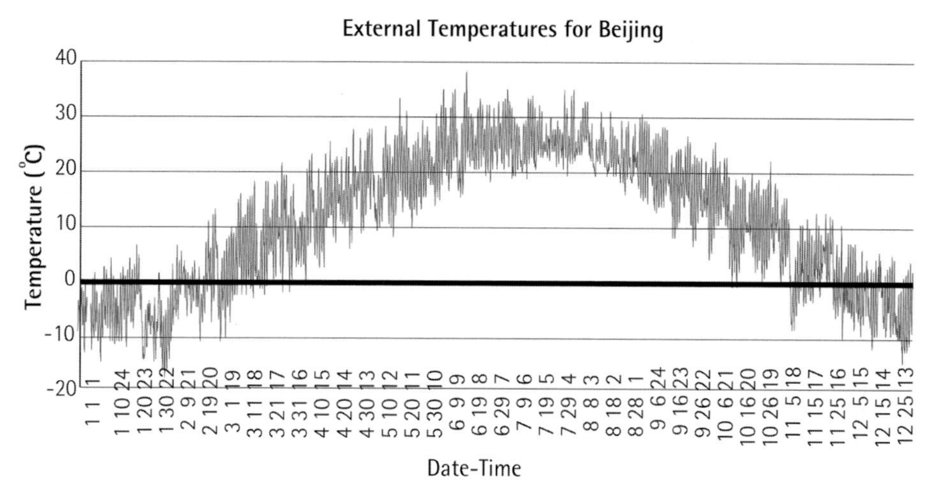

Figure 2 Hourly outdoor dry-bulb temperatures for Beijing; the date-time axis indicates month, day, and hour

	Days with Maximum Temperature above 30° C					
	Number of warm days	Day-to-night temperature variation	Average at hour of maximum temperature		Average wind speed	
			Temp.	RH (%)	(10 PM - 8 AM)	(11 AM - 5 PM)
Beijing	41 (27%)	9.7°K	32.8°C	42%	1.4 m/s	2.8 m/s
Shanghai	55 (36%)	4.6°K	32.9°C	66%	2.6 m/s	2.6 m/s

Table 1 Important climatic parameters for Beijing and Shanghai. The percentage of warm days refers to the period from 1 May to 30 September. The temperature variation (T_{Var}) is calculated by subtracting the lowest nighttime temperature (T_{Low}) from 30°C (T_{Var}=30-T_{Low}).

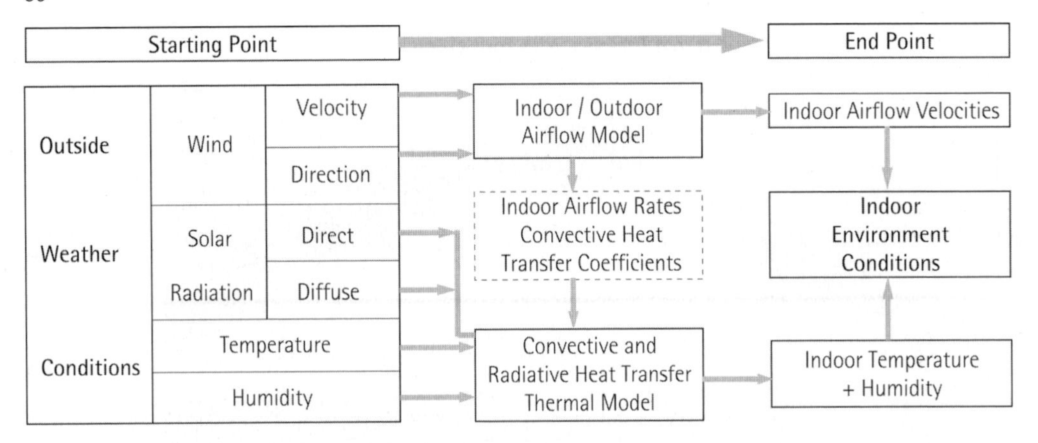

Figure 3 The structure of the coupled program (Source: Carrilho da Graça et al 1999)

was to reduce but not eliminate the use of air-conditioning. Accordingly, temperature data was not analyzed in the same manner. Average daytime highs in Shenzhen are 29°C in May and October and exceed 30°C for June through September.

STUDY 1: SIMULATION OF VENTILATIVE COOLING IN BEIJING AND SHANGHAI

We evaluated the performance of two passive-cooling strategies, daytime ventilation and night cooling, for a generic, six-story suburban residential in Beijing and Shanghai. The investigation used a coupled, transient-simulation approach to model heat transfer and airflow in the units. Wind-driven ventilation was simulated with computational fluid dynamics (CFD). Occupant thermal comfort was assessed with Fanger's comfort model (ISO 1994). The results showed that night cooling is superior to daytime ventilation. Night cooling was an acceptable alternative to air-conditioning systems for a significant part of the cooling season in Beijing, but with a high condensation risk. For Shanghai, neither of the two passive cooling strategies could be considered successful. Readers not interested in calculation details are invited to skip over the description of the methodology and resume with *Cases Studied.*

Methodology

The performance study of the two passive-cooling systems required accurate modeling of the building's thermal response to outdoor climate conditions and internal heat gains. The thermal response, in conjunction with ventilation and thermal comfort models, can determine whether there is a need for additional mechanical cooling systems. An effective passive cooling system requires little or no mechanical cooling.

Kammerud et al (1984) used a building thermal analysis program (BLAST) to study the effect on the cooling load of a night cooling strategy in a typical house for several locations in the United States. He cited two main problems encountered in his work: difficulty in modeling natural ventilation in the building and incorrect estimation of the convective heat transfer coefficients (these coefficients depend on the air velocity). This investigation used a model that addressed these two problems. PHOENICS (CHAM 1998), a CFD program, was used to calculate natural ventilation in the unit. PHOENICS calculated detailed airflow in and around the building. An experimental correlation for naturally ventilated buildings (Chandra and Kerestecioglu 1984) was used to calculate the convective heat transfer coefficients as a function of the air velocity near the walls. The detailed CFD results provided air velocity distributions near the walls, allowing for the accurate determination of the convective heat transfer coefficients. These coefficients were used as boundary conditions in the building thermal analysis. In this way, the approach used couples the thermal analysis with CFD. Figure 3 shows the coupled program structure. The starting point is the weather data (on the left). The data was required by both the ventilation calculation (upper box) and the thermal analysis (lower box). The CFD program predicted the air velocities near the walls in order to determine the convective heat transfer coefficients for the thermal analysis. Finally, the results of the airflow simulation and thermal analysis were used as inputs for the thermal-comfort model (ISO 1994).

The thermal response of the building walls and floors was calculated with an explicit finite difference method (Mills 1995) using a one-dimensional heat transfer approximation. The radiative heat transfer was calculated using the radiosity method (Mills 1995). The program separately calculated the short-wave radiation and infrared radiation. To reduce the number of simulations, the study used a fixed airflow pattern for each outside wind direction and velocity, independent of the thermal conditions. This approximation is acceptable because the work done by the thermal buoyancy forces is smaller than the momentum in cross-ventilated, single-story-building airflows.

User control of windows was simulated on the basis of the following common-sense actions:

- if the outdoor air is colder than indoors, and the indoor temperature is above 24°C, the occupants will open the windows;
- if the indoor temperature is below 24°C at the end of the day, the occupants will keep the windows closed for the night;
- for daytime ventilation, the occupants will open the windows to increase convection if the unit temperature exceeds 28°C; and
- during the day, occupants turn on ceiling fans.

Cases Studied

Figure 4 shows the north-south-oriented building and unit studied. It was a six-story residential building with ten units (two units per floor) and a ground floor used for storage. This building was isolated and located in an open suburban area. Figure 4b shows the internal layout of the units. Each unit had three bedrooms and a total floor area of 115 square meters. Access to the units was through an external corridor on the north side of the units. There was an external balcony on the south side of the living room.

Several features made this building suitable for cross ventilation. The separation of the living room from the three bedrooms allowed for high ventilation rates in the living room during the night when the occupants were in the bedrooms. This arrangement greatly enhanced nighttime heat release from the living room surfaces. The large apertures above the internal doors allowed air to flow through the bedrooms even when the doors were closed. External shading was used in all south windows to block the summer sun. When the unit was in maximum ventilation mode, both the balcony window and the north windows in the living room were fully open. During the day, ceiling fans were used in the unit to enhance convective heat transfer between the thermal mass and the occupants.

The windows were double-glazed with aluminum frames without thermal breaks. This is the typical design for this project but is not recommended for future designs because of their additional heat loss. The flooring material was ceramic tile. The living room windows were partially (50 percent) covered with light venetian blinds with a shading coefficient of 0.33. Table 2 defines further the cases studied. The internal gains, as shown in Table 3, were estimated according to ASHRAE recommendations (ASHRAE 2001).

The reference case, as shown in Table 2, had a reasonable amount of thermal mass, which is also true for most of the existing buildings in Beijing and Shanghai. Because most of these buildings and current designs do not use passive cooling strategies, this study estimates that in conventional buildings (represented by the reference case), the flow rate due to natural ventilation is 30 percent of that in our designs. The reference case did not use a ceiling fan. The daytime ventilation case had only 20 percent of the window area open during the night; 100 percent of the windows were open during the night for the night cooling case. During the daytime, cross ventilation was used whenever the indoor air temperature was higher than the outside air temperature. The daytime ventilation case used a lighter structure than the other cases.

Results

The analysis uses two criteria to evaluate the passive cooling systems: (1) the number of hours of thermal

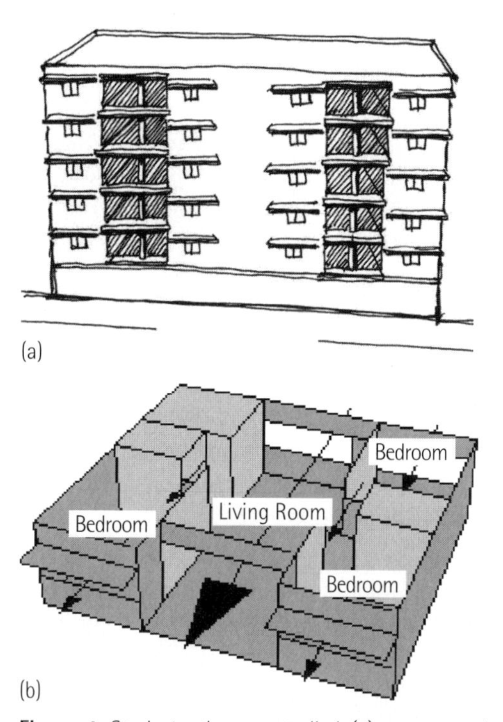

Figure 4 Study 1 - the case studied: (a) represents the six-story residential building simulated, and (b) shows the detailed layout of a unit in the building with north winds

Case	Composition of the Partitions: Floor	Composition of the Partitions: Internal Wall
Reference	10 cm concrete	10 cm concrete
Night cooling	10 cm concrete	10 cm concrete
Daytime ventilation	10 cm perforated concrete	lightweight (gypsum board)

Table 2 Study 1 - description of the cases studied

Lighting and Appliances		Occupants	
Day	Night	Day	Night
350 W	175 W	4 x 85 W	4 x 85 W

Table 3 Study 1 - internal gains used in the calculation

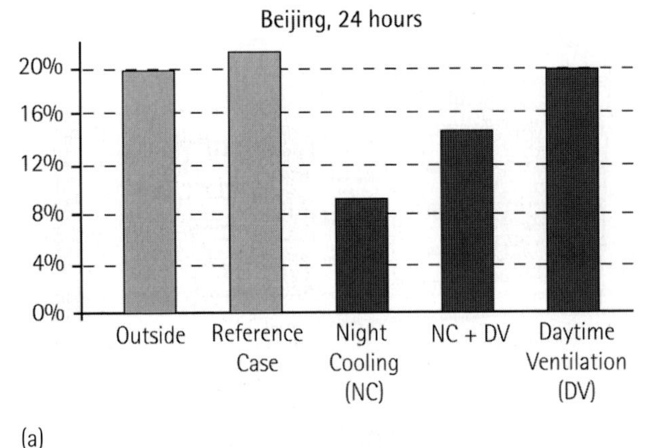

(a)

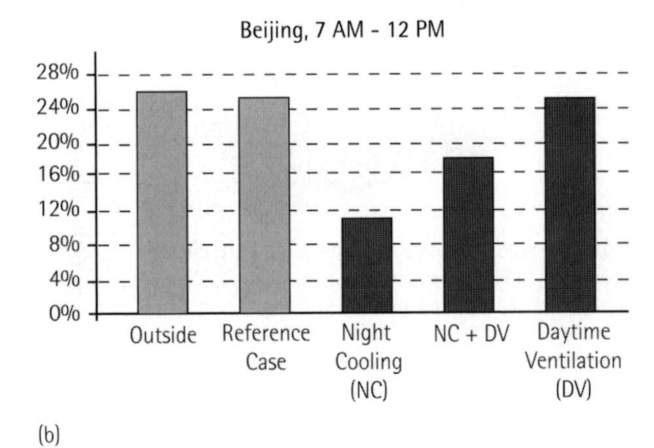

(b)

Figure 5 Study 1 - percentage of discomfort hours using different cooling strategies in Beijing (a) for the whole day, and (b) from 7 AM to 12 PM

discomfort during the warm season (from 1 May to 30 September), calculated using Fanger's thermal-comfort model (ISO 1994); and (2) the maximum indoor air temperature for each day in the warm season.

A supplementary criterion, used only for the warmest days of the season (maximum T_{out}>30°C), is the average difference between the maximum indoor and outdoor air temperatures.

The Performance of Passive Cooling Systems in Beijing Figure 5 shows the percentage of discomfort hours in the warm season. Figure 5a is for the whole day and Figure 5b from 7 AM to 12 PM. About 95 percent of the discomfort hours occur in the period between 7 AM and 12 PM. The results show that night cooling is very effective for Beijing. There are only 330 discomfort hours, 9 percent of the total number of hours in the warm season, and a reduction of 437 hours, or 57 percent, from the reference case. Daytime ventilation improves thermal comfort only slightly relative to the reference case; the number of discomfort hours is close to that under outdoor conditions. This is a consequence of two competing effects: as air moves through the unit during the day it not only removes the heat gains but it also heats the walls of the unit. The case labeled NC+DV uses the cooling control strategy during the night and the daytime ventilation control strategy during the day. For this case, the results are much better than the reference case, but still not as good as the night cooling case.

Figure 6 shows the temperature evolution over a three-day period in Beijing for night cooling. During the night period, the indoor air temperature is close to the outdoor temperature, except when the wind velocity is low. In these low-wind periods, the indoor temperature approaches the wall temperature; this is visible in the sharp peaks present in the indoor temperature profile in the second and third night periods. Figure 7 shows the temperature evolution over a similar period in Beijing for daytime comfort ventilation. One noticeable difference compared with Figure 6 is that the indoor temperature closely follows the outdoor temperature. The lower thermal mass for this case, noted in Table 2, is visible in the larger variations in the wall temperature.

Figures 8 and 9 show the maximum indoor and outdoor air temperatures in the warm season for the two cooling strategies. The air temperature in the living room with night cooling is noticeably lower than the outside air temperature. The maximum air temperature in the living room in the season is 31.6°C. On average, the indoor air temperature is 3.9°K cooler than outdoors. With daytime ventilation, the windows are open whenever the air temperature in the living room is higher than the outside air temperature. The maximum indoor air temperature exceeded 34°C on eight days. These graphs provide more evidence that night cooling is superior to daytime ventilation in Beijing.

The Performance of Passive Cooling Systems in Shanghai Shanghai is warm and humid and has a very small daily temperature variation, as shown in Table 1. The climate makes it difficult to successfully use natural ventilation for cooling. Nevertheless, Figure 10 shows that the two passive-cooling systems can help to achieve more comfortable conditions than the two reference cases. Night cooling is the best among the four cases but the improvement is modest when compared with daytime ventilation or the reference case. This is due to the small daily temperature variation. Although the maximum outdoor air temperature is almost the same as that in Beijing, the small temperature variation outdoors does not allow for significant cooling of the building thermal mass.

Even with night cooling, the maximum indoor air temperature can be as high as 34.9°C. On average, night

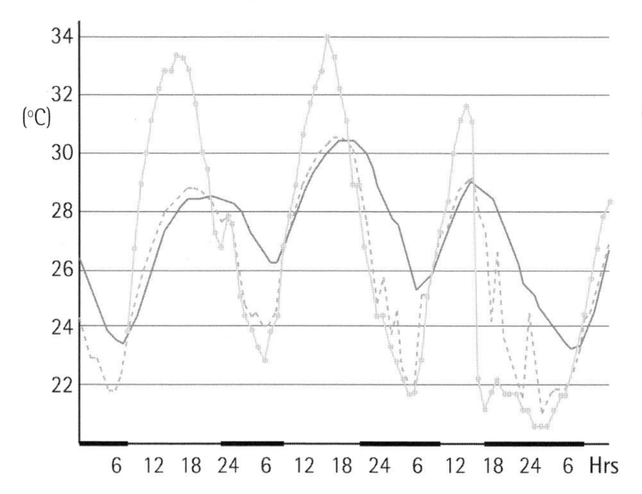

Figure 6 Study 1 - hourly temperature variation for Beijing night cooling – light gray with circles: outside temperature; medium grey (dashed): wall temperature; dark gray (solid): indoor temperature. The dark line on the horizontal axis indicates periods when the windows were fully opened.

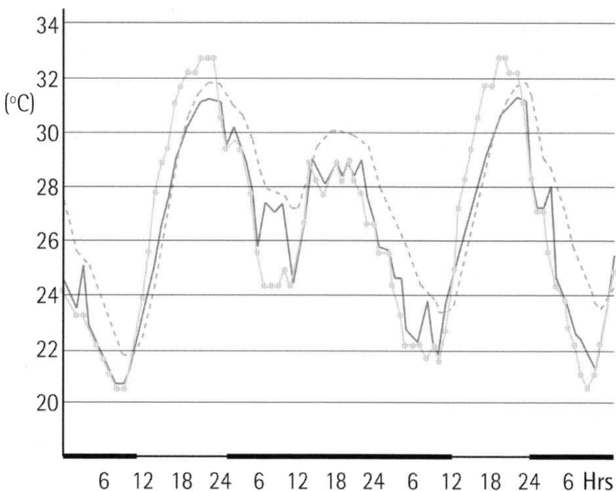

Figure 7 Study 1 - hourly temperature variation for Beijing daytime ventilation – light gray with circles: outside temperature; medium gray (dashed): wall temperature; dark gray (solid): indoor temperature. The dark line on the horizontal axis indicates periods when the windows were fully opened.

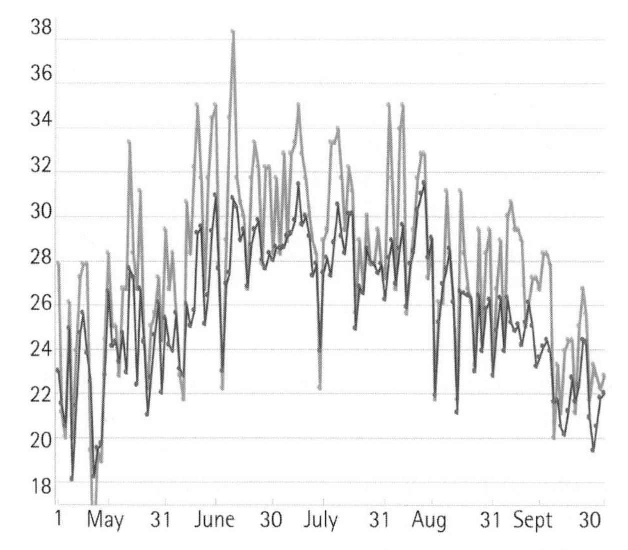

Figure 8 Study 1 - daily maximum air temperature in the living room with night cooling (dark gray) and outside (light gray) in Beijing

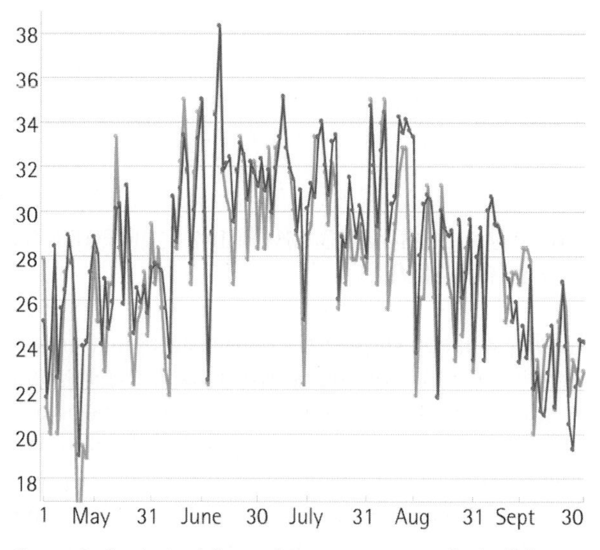

Figure 9 Study 1 - daily maximum temperature, in the living room with daytime ventilation (dark gray) and outside (light gray) in Beijing

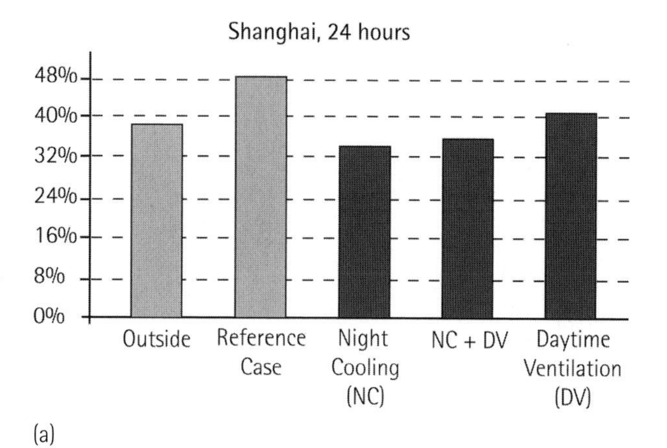

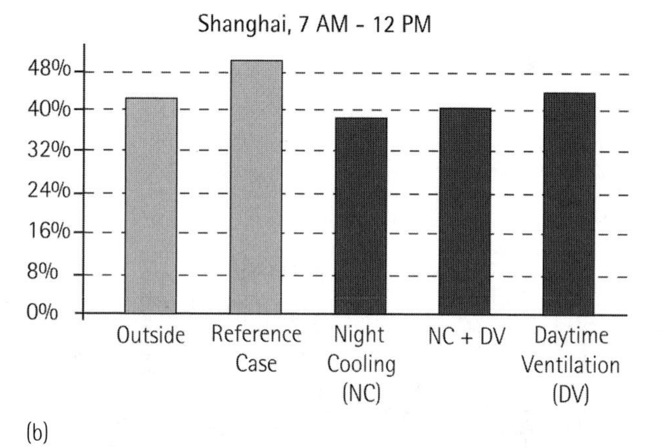

Figure 10 Study 1 - comparison of the discomfort hours using different cooling strategies in Shanghai (a) for the whole day, and (b) from 7 AM to 12 PM

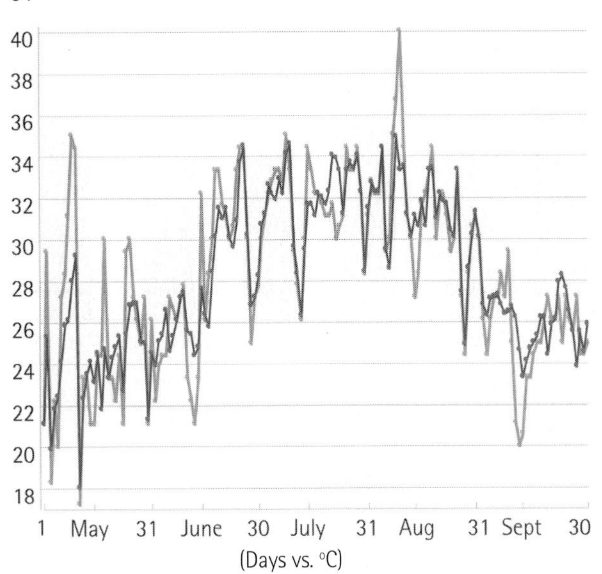

Figure 11 Study 1 - daily maximum air temperature in the living room (dark gray) and outside (light gray) in Shanghai

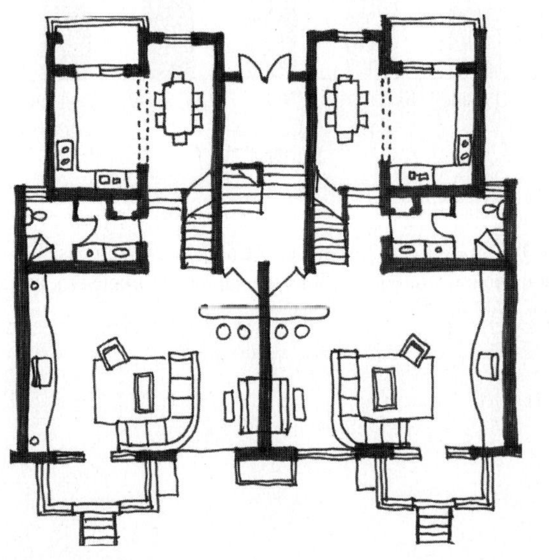

Figure 12 Study 2 - plan view of base case, an existing four-family residential building in Beijing

cooling only lowers the room air temperature by 0.9°K, as shown in Figure 11. There are 29 days in the season with a maximum daytime temperature above 32°C, and 32 days with a maximum daytime temperature between 30°C and 32°C. The indoor air is too hot to be acceptable for the occupants. The results show that using the two passive-cooling systems alone in the warm season in Shanghai is not enough to provide comfortable indoor thermal conditions.

Night cooling can cause condensation on interior surfaces by lowering their temperature below the dew point of the air brought into the building. Somewhat surprisingly, there is less risk of condensation on interior surfaces in Shanghai than in Beijing, even though the outdoor relative humidity is higher; this is because the higher outside temperatures are not as effective in lowering the mass temperature. Our calculations show the potential for only 8 hours of condensation in Shanghai under a night cooling strategy, but 60 hours in Beijing. The risk of condensation can be lowered by restricting the amount of night cooling, by reducing the size of window openings. However, this will lead to higher indoor temperatures the next day.

STUDY 2: IMPROVING THERMAL COMFORT IN BEIJING HOUSING

This investigation considered whether careful attention to architectural design and use of mechanical ventilation are sufficient to maintain thermal comfort in dwellings in Beijing without the use of vapor-compression air-conditioning equipment. The study was motivated by a desire to define strategies that would reduce building operating costs, energy demands, potential environmental impacts associated with energy usage, and potential occupant dissatisfaction with air-conditioning due to reduced outdoor-air intake and large temperature changes associated with entering or leaving a building. Results show that drastically lowering solar gains and use of large, fan-driven airflows at night are sufficient to reduce indoor peak temperatures to as low as 29°C, when the outdoor peak is 33°C. Further reductions to 26.5°C are possible as internal gains are reduced to the limit of being completely eliminated.

Description of Building and Simulation Methodology

The study was undertaken in support of the design of multi-family, low-rise housing at a development, Beijing City Garden, near the Beijing airport. The developer had already constructed most of the buildings at this site but invited researchers at MIT and Tsinghua University to develop and analyze a design that would be more energy efficient. The site and the prospective location of the low-energy building are shown in Jiang et al (1999). To help guide the design process, we simulated an existing building as a base case and then, in simulation, modified the design and operation. This structure was a 10 meter x 12 meter, five-story (with an additional basement level), four-unit building, freestanding in our simulation, but in practice likely to be attached to adjacent dwellings. Residential units in this development were considered to be luxury housing and were relatively large, about 112 square meters. Figure 12 shows one floor in this building, with living and dining rooms. Our specification of building materials, the size and orientation of windows, and wall and roof construction were based on available information about the building. While internal zoning may be important in practice, for simplicity we simulated the building as a single thermal zone.

We conducted our simulations with the DOE-2.1E simulation program (AEC 2006). As simulated, the building had no air-conditioning. Our goal was to minimize the number of hours over the year when the indoor temperature exceeded 29.4°C. This temperature value was automatically selected by the simulation program as the lower bound of very hot temperatures and corresponded to 85°F. As noted previously, this indoor temperature can be justified as thermally acceptable.

We characterized each simulation run by plots of hourly indoor and outdoor temperature over the course of a single July day, when outdoor temperatures reach 33°C. As revealed in Figure 2, this is a representative hot day, with a peak temperature exceeded on only a few days. It also matches the 33.2°C dry-bulb temperature specified for air-conditioning system design in Beijing (National Standard of PRC 1987).

We conducted two series of simulations. First, we examined reasonable alterations to the building:

- internal versus external insulation for exterior walls;
- wind-driven and mechanical ventilation of the building (7.5, 15, and 30 air changes per hour (ACH)); and
- shading of south-facing windows, with exterior, horizontal, fixed shades.

Second, we examined alterations that would be difficult to achieve in practice but which point toward substantial improvements, even if only partially realized:

- no absorption of solar radiation by the roof and exterior walls (modeled with an absorptivity of zero but achievable with shading or very high thermal resistance);
- greatly reduced transmission or absorption of solar radiation by windows (shading coefficient of 0.14); and
- reduced internal gains.

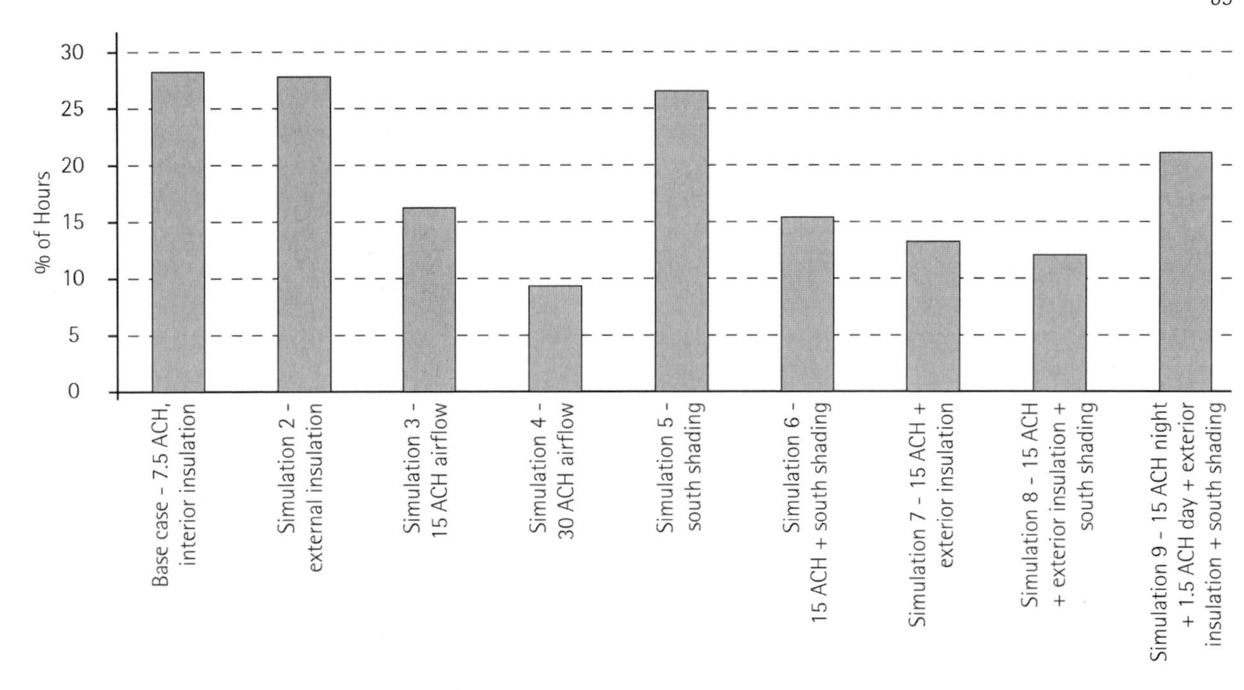

Figure 13 Study 2 - percentage of discomfort hours for variations in building construction, shading, and ventilation for Series I simulations

Finally, in conjunction with the second series of measures, we restricted ventilation to cooler night hours. As we will show, this strategy is counterproductive until solar gain is controlled.

Series I Simulations

Results for the first series of simulations are summarized in Figure 13, which presents the number of hours over the five-month simulation period when the indoor temperature falls into specified temperature bins (a total of 3,648 hours). Results for each case will be discussed next.

In addition to building-construction details, the base case is defined by the air-change rate and the schedule and magnitude of internal heat gains, both of which strongly affect indoor temperatures. We estimated a wind-driven airflow of 7.5 ACH, deriving this value from nodal airflow equations and an estimate of both open-window area and wind speed at the building façade. We used a southerly wind of 0.5 m/s, which was reduced from the annual mean meteorological value of 1.9 m/s (Chen and Cai 1994) by the screening effect of neighboring buildings (Jiang et al 1999).

Internal gains include lights, equipment, and occupants. We modeled lighting with a peak power of 8 W/m^2 and 7 hours of operation, with 2 hours in the morning and the remainder in the evening. We assigned an average of 3 W/m^2 for equipment, peaking in the

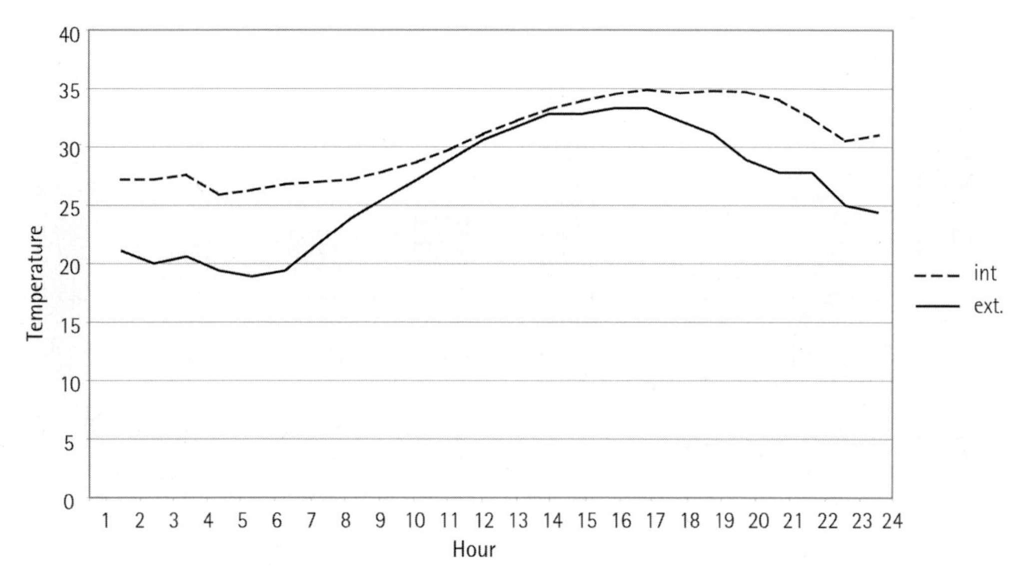

Figure 14 Study 2 -hourly indoor and outdoor dry-bulb temperatures for the base case (7.5 ACH, 16 July of TMY weather data)

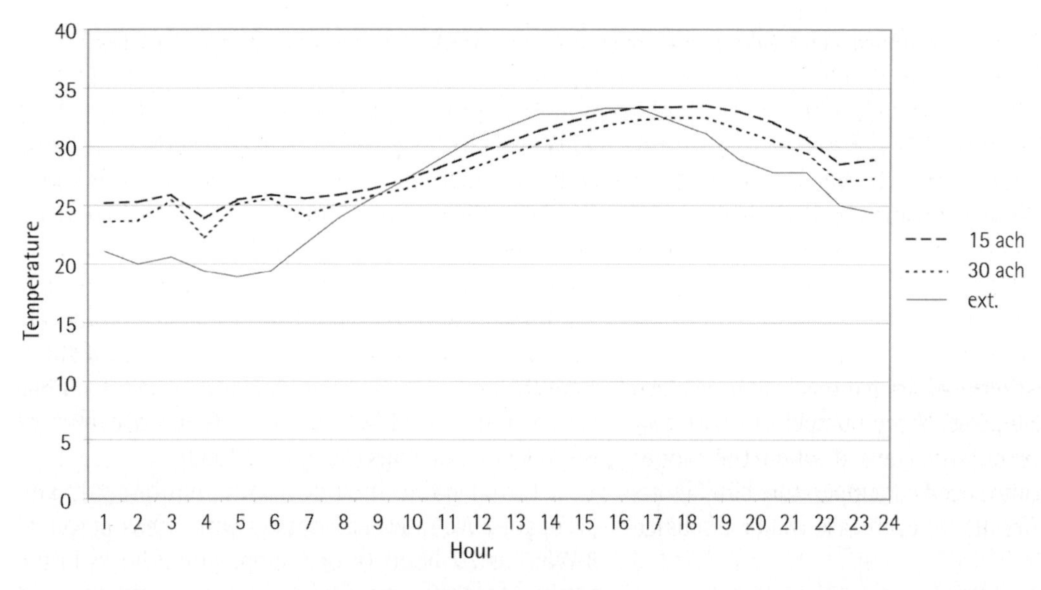

Figure 15 Study 2 - hourly indoor and outdoor dry-bulb temperatures for mechanical ventilation with 15 and 30 ACH (16 July of TMY weather data)

morning and evening, and simulated an occupancy of four people, not at home during the day. The average lighting and equipment load per unit was 470 W, appropriate for a single-family residence (ASHRAE 2001). ASHRAE notes that a lower figure of 350 W is appropriate for a unit, but that computers or other equipment may increase that value. We investigated the impact of lowering the internal loads as part of our study. The sensible heat gain from occupants was 67 W/person, appropriate for sedentary occupants (ASHRAE 2001).

Indoor temperatures are uncomfortable (> 29.4°C) 28 percent of the summer hours for the base case. In practice, the number of hot hours may be even larger, because occupants often enclose their balconies to make a sunspace, thereby eliminating the benefit of the balcony floor as a shade for windows and balcony doors below it. Figure 14 shows the diurnal variation of indoor and outdoor dry-bulb temperatures for a single July day. The indoor temperature always exceeds the outdoor temperature, peaking at 35°C later than the outdoor peak of 33°C.

Subsequent simulations shown in Figure 13 explored methods to reduce the number of uncomfortable hours. Simulation 2 placed insulation on the exterior rather than interior surface of exterior walls, to promote the flow of heat into and out of building mass. This may reduce indoor temperatures during very hot weather but will also reduce the benefit of adjusting the thermostat to lower values in winter during unoccupied or nighttime hours. Thermostat adjustments would also make sense in summer if the occupants of such a building found it necessary to use mechanical cooling. Our simulations focused only on summer months and indoor temperatures were allowed to float, so effective building thermal mass should in principle be of some benefit. Figure 13 shows that the benefit of external insulation, with no other measures, is extremely modest. For the base-case building, therefore, the location of insulation should be made on the basis of other factors, including ease of construction and the elimination of thermal bridges, which can occur where interior and exterior walls join and which we have not simulated. Exposed thermal mass can have a more beneficial impact when other changes to the building are made.

Simulation 3 focused on mechanical ventilation. Jiang et al (1999) showed that wind-driven natural ventilation was constrained at the building location in question because the demonstration building was very close to a tall building to the north and low-rise buildings to the south, both of which blocked ambient wind. To enhance low levels of natural ventilation, we proposed the use of one or more central fans to exhaust air from the units in the building. Around-the-clock ventilation at 15 ACH reduced the uncomfortable hours to 16 percent of the total. This air-change rate corresponded to a volumetric flow of about 2 m^3/s for each of the 4 units in the building and an average flow velocity of about 1 m/s through open windows. Such a ventilation rate appeared very feasible; we measured an average air speed of 2.8 m/s and a volumetric flow of about 0.7 m^3/s for a 0.25 m^2 fan that could be placed in a window. An improvement over such fans would be a single exhaust fan for each unit.

Simulation 4 increased the air-change rate to 30 ACH, reducing the uncomfortable hours to 9 percent of the cooling season. Figure 15 shows the impact of mechanical ventilation on indoor temperatures over a diurnal cycle. Note that with 15 ACH, the indoor peak relative to the base case (shown in Figure 14) has been reduced by 2°K, and that an additional doubling of the airflow rate further reduces the peak by a smaller amount, 1°K.

Simulations 5 and 6 quantified the impact of adding shading to south-facing windows that did not already have it in the base case. Most of the openings in the south façade were sliding doors that opened to balconies and were shaded by balconies on the floor above. Adding shading to the remaining south windows reduced the uncomfortable hours to 27 percent, only one percent better than the base case. Shading south-facing windows had an equally modest impact when added to a case that included mechanical ventilation: 15 percent uncomfortable hours, compared to 16 percent for simulation 3.

Simulation 7 combined exterior insulation with around-the-clock mechanical ventilation set at 15 ACH. With increased airflow to take away heat stored in exposed building mass, exterior insulation had a bigger impact than in the absence of increased ventilation, reducing the uncomfortable hours from 15 percent in simulation 3 to 13 percent.

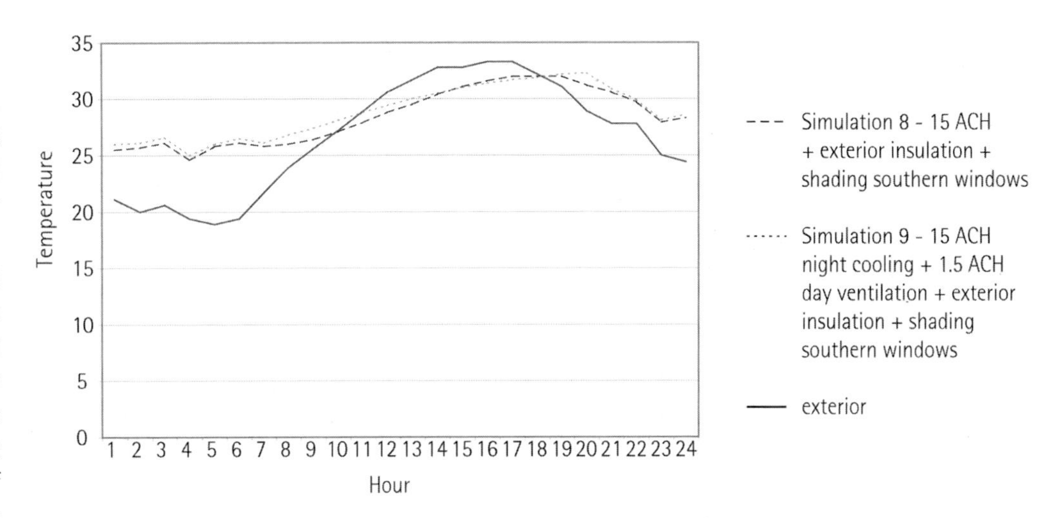

Figure 16 Study 2 -hourly indoor and outdoor dry-bulb temperatures for two cases: continuous and night-only mechanical ventilation, each with shaded southern windows and external insulation

Simulation 8 combined 15 ACH of ventilation with exterior insulation and shading of southern windows, a reasonable combination of strategies that reduced the uncomfortable hours to 12 percent. Simulation 9 reduced airflow during the day (8 AM-9 PM) to 1.5 ACH and retained 15 ACH at night. This strategy led to higher indoor temperatures because there were both solar and internal heat gains during the day. The number of uncomfortable hours increased to 21 percent. For the same day as used previously, diurnal temperatures for these two cases are plotted in Figure 16. Indoor peaks are below the outdoor peak, improving thermal comfort relative to the base case.

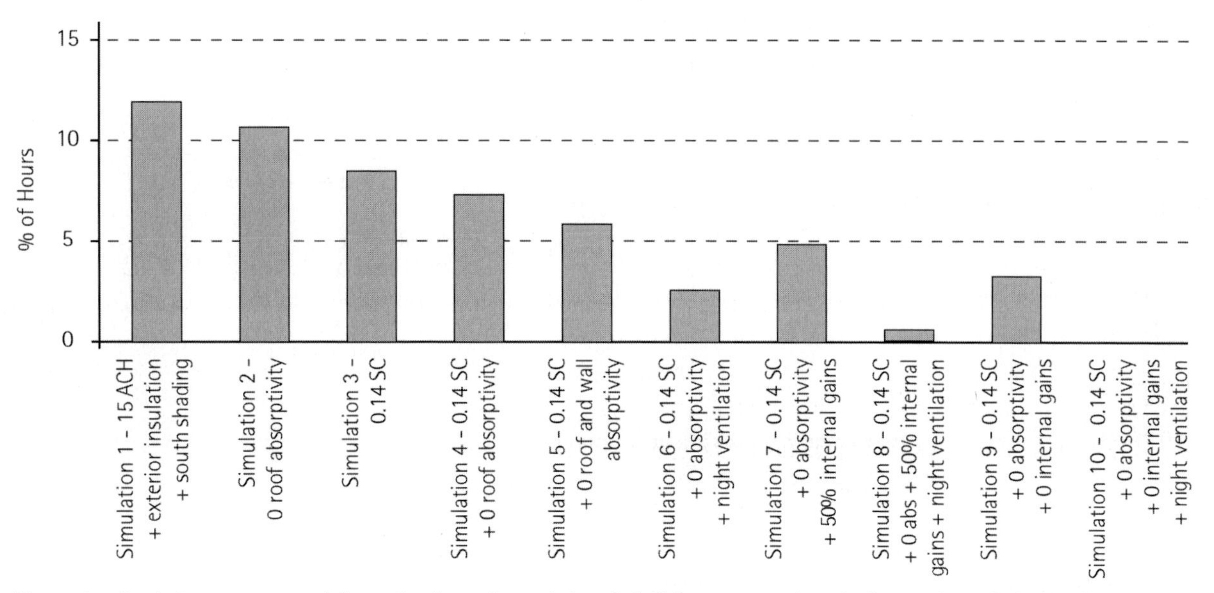

Figure 17 Study 2 - percentage of discomfort hours for variations in building construction, shading, and ventilation for Series 2 simulations

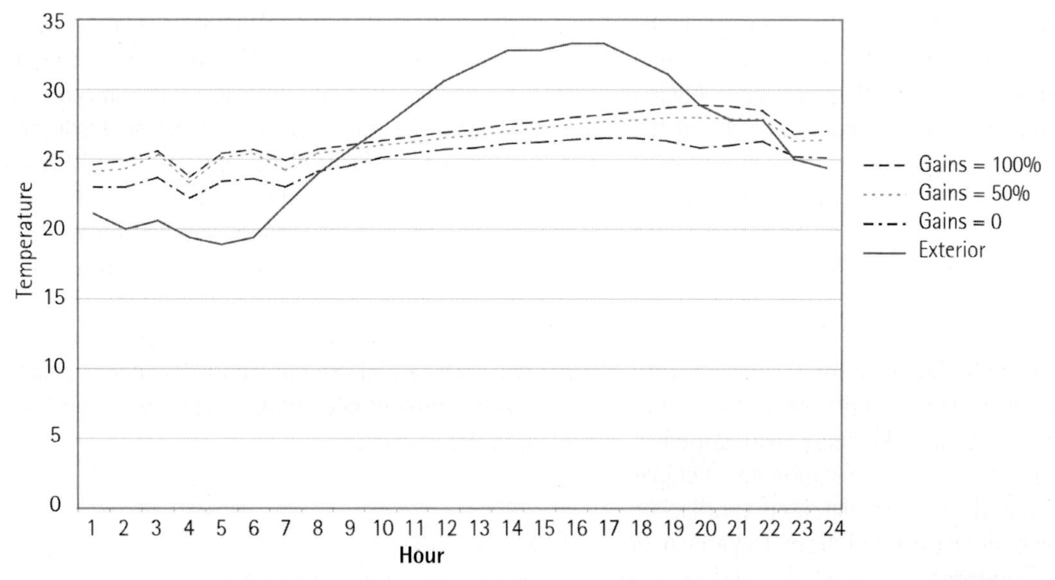

Figure 18 Study 2 - diurnal indoor and outdoor temperatures when the building is sealed during the day and ventilated at night, with reduced internal heat gains (16 July of TMY weather data)

Series 2 Simulations

The second series of simulations is summarized in Figure 17. Simulation 1 is identical to Simulation 7 from the first set of simulations and defines reasonable measures to reduce indoor overheating. Simulation 2 of the second series showed a further modest improvement by eliminating solar heat gains through the roof. This can be done in practice by heavily insulating the roof or venting a cavity under the outer roofing material and above the living space. For the simulation, we set the solar absorptivity to zero. Simulation 3 achieved a larger improvement by shielding windows rather than the roof from solar heat gains. By setting the shading coefficient for all windows to 0.14, lower than the base case value of 1.0 (single-pane, clear glass), the number of uncomfortable hours was reduced to 9 percent of the total.

Simulations 4 and 5 explored further reduction in solar heat gains, first by combining the benefits of reduced solar gain through the roof and windows, then by including a reduction of solar gain through the external walls. Reducing the absorptivity of exterior surfaces to zero is not fully achievable in practice, of course, but careful specification of materials and finishes and the use of aggressive shading, including vegetation, may substantially reduce solar heat. Shading with deciduous trees has the further benefit of allowing solar radiation to strike walls in winter. Eliminating solar gain through the roof and reducing solar gain through windows lowered the number of hot hours to seven percent of the total; by eliminating solar gain through walls as well, the uncomfortable hours were reduced to six percent.

Simulation 6 revisited a night-only ventilation strategy. Recall that such a strategy was counterproductive when solar gains were not controlled.

With no solar gains, night-only ventilation dropped the number of hot hours to only three percent of the total.

Simulations 7 through 10 explored variations of reduced internal gains and night-only ventilation, all with the drastic control of solar gains. Simulation 7 showed that reducing internal gains to 50 percent of the base case had substantially less impact than shifting the ventilation strategy from around-the-clock to night-only. Our estimate of internal gains for the base case was not grounded in measurements in Chinese buildings, and this seventh simulation could be considered an exploration of uncertainty as much as an energy-conservation strategy. Based on personal observations in China, there is some opportunity to reduce internal gains through more efficient lights and appliances. Simulation 8 combined lower internal gains and night ventilation to leave less than one percent uncomfortable hours. Simulation 9 showed that eliminating internal gains altogether but continuously ventilating the building yields three percent hot hours. Simulation 10 combined previous measures - the elimination of absorbed solar radiation, the elimination of internal heat gains, and the reduction of solar radiation through windows via transmission or absorption, with sealing the building during the day and ventilating only at night to produce the best results. The hot hours were entirely eliminated.

Figure 18 shows simulated hourly temperatures for the three runs that incorporated effective night-only ventilation, those with internal gains at the base-case level, at 50 percent of the base case, and entirely eliminated. The resulting indoor temperature was remarkably flat and never exceeded 29°C, even when internal gains were at the base-case level. The diurnal-average indoor temperature was only 0.1°K higher than the outdoor average of 26.6°C. With internal gains reduced by 50 percent, the peak indoor temperature was 28°C; in the limit, with no internal gains, the peak indoor temperature was 26.5°C. In this limit, the diurnal-average indoor temperature was 1.6°K lower than the outdoor average, a reduction achieved by selectively ventilating the building only at night, when outdoor temperatures are low.

The elimination of the need for mechanical cooling in this limiting case has a value that can be estimated by comparison with the base case. Base-case simulations showed that 11.1 MWh of thermal energy would be required to keep indoor temperatures from exceeding 29.4°C over a summer. Vapor-compression cooling equipment with a coefficient-of-performance (COP) of 3 would use 3.7 MWh of electrical energy to meet this cooling load, at an approximate cost of US$220, or 1,840 RMB, per year for the test building, which included six units. If installed air conditioners were used to reduce indoor temperatures to below 29.4°C, as is likely, energy costs would increase. In addition, occupants who choose to rely solely on ventilative cooling will reduce first costs by buying a fan rather than an air conditioner.

Net energy savings must account for fan power needed for ventilation. We did not estimate fan energy usage in our simulations but instead simply set airflow across the building envelope at a specified amount. An airflow of 15 ACH corresponds to 6 m^3/s for the test building. If airflow resistance is, to first order, limited to two orifices representing inlet windows and an outlet exhaust duct, the latter of 1 m^2, fan power of about 1 kW would be required. Were such a fan to be used for the 1,032 hours when indoor temperatures exceeded 29.4°C in the base-case design, fan energy would total about 1 MWh, or 27 percent of the energy required by an air conditioner. Fan energy could be substantially lowered by using a variable-speed drive and reducing fan speed and airflow at off-peak conditions, although it is likely that occupants would prefer more airflow and lower temperatures.

For housing marketed to the more affluent, as is the case with the development we studied, it would be necessary to eliminate all or nearly all uncomfortably warm indoor hours to give residents a reasonable alternative to vapor-compression cooling. It appears possible to do so in Beijing, albeit not easily. Simulations such as we performed can be extended to consider specific construction proposals to reduce solar gains, for example. They can and should be tested by comparison of predictions with measurements in occupied buildings. Further, the simulations can form the foundation of an owner's manual that would describe to residents the importance of controlling solar and internal gains and properly operating ventilation fans.

STUDY 3: LOW-ENERGY BUILDING DESIGN IN SHENZHEN

This study focused on a new residential development in Shenzhen. The issues addressed included siting of the buildings, shading of windows, and control of noise. The study produced a set of recommendations aimed toward reducing energy consumption and noise levels in the proposed houses.

Figure 19 shows the site as modeled for an airflow study. The existing buildings are at the top of the figure to the east of the proposed houses. Note that the proposed buildings are very strongly aligned on east-west axes, to reduce unwanted morning and afternoon solar gain, when the sun is lower in the sky and windows are harder to shade. The proposed buildings are generally low-rise but increase in height on the northern edge, to shield the development from noise from trains running on tracks adjacent to the site.

Noise control, while tangential to the central issues of this book, strongly influenced our proposal to the developer. Recommendations focused on the dwellings adjacent to the rail line and included heavy wall construction on the side (north) facing the tracks, small, sealed windows, no ventilation through the north side, and careful sealing of cracks.

Energy consumption was studied in a series of simulations. Table 4 shows the characteristics for a base-case house, derived from typical practice. An assessment of a base-case dwelling showed that solar heat through windows dominated summer heat gains (Figure 20). Based on these loads, strategies to reduce cooling loads were recommended; these are covered in the next four subsections.

Window Orientation, Shades, and Overhangs

Solar gain through windows should be minimized by orienting buildings on an east-west axis as shown in Figure 19, with as much surface area as possible facing north and south. Designers should use overhangs and shades, and select glazing with low solar heat gain coefficients. Examples of shading devices are shown in Figure 21.

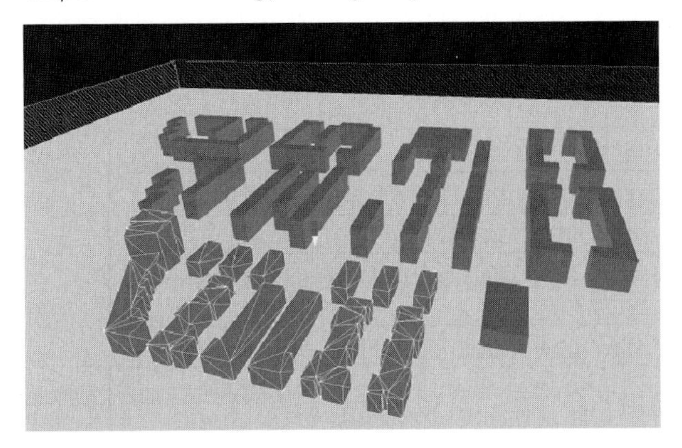

Figure 19 Study 3 - site diagram of the Shenzhen Wonderland housing development under investigation (north is to the left)

Building Component	Specification
Floor area	6 m x 12 m
Orientation	North/south facing façades
Wall type	Concrete block, no insulation
Window type	Clear single-pane, aluminum frame
Window area	40% window-to-wall area
Overhang type	No overhang
Electric heat pump	10 SEER / 7.8 HSPF
Heating set point	20°C
Cooling set point	24°C
Ventilation	No ventilation

Table 4 Study 3 - specifications for base case simulated building

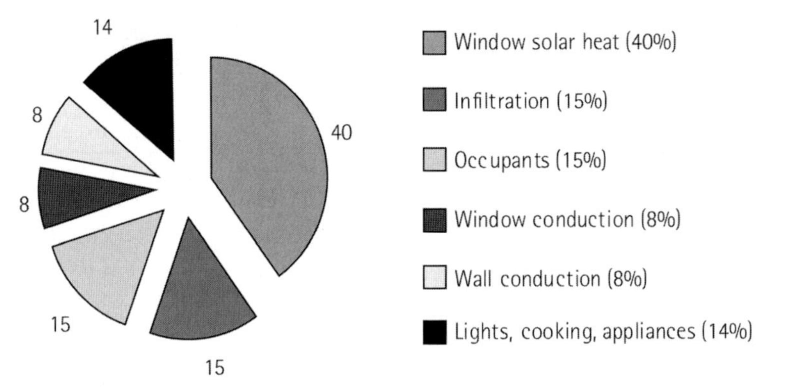

Figure 20 Study 3 - Breakdown of summer heat gains in base case dwelling in Shenzhen

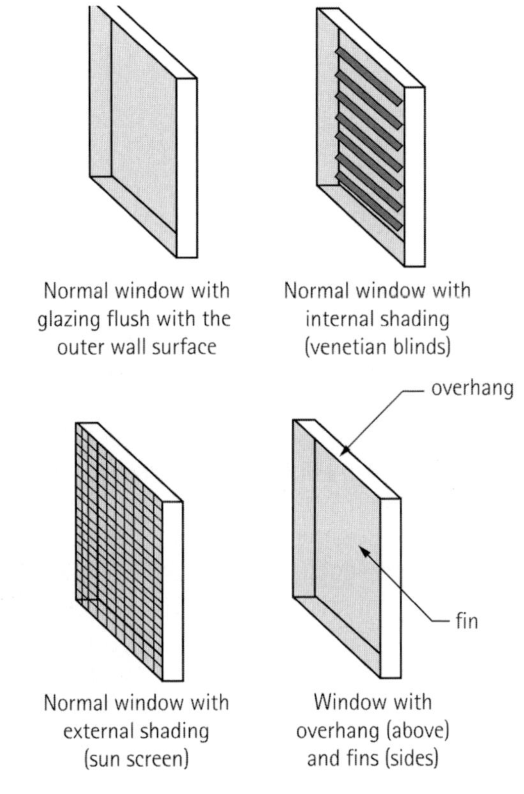

Figure 21 Window shading options

Daily solar-energy flux for windows was assessed through detailed simulation studies, listed in Table 5; results are shown in Figures 22 through 24. For east- and west-facing windows, the summer solar heat gains are in three distinct groups, each with three simulations. The base case, with a single layer of clear glass and no shading, yields the largest daily solar-energy flux for each month of the year. Vertical fins with a projection equal to 30 percent of the window width, Case 7, reduce the solar heat gain a modest amount, but little in the summer, when the reduction is most needed. Case 8, with larger vertical fins, provides little additional improvement. The next group offers summer solar heat gains about 40 percent below the base case. Case 1 reduces the solar gain via venetian blinds. The same effect in summer can be achieved with an overhang that extends a distance equal to 30 percent of the window height, as shown in Case 3. Adding vertical fins with a projection equal to 30 percent of the window width, as shown in Case 4, offers no additional benefit in summer. The third group drops the daily solar gain in summer to 20–30 percent of the base case. Case 6, with large overhang and fins, minimizes solar heat in summer, when desired, but also in winter, when more light and heat would be welcome. It also restricts views through the window. Case 5, with full overhang (100 percent of the height of the window) and no fins, does nearly as well in summer and brings in more energy in winter. A reflective coating, Case 2, matches Case 5 in summer. The performance offered by Cases 2 and 5 are a good target, reducing solar gain in summer to less than 30 percent of the base case (Figure 22).

Case	Shading Coefficient	Overhang %	Fin %	Description
0 (base)	1.00	0	0	3 mm clear, single-glazed
1	0.58	0	0	3 mm clear, single-glazed with venetian blinds
2	0.28	0	0	3 mm clear, single-glazed with exterior shade or 6 mm reflective double-glazed
3	1.00	30	0	Base case with overhang extending 30% of window height
4	1.00	30	30	Base case with overhang extending 30% of window height and fin extending 30% of window height
5	1.00	100	0	Base case with overhang extending 100% of window height
6	1.00	100	100	Base case with overhang extending 100% of window height and fin extending 100% of window height
7	1.00	0	30	Base case with fin overhang extending 30% of window width
8	1.00	0	100	Base case with fin overhang extending 100% of window width

Table 5 Study 3 - description of window simulation studies

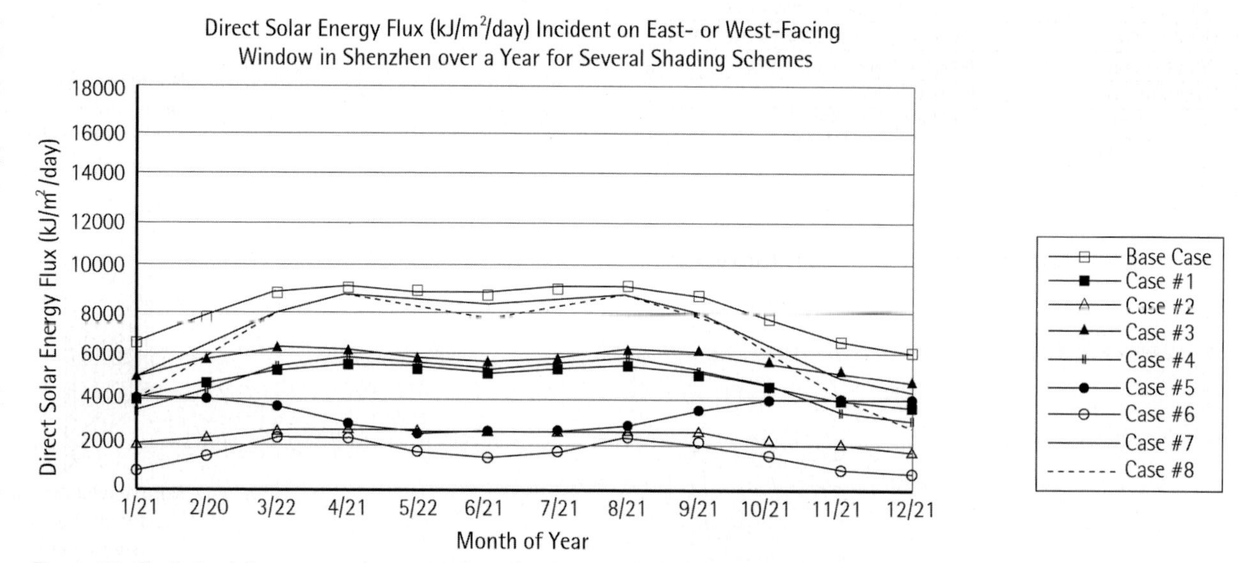

Figure 22 Study 3 - daily average solar energy through east- or west-facing windows, for each month and for a variety of window types

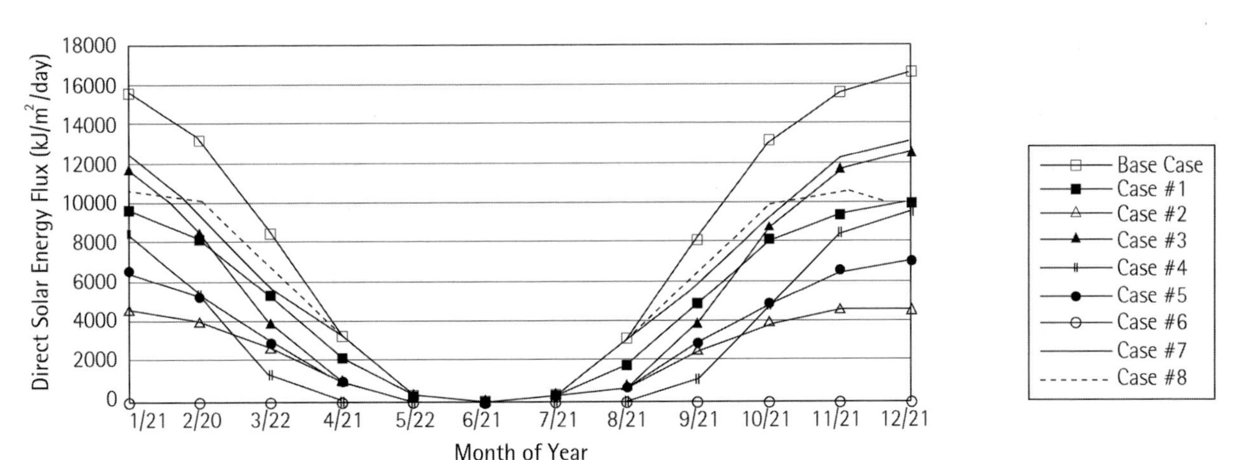

Figure 23 Daily average solar energy through south-facing windows, for each month and for a variety of window types

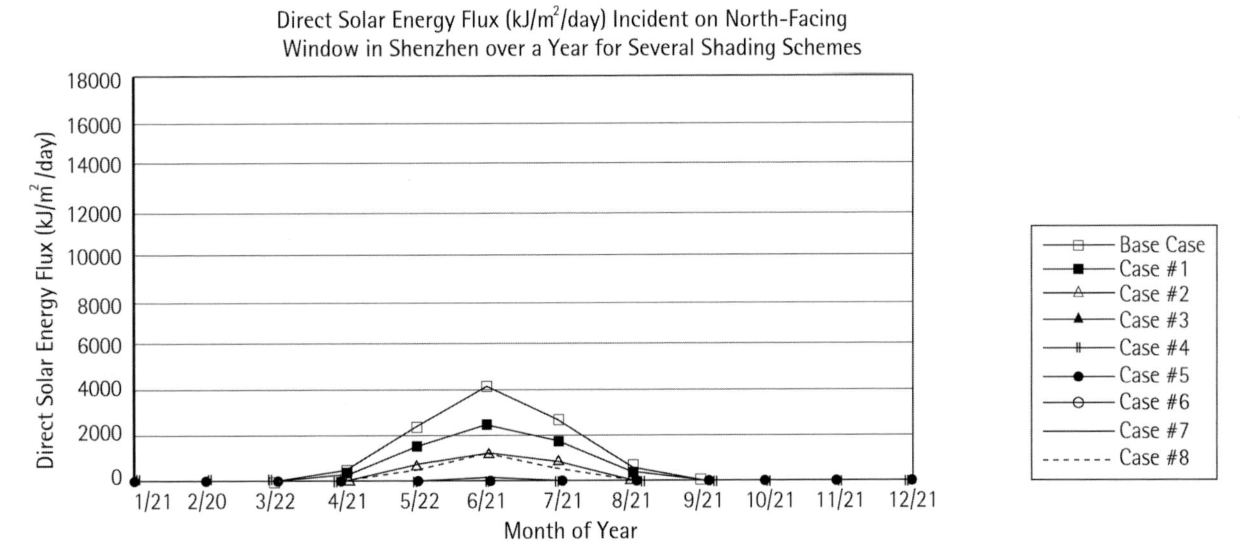

Figure 24 Daily average solar energy through north-facing windows, for each month and for a variety of window types

There is essentially no solar gain through south-facing windows in June for buildings in Shenzhen's latitude, regardless of window glazing or shading. This statement also applies, with little error, to May and July as well. However, window treatment makes a dramatic difference in late summer months, August and September, and in the matching spring months. The base case gives the highest solar gain and Case 6, with full overhang and fins, gives essentially no solar gain throughout the year. Vertical fins provide little reduction in solar gain and are not appropriate for south-facing windows, as Cases 7 and 8 indicate.

Better choices, in order of increasingly lower solar gains in September, are Case 1, venetian blinds; Case 3, a window with an overhang extending 30 percent of the height of the window; Case 5, a window with a 100 percent overhang; Case 2, reflective glass; and Case 4, a window with a 30 percent overhang and vertical fins that project a distance equal to 30 percent the width of the window. Case 4 is especially good at reducing solar heat gain in warm weather and allowing some heat in winter (Figure 23). Combining a shallow overhang with occupant-adjustable blinds, not simulated, would also be a good choice, allowing the user to reduce as needed the solar gains associated with a window with an overhang alone.

Solar heat gains for north-facing glass peak in the summer. For a single layer of unshaded, clear glass, the daily solar heat gain in June is significant: over 40 percent of the value for an east- or west-facing window. Shallow vertical fins effectively block early-morning and late-afternoon summer sun (Figure 24).

Insulation

There was little benefit from insulating walls, although insulation in units with east- and west-facing walls reduced the impact of absorbed solar radiation. As shown in Figure 25, insulation in the ceiling of top-floor units reduces total energy use by as much as 6 percent; attic ventilation provides additional savings.

Ventilation and Infiltration

Ventilation was found to reduce annual cooling energy consumption by about 24 percent, via use of outdoor air when conditions are considered comfortable. Figure 26 shows the monthly impact of ventilation at 10 ACH. For each month, savings are given for 10-hour-per-day (night) and 24-hour ventilation; the sum of the monthly savings gives the annual savings. From January through March and again in November and December, most of the savings are during the day, when heat loads can be removed by cool outdoor air. In April through June and again from September through October, when outdoor temperatures during the day are warmer, most of the savings are due to ventilation at night. Infiltration contributes to the building cooling load in hot weather, when outdoor temperatures exceed indoor temperatures. Fans would be needed when winds are not adequate to provide the desired airflow.

Equipment Efficiency and Occupant Habits

Simulations showed that further savings could be realized if high-performance air conditioners or heat pumps were installed. Simulations did not consider solar hot-water heating nor the benefits associated with an expanded thermal-comfort range.

Figure 27 shows the savings associated with a number of individual actions, including orientation, insulation, ventilation, and window shading. Note that a building with east-west orientation uses about 15 percent more

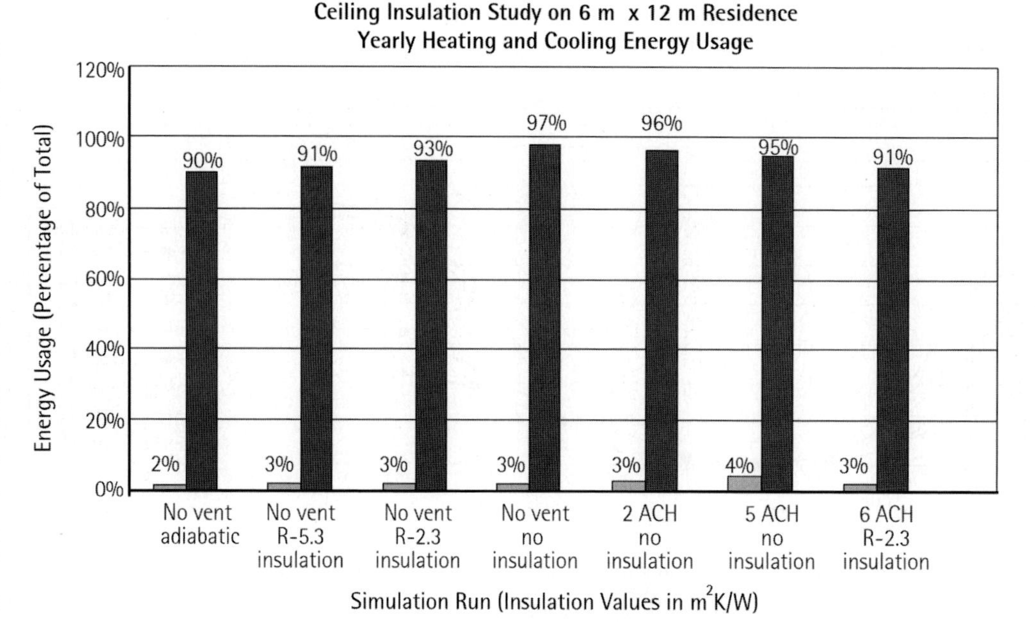

Figure 25 Impact of attic insulation and ventilation on total energy use

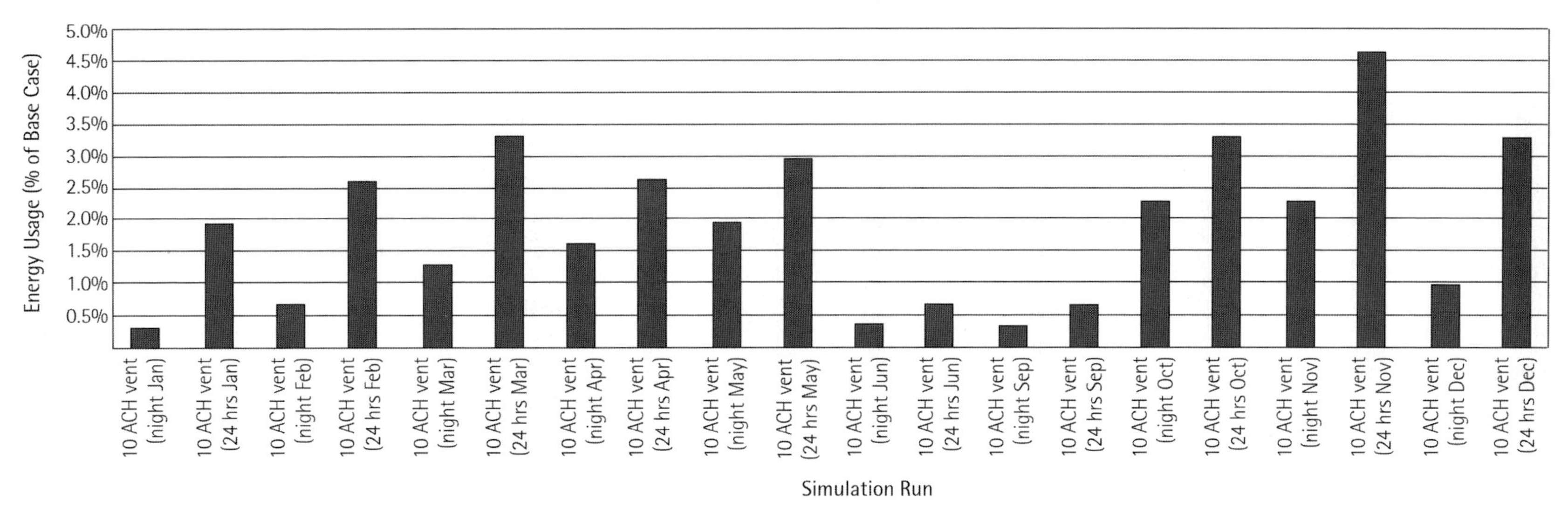

Figure 26 Monthly cooling-energy savings associated with ventilation

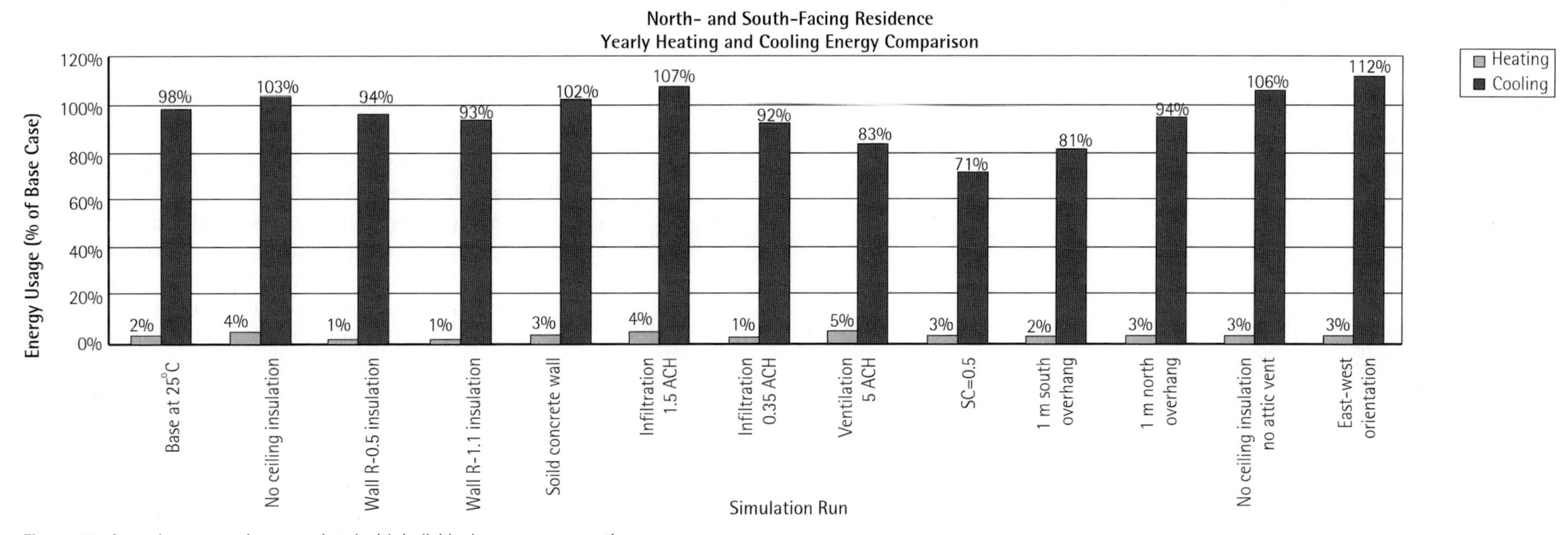

Figure 27 Annual energy savings associated with individual energy-conservation measures

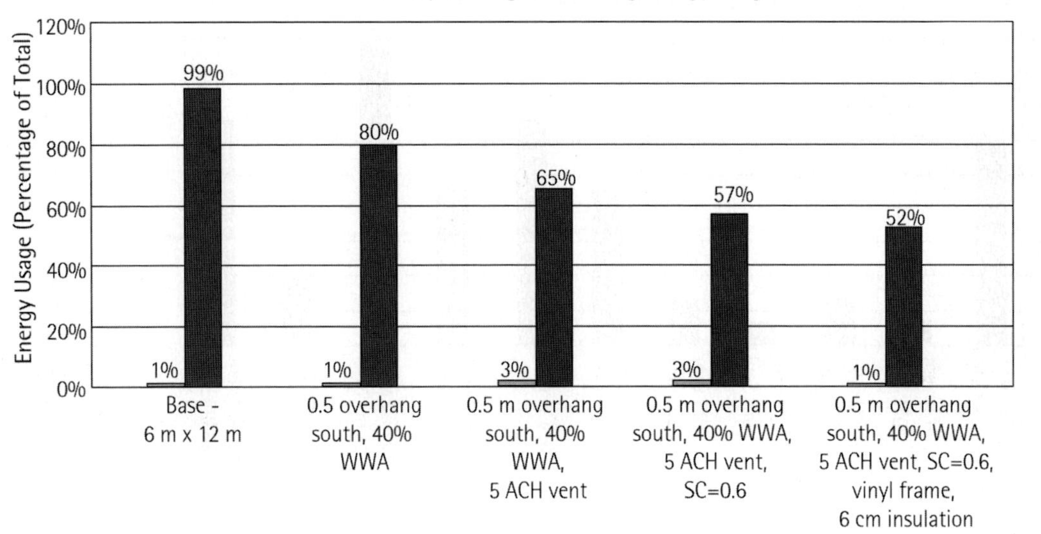

Figure 28 Annual energy savings associated with combinations of measures

heating and cooling energy than the same building oriented north and south. Figure 28 combines measures; the best set of options for window overhang, house ventilation and roof insulation cut annual cooling energy use by 48 percent. Energy savings translate into reduced operating costs. Orienting a building to avoid strong morning and afternoon sun reduced estimated annual energy costs for heating and cooling from 3,400 to 3,000 RMB. The addition of window shades, window overhangs and ventilation dropped the cost to 1,750 RMB.

SUMMARY

The three simulation studies described in this chapter point toward significant opportunities to reduce the demand for mechanical cooling in dwellings in Beijing, Shanghai, and Shenzhen. The first study evaluated two entirely passive cooling systems, daytime ventilation and night cooling, for a generic, six-story unit building in Beijing and Shanghai. Results showed that night cooling is superior to daytime ventilation in both cities. Night cooling may replace air-conditioning systems for a major part of the cooling season in Beijing (90 percent of the time). On average, night cooling reduced the maximum daytime indoor air temperature to a level 3.9°K cooler than the maximum outdoor air temperature. Daytime ventilation helped to provide a better comfort level indoors, but the improvement was not very significant. The two passive-cooling systems do not work as effectively in Shanghai, where comfort conditions were only assured in 66 percent of the warm period. This was due to the small daily temperature variation and high air temperature and humidity in Shanghai. Condensation on inside wall surfaces associated with reducing wall temperatures via night cooling was found to be a potential problem in Beijing. The study found that condensation could occur for 60 hours in the warm period. Condensation was negligible in Shanghai, even though it is a hot and humid climate, because night cooling was not as effective in lowering inside surface temperatures.

The second study complemented the first by examining the benefit of reducing building loads and using mechanical ventilation in a four-unit building proposed for a site in Beijing. Mechanical ventilation guaranteed a specified airflow, necessary in this case because buildings already constructed on the densely developed site screened the target building from the wind. The base-case building was comfortable about 72 percent of summer hours. By limiting solar gain, comfortable conditions could be maintained throughout the summer. Means to accomplish this included providing shading using architectural features (window overhangs), vegetation, and heavy insulation that minimized the impact of absorbed solar energy; preventing internal gains, which could be approached via careful selection and operation of appliances; and nighttime ventilation.

The third study focused on a housing development in Shenzhen. Building orientation and window shading were critical in reducing the use of cooling equipment. Careful window shading combined with ventilation at appropriate times and roof insulation reduced cooling energy consumption by 47 percent.

ACKNOWLEDGMENTS

Portions of this chapter are reproduced with permission from Elsevier from the paper: Carrilho da Graça, G., Q. Chen, L. Glicksman and L. Norford. 2002. Simulation of Wind-Driven Ventilative Cooling Systems for an Apartment Building in Beijing and Shanghai. *Energy and Buildings* 34(1): 1-11. The author would also like to acknowledge use of the following paper in this chapter: Norford, L., L. Caldas, J. Kaufman, L. Glicksman, A. Scott and Q. Chen. 1999. Opportunities to Maintain Thermal Comfort in Beijing Housing without Vapor Compression Cooling Equipment. *Proceedings of the 3rd International Symposium on HVAC*, 17-19 November 1999, vol. 1: 226-36. Shenzhen.

REFERENCES

AEC. 2006. *VisualDOE 4.0.* Architectural Energy Corporation. <http://www.archenergy.com/products/visualdoe>

ASHRAE. 2001. *Handbook of Fundamentals.* American Society of Heating, Refrigerating and Air-Conditioning Engineers. Atlanta, GA.

ASHRAE. 2004. *ASHRAE Standard 55-2004. Thermal Environmental Conditions for Human Occupancy.* American Society of Heating, Refrigerating and Air-Conditioning Engineers. Atlanta, GA.

Brager, G. and R. de Dear. 2000. A Standard for Natural Ventilation. *ASHRAE Journal*, 42(10): 21-28.

Busch, F. 1992. A Tale of Two Populations: Thermal Comfort in Air-Conditioned and Naturally Ventilated Offices in Thailand. *Energy & Buildings*, 18: 235-49.

Carrilho da Graça, G., Q. Chen, L. Glicksman and L. Norford. 1999. Simulation of Wind Driven Ventilative Cooling in an Apartment Building in Beijing and Shanghai. *Proceedings of the 3rd International Symposium on HVAC*, 17-19 November 1999, Vol. 2: 648-658. Shenzhen.

Carrilho da Graça, G., Q. Chen, L. Glicksman and L. Norford. 2002. Simulation of Wind-Driven Ventilative Cooling Systems for an Apartment Building in Beijing and Shanghai. *Energy and Buildings* 34(1): 1-11.

CHAM. 1998. *PHOENICS Version 3.1.* Software. CHAM Ltd. London.

Chandra, S. and A. Kerestecioglu. 1984. Heat Transfer in Naturally Ventilated Rooms, Data from Full-Scale Measurements. *American Society of Heating, Refrigerating and Air-Conditioning Engineers Transactions*, 90(2): 211-224.

Chen, D and J. Cai. 1994. *Documents Collection for Building Design Vol. 1* (in Chinese). Chinese Construction Industry Publications Inc., Beijing.

de Dear, R. 1998. A Global Database of Thermal Comfort Field Experiments. *American Society of Heating, Refrigerating and Air-Conditioning Engineers Transactions*, 104(1B): 1141-52.

de Dear, R. and G. Brager. 1998. Developing an Adaptive Model of Thermal Comfort and Preference. *American Society of Heating, Refrigerating and Air-Conditioning Engineers Transactions*, 104(1A): 145-67.

DeST. 2006. *"Meteorological data producer for HVAC analysis."* Available from Designer's Simulation Toolkit at <http://www.hvacr.com.cn/technology/software/dest-e/medpha.html>.

Givoni, B. 1994. *Passive and Low-Energy Cooling of Buildings.* Van Nostrand Reinhold, New York.

Givoni, B. 1998. *Climate Considerations in Building and Urban Design.* Van Nostrand Reinhold, New York.

Humphreys, M. and J. Nicol. 1998. Understanding the Adaptive Approach to Thermal Comfort. *American Society of Heating, Refrigerating and Air-Conditioning Engineers Transactions*, 104(1B): 991-1004.

ISO. 1994. Moderate Thermal Environments - Determination of the PMV and PPD Indices and Specifications for Thermal Comfort. *International Standard 7730.* ISO, Geneva.

Jiang, Y., H. Xing, C. Straub, Q. Chen, A. Scott, L. Glicksman and L. Norford. 1999. Design of Natural Ventilation and Outdoor Comfort by a Team of Architects and Engineers with the CFD Technique. *Proceedings of the 3rd International Symposium on HVAC*, vol. 2: 591-601. Shenzhen.

Jitkhajorwanich, K., A. Pitts, A. Malama and S. Sharples. 1998. Thermal Comfort in Transitional Spaces in the Cool Season of Bangkok. *American Society of Heating, Refrigerating and Air-Conditioning Engineers Transactions*, 104(1B): 1181-93.

Kammerud R., E. Ceballos, B. Curtis, W. Place and B. Anderson. 1984. Ventilation Cooling of Residential Buildings. *American Society of Heating, Refrigerating and Air-Conditioning Engineers Transactions*, 90(2): 226-251.

Khedari J., N. Yamtraipat, N. Pratintong and J. Hirunlabh. 2000. Thailand Ventilation Comfort Chart. *Energy and Buildings*, 32: 245-249.

Kwok, A. 1998. Thermal Comfort in Tropical Classrooms. *American Society of Heating, Refrigerating and Air-Conditioning Engineers Transactions*, 104(1B): 1031-47.

Lechner, N. 2001. *Heating, Cooling, Lighting: Design Methods for Architects.* John Wiley & Sons, New York.

Malama, A., S. Sharples, A. Pitts and K. Jitkhajornwanich. 1998. An Investigation of the Thermal Comfort Adaptive Model in a Tropical Upland Climate. *American Society of Heating, Refrigerating and Air-Conditioning Engineers Transactions*, 104(1B): 1194-1203.

Mills, A. 1995. *Basic Heat and Mass Transfer.* Prentice Hall, New Jersey.

National Standard of PRC. 1987. Design Code for Heating, Ventilating and Air Conditioning, GBJ 19-87. China Planning Press, Beijing. Available from China

Architecture & Building Press <www.china-abp.com.cn>.

Nicol, J. and M. Kessler. 1998. Perception of Comfort in Relation to Weather and Indoor Adaptive Opportunities. *American Society of Heating, Refrigerating and Air-Conditioning Engineers Transactions*, 104(1B): 1005-17.

Nicol, J. , M. Humphreys, O. Sykes and S. Roaf, eds. 1995. *Standards for Thermal Comfort.* E & F Spon, London.

Norford, L., L. Caldas, J. Kaufman, L. Glicksman, A. Scott and Q. Chen. 1999. Opportunities to Maintain Thermal Comfort in Beijing Housing without Vapor Compression Cooling Equipment. *Proceedings of the 3rd International Symposium on HVAC*, 17-19 November 1999, vol. 1: 226-36. Shenzhen.

Oseland, N. 1998. Acceptable Temperature Ranges in Naturally Ventilated and Air-Conditioned Offices. *American Society of Heating, Refrigerating and Air-Conditioning Engineers Transactions*, 104(1B): 1018-30.

Xia, Y. and R. Zhao. 1999. Effects of Air Turbulence on Human Thermal Sensation in Warm Isothermal Environment. *Proceedings of the 3rd International Symposium on HVAC*, 17-19 November 1999, vol. 1: 147-53. Shenzhen.

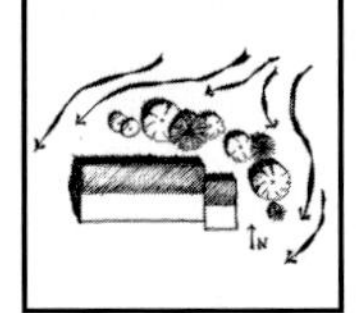

CHAPTER SIX

WIND IN BUILDING ENVIRONMENT DESIGN

Qingyan Chen

L. Glicksman and J. Lin (eds), Sustainable Urban Housing in China, 100-115

INTRODUCTION

Wind can be a building's "friend" because it can naturally ventilate the building, providing a comfortable and healthy indoor environment, as well as saving energy. Conventional design approaches often ignore opportunities for innovations with wind that could condition buildings at a lower cost, while providing higher air quality and an acceptable thermal-comfort level by means of passive cooling or natural ventilation. Natural ventilation can be used for cooling in the spring and autumn for a moderate climate (e.g., Nashville, TN), the spring for a hot and dry climate (e.g., Phoenix, AZ), the summer for a cold climate (e.g., Portland, ME), and the spring and summer for a mild climate (e.g., Seattle, WA). Natural ventilation can also be used to cool environments in a hot and humid climate during part of the year (e.g., New Orleans, LA) (Lechner 2000).

On the other hand, wind can be a building's "enemy" when it causes discomfort to pedestrians, usually as a result of high wind speed around the building. Table 1 summarizes the effects of wind on people. The wind speed is normally referred to as the speed of wind at ten meters above an open terrain. The wind speed at pedestrian level is roughly 70 percent of the tabulated values. Visser (1980) proposed comfort criteria with different activities versus the frequency of wind speed higher than five meters per second, as shown in Table 2.

For example, in an area where the number of days with an average wind speed higher than 5 m/s is 150 days per year (or the frequency of wind with a speed higher than 5 m/s is 150 days/365 days x 100 percent = 41 percent), people who walk fast would feel unpleasant. Clearly, wind speeds greater than five meters per second are considered uncomfortable for most activities. Therefore, it is essential to reduce the wind speed around buildings.

In addition, in a mild, moderate, and cold climate, it is very important to minimize infiltration of cold air into the building during the winter to reduce wind speed around buildings. The reduction of wind speed can be achieved by: avoiding windy locations such as hilltops; using wind barriers like evergreen vegetation; clustering buildings for mutual wind protection; and designing buildings with streamlined shapes and rounded corners to both deflect the wind and minimize the surface-to-volume ratio (Lechner 2000).

For small-scale buildings, there are established guidelines for passive solar heating. However, natural ventilation and outdoor thermal comfort are very difficult to design, even in simple cases. The purpose of the present chapter is to demonstrate, with the help of the computational fluid dynamics (CFD) technique, how architects can work with engineers to design naturally ventilated buildings and comfortable outdoor environments around buildings.

Beaufort Number	Description	Wind Speed	Wind Effect
2	Light breeze	1.6 - 3.3	Wind felt on face
3	Gentle breeze	3.4 - 5.4	Hair disturbed; clothing flaps; newspaper difficult to read
4	Moderate breeze	5.5 - 7.9	Raises dust and loose paper; hair disarranged
5	Fresh breeze	8.0 - 10.7	Wind force felt by body; possible stumbling when entering a windy zone
6	Strong breeze	10.8 - 13.8	Umbrellas used with difficulty; hair blown straight; difficult in walking steadily; wind noise on ears unpleasant
7	Near gale	13.9 - 17.1	Inconvenience felt when walking
8	Gale	17.2 - 20.7	Generally impedes progress; great difficulty with balance in gusts
9	Strong gale	20.8 - 24.4	People blown over

Table 1 Effects of wind on people. Beaufort number classifies wind as 0 (calm) to 12 (hurricane) (Source: Bottema 1993)

Activities	*Acceptable*	*Unpleasant*	*Intolerable*
Walking fast: carpark, sidewalk, road, cycle-track	<35%	35% - 75%	> 75%
Strolling: park, shop center, footpath, building entrance, bus station	<5 %	5% - 35%	>35%
Sitting/standing short: shop center, square, playground	<0.1%	0.1% - 5%	> 5%
Sitting/standing long: terrace, swimming pool, open-air theater	0%	0% - 0.1%	> 0.1%

Table 2 Comfort criteria for different frequency (day/year) when wind speed is higher than 5 m/s (Source: Visser 1980)

WIND DATA

To design naturally ventilated buildings and/or comfortable outdoor environments, the first step is to obtain reliable wind information, such as wind speed and direction. For example, the National Renewable Energy Laboratory derived a set of typical meteorological weather data for 229 stations throughout the United

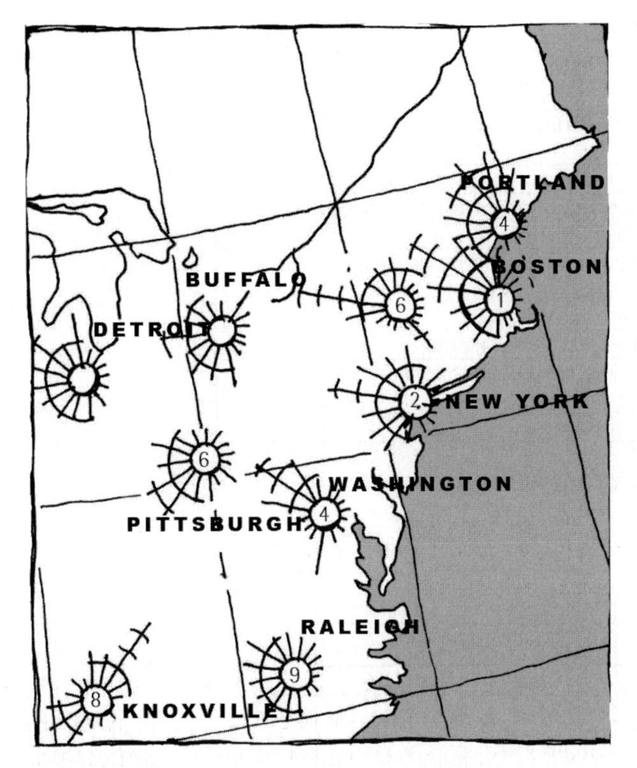

Figure 1 Surface wind roses in January for northeastern U.S. (Source: adapted from NOAA 1983)

Figure 2 Diurnal and nocturnal air movements near a large body of water (Source: adapted from Moore 1993)

States and its territories (Marion and Urban 1995). The database provides hourly wind speed and directions that can be used directly in natural ventilation and outdoor comfort design. Rather than accounting for every hour in a full reference year, a designer should analyze the data and divide it into eight directions (N, NE, E, SE, S, SW, W, and NW) for several wind speeds (e.g., Beaufort number <2, 3-4, 5-6, >7, where Beaufort number classifies wind as 0 for calm to 12 for hurricane). The weather data can also be used to determine the percentage of wind for each direction and speed combination (32 in total). The total number can be reduced, eliminating those with very low probabilities.

For countries where typical meteorological weather data are not available, wind roses can be used. Figure 1 shows a part of the wind rose map for northeastern United States in January. The wind roses give the wind direction and percentage. The number inside the wind rose stands for the percentage of calm period. NOAA (1983) uses another figure to provide the monthly average wind speed over a year.

Note that the wind data from weather databases, wind roses, or a weather station is for an open terrain. Numerous factors could have a significant impact on the local climactic conditions. For example, a large water body, such as a lake, can create a local wind from the water body to the land during the day and a local wind from the land to water during the night (Figure 2). This is because water has a higher effective thermal mass than that of land. Under the sun, the surface of land is heated much faster than water. The warmer air above the land goes up due to the buoyancy effect, creating an air pressure differential from the water to land. During the night, land cools faster than water due to thermal radiation. It is again the thermal buoyancy in the air that forms a land-to-water wind. Other factors include valleys, mountain ranges, and even large building blocks.

DESIGN TOOLS

Traditionally, many architects predict the airflow in and around buildings by using "smart arrows," as shown in Figure 3. Drawing the airflow correctly requires a rich knowledge of fluid mechanics. Unfortunately in many cases, the "predicted" airflow pattern can be completely different from that in reality. Furthermore, the smart arrows cannot give the wind speed, or at least the reliable air speed, which is an important parameter for evaluating the benefits of natural ventilation and outdoor comfort. Chandra, Fairey and Houston (1983) developed a simple model for calculating the air exchange rate for natural cross ventilation. However, it is limited in that it can only be applied to buildings with simple geometry and surroundings.

Many empirical and analytical tools have also been developed for manual prediction of natural ventilation in buildings and outdoor thermal comfort, as documented by Allard (1998), Awbi (1996), CIBSE (1997), and Linden (1999). These manual methods are generally very simple and can be expressed by algebraic equations and spreadsheets. Despite being useful, these empirical and analytical tools have great uncertainties when used for complex buildings.

As a result, most traditional studies use wind tunnels to simulate and measure the airflow around buildings for outdoor thermal comfort and a full-scale mock-up room to determine natural ventilation. Figure 4 shows a site model placed in a wind tunnel. By rotating the site model disc and by changing the fan speed, different wind directions and speeds can be simulated.

When the buoyancy effect is not strong, such as during natural cross ventilation, the wind tunnel, together with modeling theory, can also be used to study natural ventilation. For buoyancy-dominant natural ventilation, such as single-sided natural ventilation, ideally a full-scale mockup is needed in order to satisfy both the Reynolds number that represents inertial force from the wind and the Grashov number that represents the buoyancy force. The experiment usually measures wind speed in the wind tunnel, and wind speed and temperature in the mockup room. Rarely is the wind direction also measured. Although the experimental approaches provide reliable information concerning airflow in and around buildings, the available data is generally limited due to the expensive experimental rigs and processes. Moreover, the approach is not practical for a designer who wishes to optimize his or her designs because the experimental method is very time-consuming. Alternately, another fluid such as heavy refrigerant vapor (Olson, Glicksman and Ferm 1990) or water (Linden 1999) can be used for modeling. These fluids allow the model size to be substantially reduced. Whole buildings can be simulated with the water models. Also, by relaxing some of the modeling criteria, such as matching the Reynolds number, small-scale air models can also be employed.

Numerical simulation has become a new trend for determining natural ventilation and outdoor thermal comfort. Two numerical methods are available for predicting natural ventilation. The first one is the zonal method, which calculates inter-zonal airflow using the Bernoulli equation along with experimental correlations of flow resistance through doorways, windows, and other orifices.

The prediction of the inter-zonal airflow relies on the external pressure distribution caused either by wind or the buoyancy effect. However, the determination of the external pressure is very complex, since the pressure distribution depends on incoming wind speed and direction, building size and shape, and the size and location of the building's interior opening (Vickery and Karakatsanis 1987). Therefore, the accuracy of the zonal method depends on the accuracy of the pressure distribution. Furthermore, the zonal model is incapable of determining thermal comfort around a building, because it does not provide wind velocity information. However, such a method can supply good preliminary estimates of air flow and temperature levels within a building if reasonable estimates of external wind pressure distributions can be made.

The other numerical method, CFD, calculates the airflow distribution for both indoor and outdoor thermal comfort. The CFD technique numerically solves a set of partial differential equations for the conservation of mass, momentum (Navier-Stokes equations), energy, species concentrations, and turbulence quantities. The solution provides the field distribution of pressure, air velocity, temperature, concentrations of water vapor (relative humidity), contaminants, and turbulence. Refer to Chen and Glicksman (2000) for a more detailed description of the CFD technique. Despite having some uncertainties and requiring an engineer with sufficient knowledge of fluid mechanics and a high-capacity computer, the CFD method has been successfully used to predict airflow in and around buildings (Chen 1997, Murakami 1998). With the rapid increase in computer capacity and the development of new CFD program interfaces, the CFD technique is becoming very popular.

The following sections will discuss the applications of CFD to outdoor thermal comfort and natural ventilation design. CFD generally includes large eddy simulation and Reynolds averaged Navier-Stokes equation modeling. Large eddy simulation, as reviewed by Murakami (1998), can give more detailed results, such as an instantaneous airflow field, but it requires more computing time than that of the Reynolds averaged Navier-Stokes equation modeling. Large eddy simulation has started appearing in building environment research, but has yet to be applied as a design tool. Therefore, this

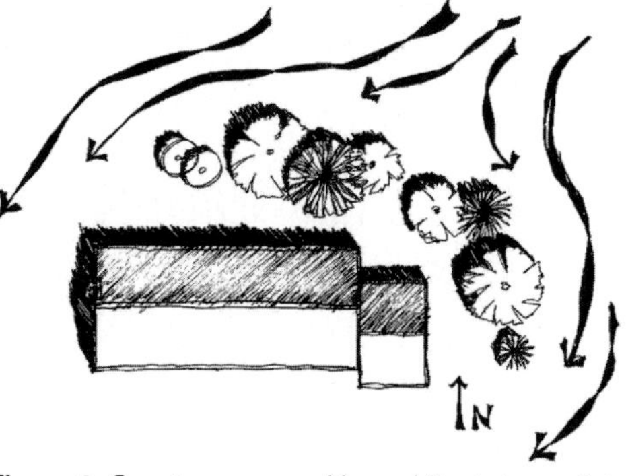

Figure 3 Smart arrows used by architects to predict airflow in and around buildings (Source: adapted from Moore 1993)

Figure 4 A building site model within a wind tunnel

(a)

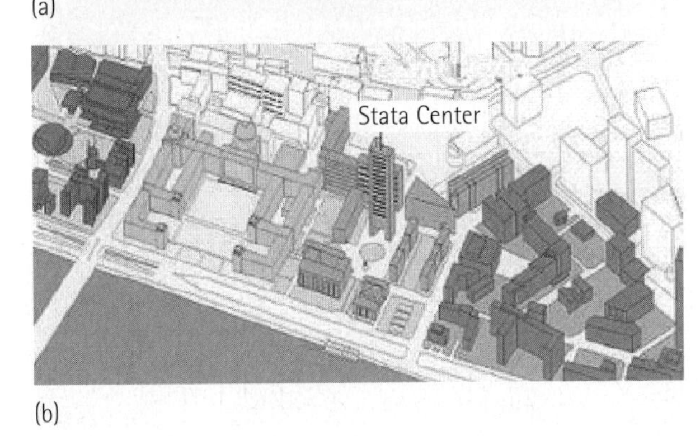

(b)

Figure 5 Stata Center: (a) model shown without glass roof and (b) surroundings

chapter focuses on Reynolds averaged Navier-Stokes equation modeling. Many commercial CFD programs based on the Reynolds averaged Navier-Stokes equation modeling are available on market and are rather similar to each other. This investigation uses the PHOENICS program (CHAM 2000).

OUTDOOR THERMAL COMFORT STUDIES

Outdoor thermal comfort design will be illustrated by two application examples. The first example concerns the design of the Stata Center at the Massachusetts Institute of Technology, and the second is a high-rise residential building complex in Beijing.

Stata Center

Figure 5a shows a model of the Stata Center and its surroundings designed by Frank O. Gehry and Associates. Since this campus building has windy surroundings, the architect was concerned about the outdoor thermal comfort in the plaza (the front part of Figure 5a). At one time, the architect wanted to add a glass roof that would provide a wind shield over the plaza. Since the glazed roof would cost several million dollars, the architect initiated a study of the wind distribution around the Stata Center, which was researched by the author.

This investigation used a commercial CFD program (CHAM 2000) for the study. The CFD program allows one to read data from an AutoCAD file. This feature is very important because of the complicated geometry of the buildings. Similar to a wind tunnel, CFD requires detailed information on the surroundings of the Stata Center in order to calculate the airflow. The surrounding buildings can either block or enhance the wind speed around the center. The computational domain for the building and surroundings is shown in Figure 5b. The domain length is about five times that of Stata Center in the four horizontal directions (or 100 times the Stata Center area size). The wind distributions around Stata Center were calculated for the north, east, south, and west wind directions with a typical wind speed for each direction. Figure 6 shows the wind distribution around Stata Center with an east wind. This study used about one million grid points; the study required three days of computing time on a Pentium II 450 PC with 512 MB of memory. That PC was considered to be high-end in 1999. Obviously, the grid number was too coarse so the wind information was not sufficiently detailed.

Therefore, the investigation used a zoom-in approach to study the details of the wind distribution. The zoom-in approach used the wind information computed (Figure 6) as boundary conditions in calculating the wind speed distribution just around Stata Center, as shown in Figure 7. With the zoom-in approach, the CFD results provided very detailed wind speed information. For example, the wind speed was found to be almost identical around Stata Center with or without the glass roof. Hence, the glass roof was not necessary.

A High-Rise Residential Building Complex in Beijing

In the past, a good living environment in China implied ample space between buildings filled with trees and grass. High-rise buildings have been regarded as a symbol of modernity and luxury. A typical building consisting of such residential units is shown in Figure 8. Jiang et al (1999) made a detailed analysis on the design and found that such a design is not sustainable in terms of energy efficiency and Chinese culture. The study showed that the best design would be made up of low-rise buildings with varying-sized courtyards. This would avoid a harsh winter wind, let the winter sun in, and promote the use of natural ventilation.

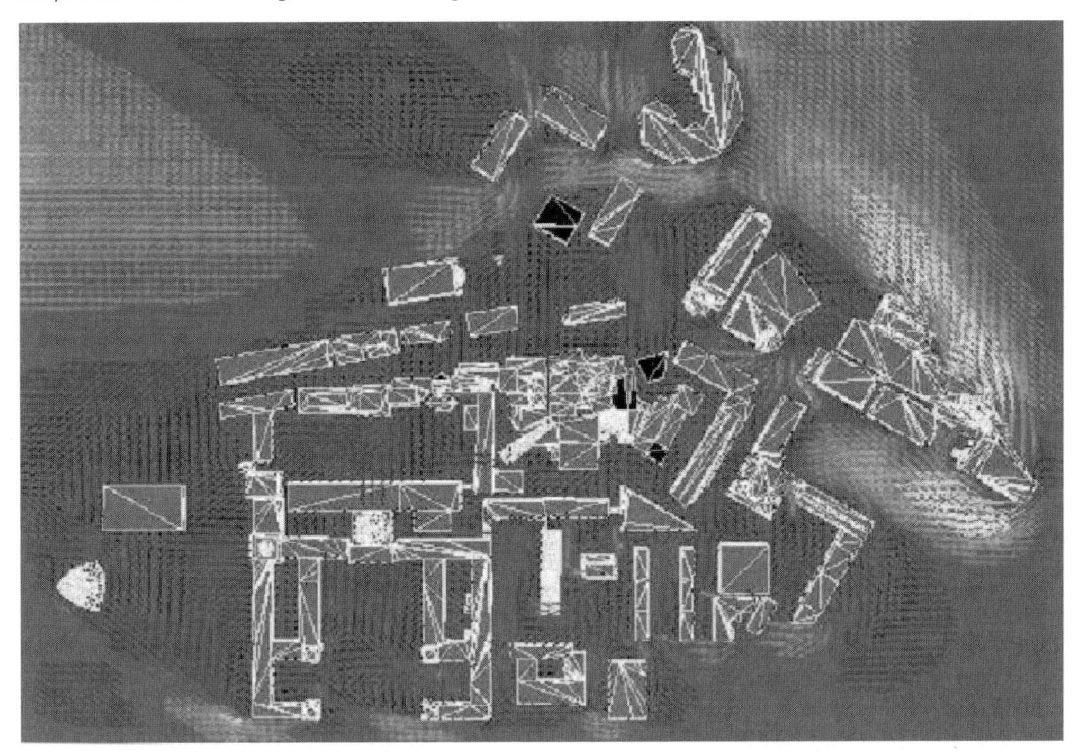

Figure 6 Wind distribution around Stata Center at the ground level (dark - low velocity and light - high velocity)

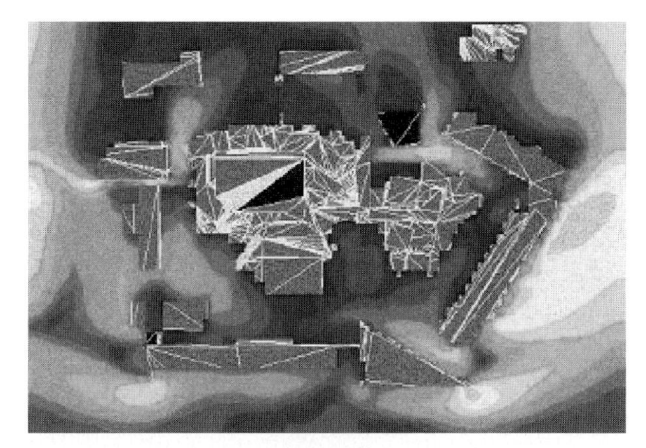

(a)

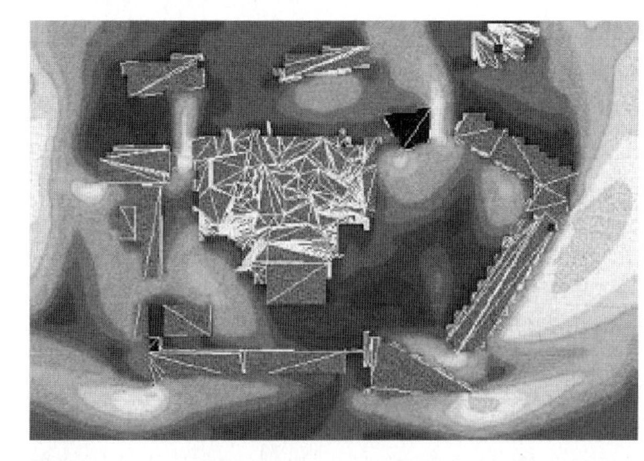

(b)

Figure 7 Wind distribution around Stata Center (zoom-in): (a) with a glass roof and (b) without a glass roof (dark - low velocity and light - high velocity)

Figure 8 Beijing Star Garden - a high-rise residential development

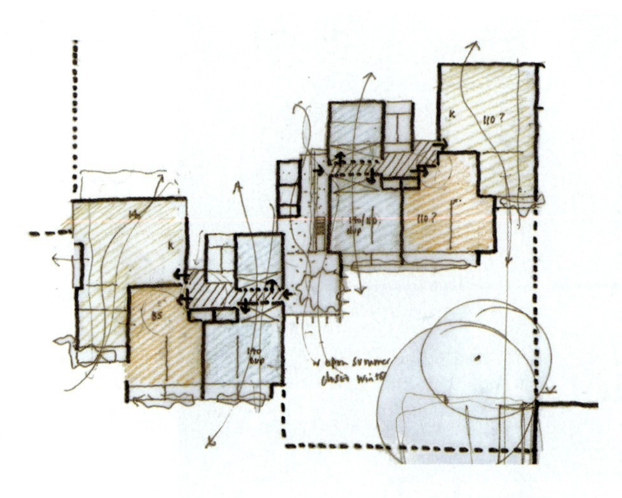

Figure 10 The unit layout in the four new towers allows more effective natural ventilation in the summer (north is up)

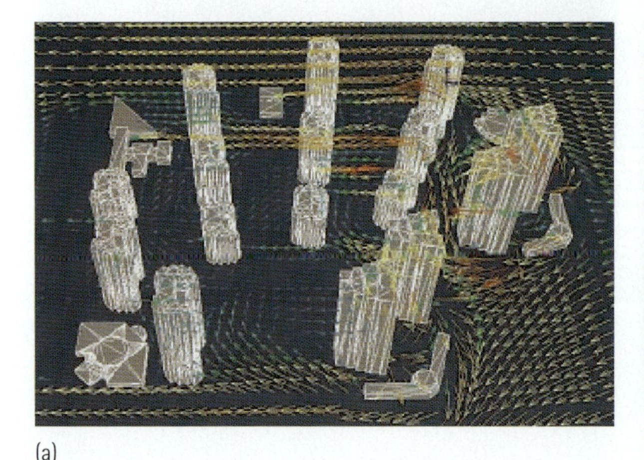

(a)

(b)

Figure 9 Wind distribution on the building site: (a) original design, and (b) design with four proposed towers to the north (to the right) (red indicates high velocity, yellow - moderate velocity, and blue - low velocity)

As is the case with many downtown areas with skyscrapers, high-rise buildings sometimes create a wind tunnel effect that is very uncomfortable to pedestrians. The proposed design for Beijing Star Garden forms a wind tunnel effect on the site with prevailing winter winds from the north. Figure 9a shows the wind distribution on the site with a north wind from the right. There are a few places that have very high wind speeds (see red arrows in Figure 9a). The developers did not adopt our suggestion of lowering the building height and creating courts to eliminate "wind tunnel" problems and enhance contact between neighbors. Instead, they sought to change the shape of the four towers in the north to eliminate high wind spots. The new design used a different building shape to deflect the wind to the westward direction. Figure 9b shows the airflow distribution with a north wind under the new design of the four towers that reduces areas of high winds.

Of course, wind is not the only factor in producing an energy-efficient building design. Changing the tower shape may have an impact on the desire to have south-facing windows. This can be achieved through architectural design, as shown in Figure 10. The thin structure also allows the use of natural ventilation in the summer. See *Chapter 11, Case Study Two – Beijing Star Garden* for more information.

NATURAL VENTILATION STUDIES

The last ten years have seen a significant shift in the development and integration of environmental, ecological, and energy issues into the architectural design of buildings. Energy-efficient buildings address not only the issues of consumption and performance, but also the development and integration of a series of design and system technologies. Buildings should provide the basic amenities of shelter and yet practice responsible use of resources. Whatever the climate zone, energy-conscious design utilizes strategies that optimize the passive environmental systems in reference to active "sealed system" strategies. This leads to the use of natural ventilation and the maximization of daylighting wherever reasonable. Even under unfavorable outdoor climate conditions, passive-based technology can be combined with active systems during shoulder seasons and sometimes for night cooling in conjunction with adequate thermal mass.

Leading architects of this generation in the United Kingdom, Germany, France, Switzerland, and Scandinavia have turned their attention to a more sustainable form of practice in both building system technologies and building typologies. This approach can be witnessed in the work of architects such as Foster and Partners, Renzo Piano, Alan Short, Thomas Herzog, Michael Hopkins, Edward Cullinan, and Kiessler and Partners. In their designs, the issue of resources and the environment is at the heart of making intelligent and well-crafted architecture. Their buildings provide an interaction between the enclosure systems and the environmental and mechanical strategies for the internal space. The buildings are reputed to save a considerable amount of energy while improving indoor air quality and comfort.

Table 3 shows the potential of using natural ventilation in the United States for residential buildings. With proper design of building orientation, location, shape, and openings, daytime natural ventilation and/or night cooling can provide a thermally comfortable indoor environment for a long period in most all U.S. climates. Even if it is not possible to avoid the use of air-conditioning in the summer, air-conditioning units can be much smaller with natural ventilation, reducing first and operating costs.

Climate Region and Reference City	Periods Suitable for Natural Ventilation (NV) and when Air-Conditioning (AC) or Heating (H) is Needed for Residential Buildings											
Month	Jan	Feb	Mar	Apr	May	Jun	Jul	Aug	Sep	Oct	Nov	Dec
1. Hartford, CT	H	H	H	H	NV	NV	NV	NV	NV	H	H	H
2. Madison, WI	H	H	H	H	NV	NV	NV	NV	NV	H	H	H
3. Indianapolis, IN	H	H	H	NV	NV	NV	AC	AC	NV	NV	H	H
4. Salt Lake City, UT	H	H	H	H	NV	NV	AC	AC	NV	NV	H	H
5. Ely, NV	H	H	H	H	NV	NV	NV	NV	NV	NV	H	H
6. Medford, OR	H	H	H	NV	NV	NV	NV	NV	NV	NV	H	H
7. Fresno, CA	H	H	NV	NV	NV	NV	NV	NV	NV	NV	H	H
8. Charleston, SC	H	H	NV	NV	NV	AC	AC	AC	AC	NV	NV	H
9. Little Rock, AR	H	H	H	NV	NV	AC	AC	AC	NV	NV	NV	H
10. Knoxville, TN	H	H	H	NV	NV	AC	AC	AC	NV	NV	NV	H
11. Phoenix, AZ	H	NV	NV	NV	AC	AC	AC	AC	AC	NV	NV	H
12. Midland, TX	H	H	NV	NV	NV	AC	AC	AC	AC	NV	NV	H
13. Fort Worth, TX	H	NV	NV	NV	AC	AC	AC	AC	AC	NV	NV	H
14. New Orleans, LA	H	H	NV	NV	NV	AC	AC	AC	AC	NV	NV	H
15. Houston, TX	H	NV	NV	NV	AC	AC	AC	AC	AC	NV	NV	H
16. Miami, FL	NV	NV	NV	NV	AC	AC	AC	AC	AC	AC	NV	NV
17. Los Angeles, CA	H	H	NV	NV	NV	NV	AC	NV	NV	NV	NV	H

Table 3 The potential for natural ventilation in the U.S. (Source: Lechner 2000)

Age	< 19		20-40		40-60		> 60	
Sex	M	F	M	F	M	F	M	F
Like AC (%)	43	52	48	35	38	37	22	30
Neutral (%)	43	32	43	53	37	37	37	33
Dislike AC (%)	14	16	9	12	25	26	41	37

Table 4 A survey conducted in Beijing with respect to the use of air-conditioning in homes (Source: Jiang 1999)

Furthermore, it is very interesting to see the survey conducted in Beijing by Jiang (1999) regarding the acceptability of air-conditioning systems. Table 4 shows the survey results separated into categories according to age and sex. People who like air-conditioning think that it provides a cool temperature (40 percent), represents a modern technology (34 percent), and offers an ability to control the climate (23 percent). On the other hand, those who dislike air-conditioning believe that air-conditioning separates them from nature (47 percent), leads to draft and a high-noise environment (26 percent), and causes high electricity and first costs (23 percent). In general, younger people tend to like air-conditioning more than elderly people do. The results show a great potential for using natural ventilation in Beijing.

It should also be noted that natural ventilation has its shortcomings; these include issues of humidity control, noise control (10 dB deduction for an open window versus 30 dB deduction for a sealed window), heat recovery, security concerns, and rain. In addition, in areas with high outdoor pollution, natural ventilation has difficulty in controlling air quality. One solution could be the use of night cooling that closes the window during daytime, as illustrated by Carrilho da Graça et al (2002) and as described in chapter 5.

According to CIBSE (1997), natural ventilation can be classified as:

- cross ventilation;
- single-sided ventilation;
- stack ventilation; and
- mechanically assisted ventilation.

Cross ventilation occurs where an indoor space has ventilation openings on both sides. Air flows from one side of the building to the other due to a pressure difference built up by wind. Single-sided ventilation implies that an indoor space has all of the openings on one side. Stack ventilation makes use of density differences due to buoyancy in promoting an outflow from a part of a building (e.g., roof) and drawing in fresh and cool air from another part of the building (e.g., windows and doors). Mechanically assisted ventilation uses mechanical ventilation to increase the airflow in any of the above-mentioned systems. A building may have more than one of the ventilation systems described above.

This section will describe the applications of CFD to design cross ventilation and single-sided ventilation in buildings. The method can be used for other ventilation systems as well.

Cross Ventilation in a Building

The design team was requested to design three mid-rise buildings for a residential building development in Shanghai (Figure 11).

Since wind around the buildings is the driving force in cross ventilation, this investigation involves the simulation of indoor and outdoor airflow by CFD. In order to study the impact of surrounding buildings, the computational domain for outdoor airflow should be sufficiently large (e.g., an area of tens of thousands to a million square meters). Due to the limitation in current computer capacity and speed, the grid size used cannot be very small (it can be a few meters). On the other hand, the grid size for indoor airflow simulation should be small enough (in terms of a few centimeters) for one to see the details. Therefore, the indoor and outdoor airflow should be separately simulated. For natural ventilation design, the outdoor airflow simulation can provide flow information as boundary conditions for the indoor airflow simulation. Zhai et al (2000) have discussed a few methods to provide the flow information.

For simplicity, this investigation used a CFD program to calculate the pressure difference around the buildings and uses it as the boundary conditions for indoor airflow simulation. Ideally, the calculation should be performed for different wind directions under various wind speeds in a period suitable for natural ventilation, such as summer. Figure 12 illustrates the pressure distribution under the prevailing wind direction (southeast) and speed (3 m/s). In order to correctly take the impact of the surrounding buildings into account, the computational domain is much larger than the one shown in the figure. Clearly, the pressure difference is the highest between the northern and southern façades. It is also interesting to note that the highest pressure difference is neither at the top floor nor at the bottom floor, but somewhere near the top, as shown in Figure 12b.

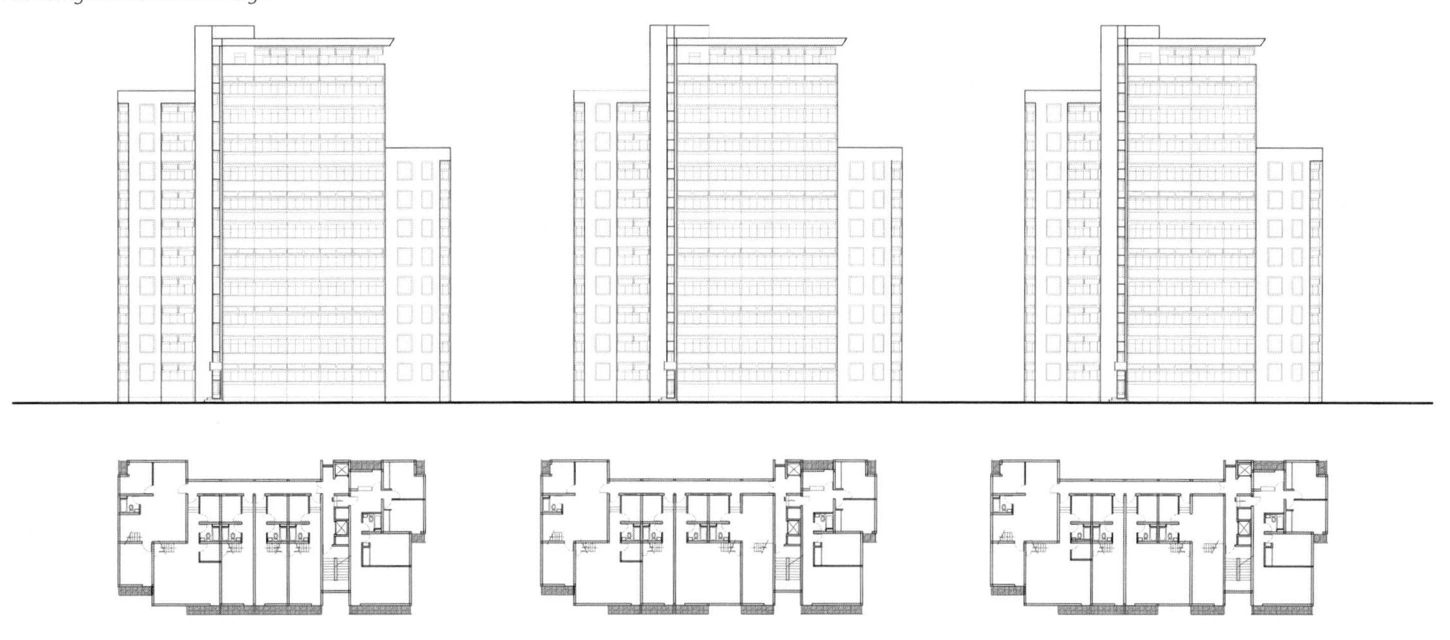

Figure 11 Architectural elevation and plans of final design for Shanghai Taidong Residential Quarter

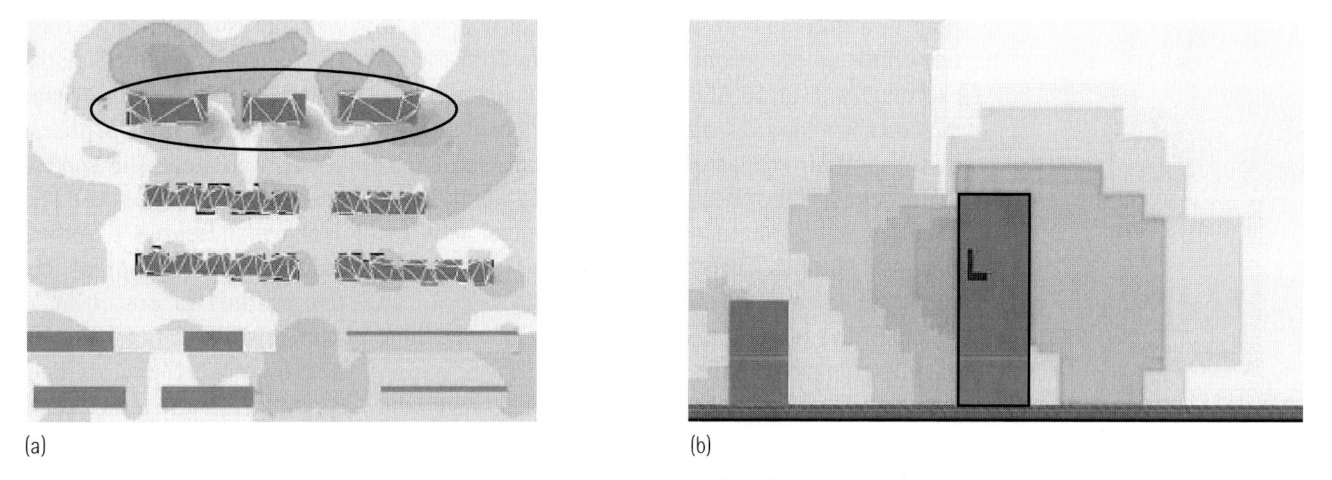

Figure 12 Pressure distribution around the buildings proposed (shown circled) for Shanghai Taidong Residential Quarter, due to prevailing winds, from the southeast at 3 m/s (dark gray - high pressure, light gray - low pressure): (a) is the plan (north is up) and (b) is the section looking west (see color version of Figure 12b in chapter 12, Figure 34)

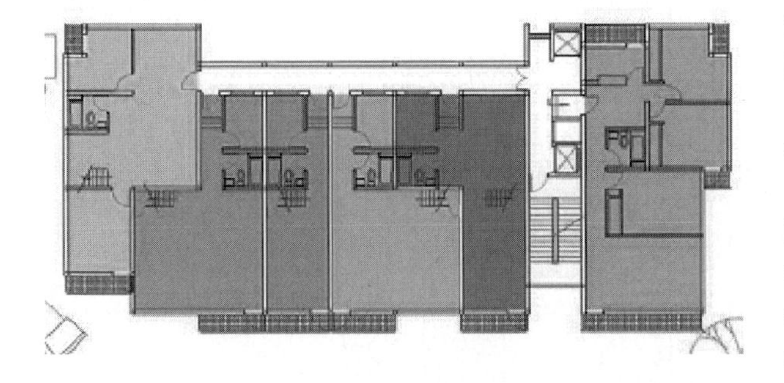

Figure 13 Building floor plan - unit G (the unit farthest to the right) was analyzed for interior CFD studies

By working together with the architects, the design team evaluated the ventilation performance for the buildings. Unit G in the middle building was used (Figure 13) as an example to illustrate the evaluation of cross-ventilation design.

With the unit layout in Figure 13, a CFD model can be established, as shown in Figure 14a. With the pressure distribution from Figure 12, the CFD program can calculate the distributions of airflow, air temperature, relative humidity, predicted percentage dissatisfied (PPD), and the mean age of air, as shown in Figure 14. CFD uses the humidity ratio and air temperature to determine the relative humidity. The PPD is determined by using the air velocity, temperature, humidity ratio, and environmental temperature. The results shown in Figure 14 are with an outside air temperature of 24°C and a relative humidity of 70 percent.

The computed results by CFD indicate that the maximum air velocity in the unit is less than one meter per second - a comfortable value for cross ventilation. The air exchange rate varies from 16 ACH on the first floor to a maximum of 40 air changes per hour (ACH) two-thirds up the height of the building. With the air exchange rate of 16 ACH, the indoor air temperature increases less than 1°K, although there are heat sources in the unit. The relative humidity is around 65 to 70 percent, a value close to that of the outdoors. Since the air exchange rate is high, the mean age of air is less than 120 seconds. Therefore, the air quality would be very good when outdoor air quality is high.

Since the air exchange rate is a very important parameter in cross ventilation design, this investigation indicates the design to be very successful. However, the wind is not always at the prevailing speed and direction, and the outdoor air temperature varies over time. A more complete evaluation of the design should be combined with an energy analysis of the building, as described in chapter 5. Carrilho da Graça et al (2002) have shown how to combine the information from flow and energy analysis for such a building. The paper also emphasizes the importance in using different control strategies. For example, in Shanghai it is more appropriate to use night cooling and minimum daytime ventilation to achieve an indoor air temperature lower than that of the outdoors. This is superior to ventilating buildings twenty-four hours a day. See *Chapter 12, Case Study Three- Shanghai Taidong Residential Quarter* for more information.

Single-Sided Ventilation in a Building

The building studied is a student dormitory in Cambridge, Massachusetts. Single-sided ventilation was evaluated for a typical room that is 4.7 meters long, 2.9 meters wide, and 2.8 meters high. The general room model used throughout the study is shown in Figure 15. The furniture within this room consisted of a bed, desk, closet, and bookcase. The heat sources were a computer (300 W), a TV set (300 W), and one occupant (100 W). For each of the heat sources, convective and radiative heat transfer was approximately equal. The surrounding walls, ceiling, and floor absorbed the radiative component and released it back to the room air by convection. Solar gains were not included for purposes of the time-averaged ventilation study. The window designed for the room consisted of an upper and lower window (0.4 m^2 each), as shown in Figure 15. The outdoor air temperature was maintained constant at 25.5°C, the average noon temperature for Boston in July. The intention was to analyze the results for this fixed outdoor temperature, and then apply them to a range of outdoor temperature conditions to develop trends.

This study stacked three identical dormitory rooms vertically above one another to evaluate the effect along a building's height. This three-story setup was placed within a larger outside domain (Figure 15b). The extension

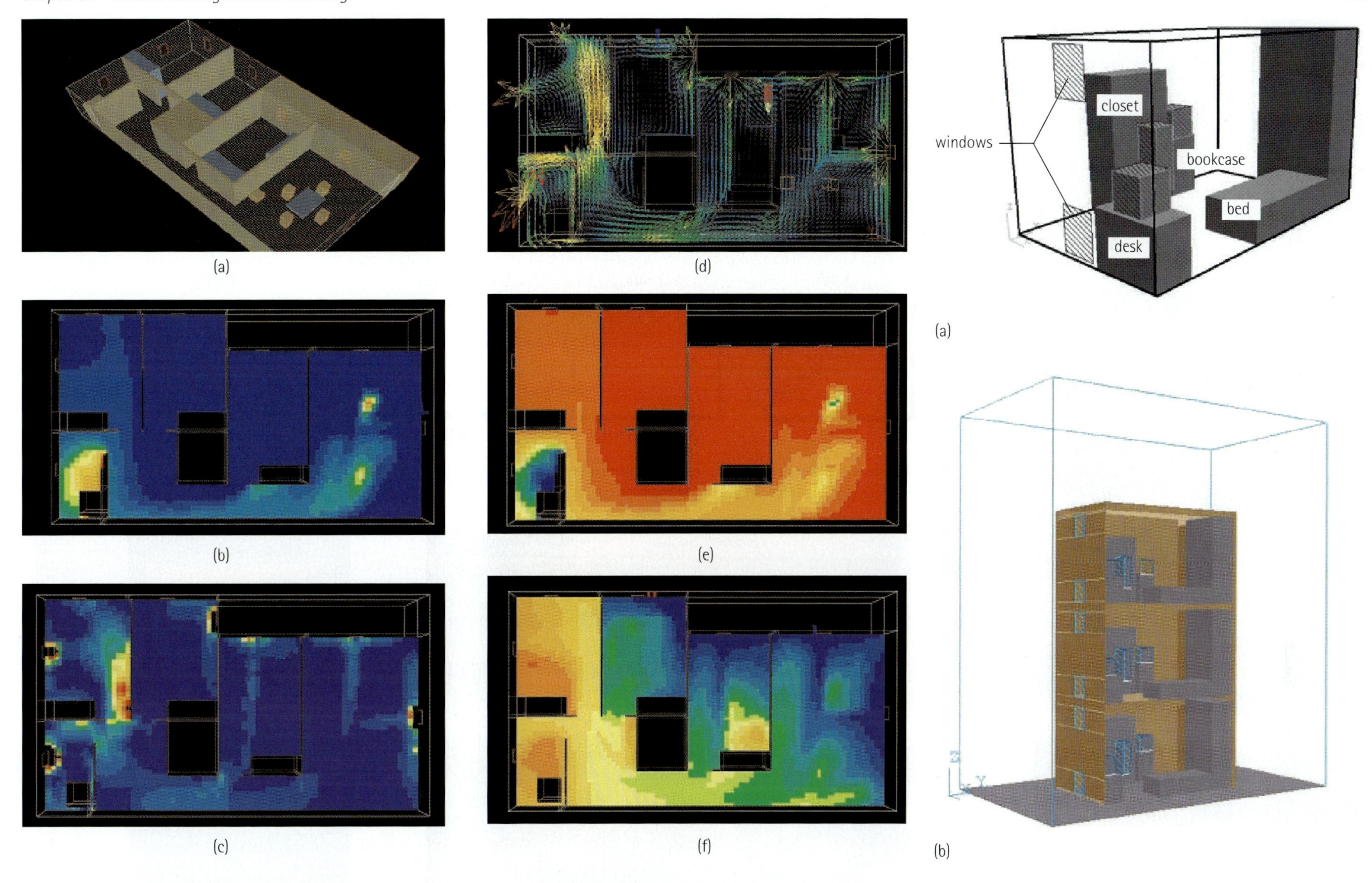

Figure 14 Cross ventilation performance analysis of Beijing unit interior (red – high, yellow – moderate high, green – moderate low, blue – low): (a) unit model, (b) air temperature, (c) predicted percentage of dissatisfied people (PPD) due to thermal comfort, (d) air velocity, (e) relative humidity, and (f) mean age of air

Figure 15 The CFD model used to study single-sided ventilation in an MIT dormitory room: (a) the room model and (b) the building and environment model

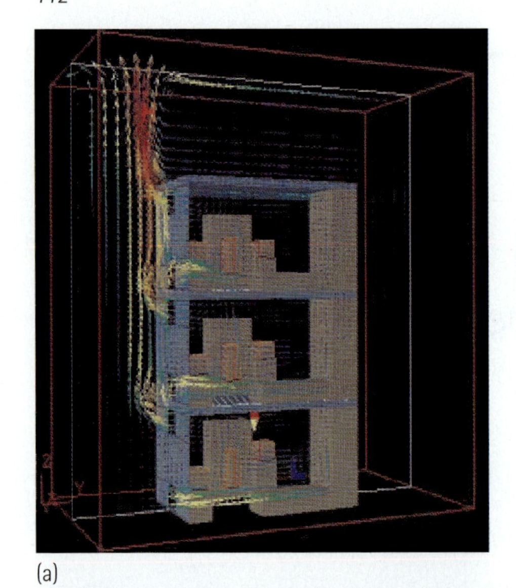

(a)

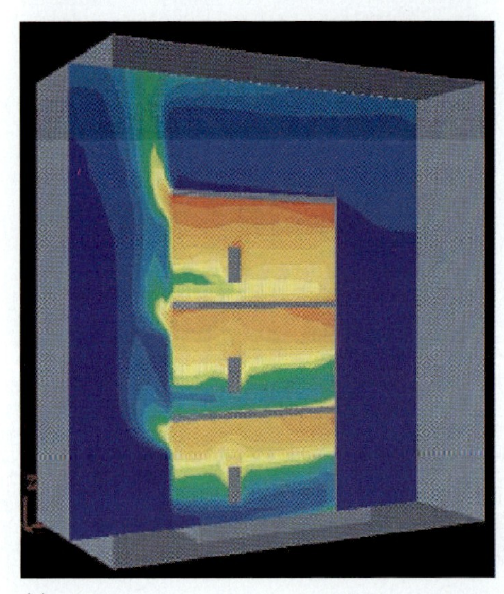

(b)

Figure 16 CFD results in the center of the rooms with stack effect (red – high, yellow – moderate high, green – moderate low, blue – low): (a) air velocity distribution and (b) air temperature distribution

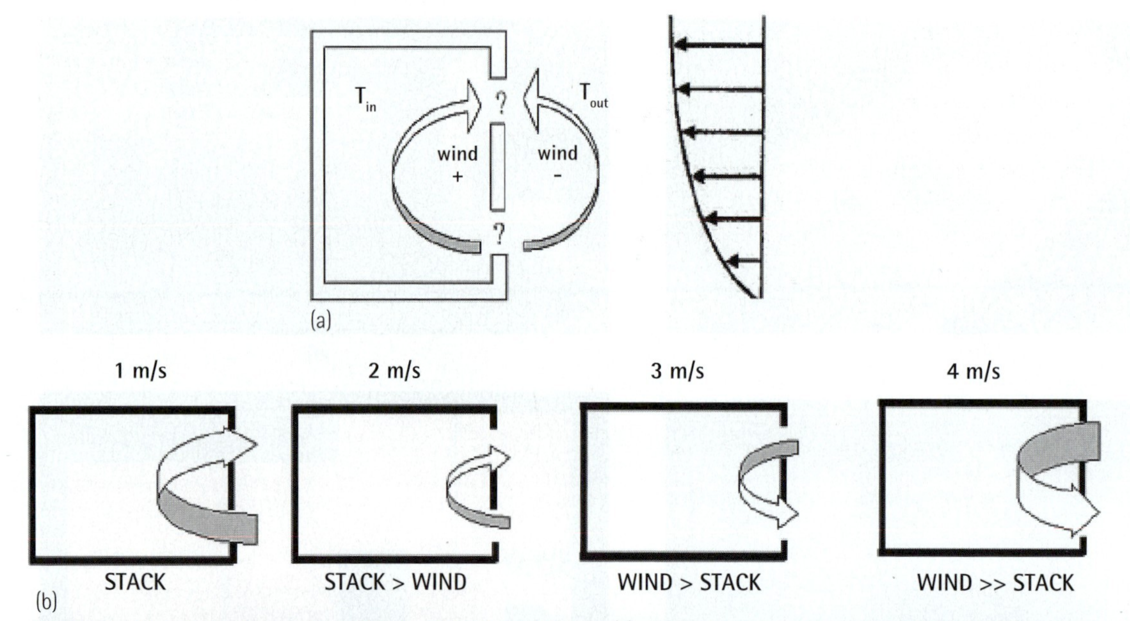

Figure 17 Uncertain effects of combined wind and stack forces: (a) reinforcing vs. counteracting effect and (b) depiction of counteracting wind and stack effects over a progression of wind speeds

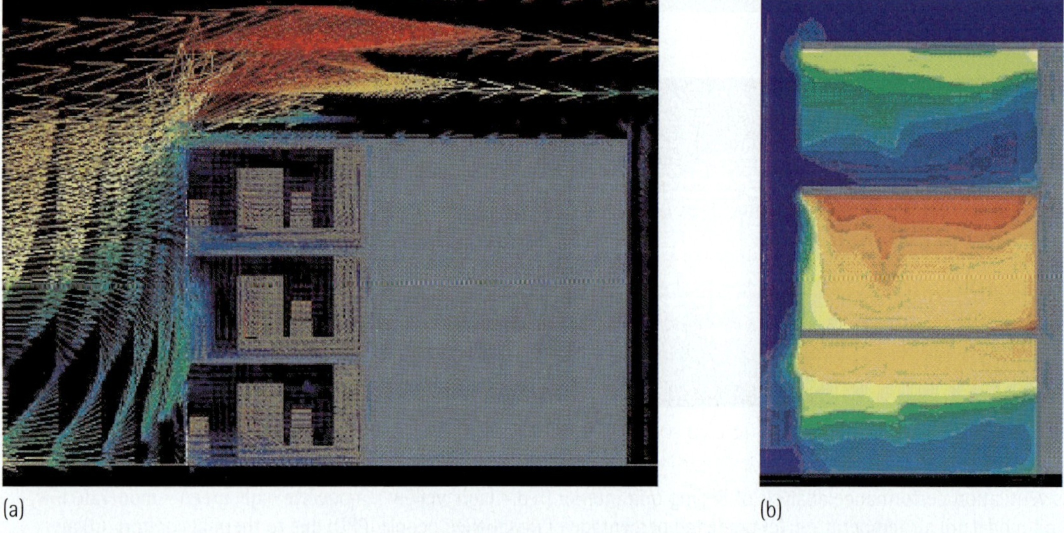

(a) (b)

Figure 18 CFD results in the center of the rooms with the combined wind and stack effects (red – high, yellow – moderate high, green – moderate low, blue – low): (a) air velocity distribution and (b) air temperature distribution

of the flow domain to the outdoors allows us to consider the vertical (hydrostatic) pressure distribution.

Under a buoyancy-driven scenario (windless condition), the temperatures in each space increased with height due to the outside thermal plume from one room entering the room above (as shown in Figure 16) despite the fact that the spaces were physically and thermally isolated from one another. This can clearly be seen from the shifts in the graph of indoor temperature versus height, as shown in Figure 16b (Allocca, Chen and Glicksman 2003). This type of effect seems plausible due to the small distance between the upper openings of one space and the lower openings of the space above. Analytical solutions or experimental measurements do not easily discover such a phenomenon found in the CFD simulation.

Although the study of pure buoyancy effects on single-sided ventilation is interesting, it is more useful to examine the effects of combined wind and buoyancy on the ventilation. The experiments (Phaff et al 1980) showed that, for a particular tested room, wind and stack flow reinforced each other. Our study found that wind and stack forces did not always reinforce each other. In fact, they opposed each other in some instances. This ambiguity is illustrated in Figure 17a. An example of the counteracting wind and stack effect during an increasing progression of wind speeds is also shown in Figure 17b.

A counteracting wind and stack flow took place in the middle unit. Figure 18a shows the airflow at the mid-section in the room. The wind force at the upper opening was stronger than the buoyancy force, thereby forcing a clockwise flow into the unit through the upper opening and out through the lower opening. Since the two forces opposed each other, the ventilation was reduced. As a result, the temperature in the middle unit was the highest, as shown in Figure 18b. However, in the upper unit, the wind aided the buoyancy effects by driving air in through the lower opening and out through the upper opening. The room air temperature was the lowest in the building. In the lower unit, the buoyancy effects were stronger than the opposing wind effects. The air still flowed in from the lower opening and out through the upper opening. The corresponding temperature in the unit was moderate since the wind velocity was lower near the base of the building.

There are no guiding rules to determine where counteracting wind and stack effect will occur. Ordinary design guides may not provide useful information, unless detailed air velocity distributions near the openings are known. It seems that CFD analysis can provide detailed information to a designer, ensuring a successful design of natural ventilation systems.

SUMMARY

The results in this chapter show that wind can have many positive attributes in an architectural environment such as providing a comfortable and healthy indoor environment that can also save energy by means of passive cooling or natural ventilation. However, this chapter illustrates that wind can also cause discomfort to pedestrians if its speed around a building is too high, and it can also increase energy loss in the winter.

This chapter has discussed different methods available for wind design. Among the methods studied, CFD seems to be attractive for building environment design, since it is the most affordable, accurate, and informative when it is properly applied. This chapter has also illustrated a number of architectural indoor and outdoor environment designs that have utilized CFD. These include:

- airflow around a building complex;
- using building shape to prevent a cold wind tunnel in a high-rise building site;
- cross natural ventilation in a building; and
- single-sided natural ventilation design.

This chapter discusses the available methods for designing a building to take advantage of the wind, such as model mockup, wind tunnel, nodal/zonal models, and CFD. CFD can provide detailed and useful information and is becoming an attractive and popular design tool. The examples illustrate how to collect wind information, develop different strategies for outdoor and indoor environment design, and use a CFD program to conduct the design. The outdoor design aims at developing pedestrian thermal comfort by varying building shape. The indoor design focuses on promoting natural ventilation, a good measure for reducing energy use in buildings and providing better indoor air quality.

The results show that CFD can be an effective tool for identifying whether outdoor wind is too strong in a building site. If it is, one can eliminate the strong wind by changing the building shape with the help of CFD. The example of the cross natural ventilation study shows that CFD can provide a lot of information concerning thermal comfort and indoor air quality. CFD can also identify some basic flow phenomena that may not be easily found with other methods of analysis as demonstrated in the single-sided natural ventilation design.

ACKNOWLEDGMENTS

This chapter is mainly reproduced with permission from Elsevier from the paper: Chen, Q. 2004. Using computational tools to factor wind into architectural environment design. *Energy and Buildings*, 36: 1197-1209. The author would also like to acknowledge the direct use of the following paper: Allocca, C., Chen, Q. and Glicksman, L.R. 2003. Design Analysis of Single-Sided Natural Ventilation. *Energy and Buildings*, 35(8): 785-795 in this chapter.

The author wishes to thank his former team members, C. Allocca, L. Glicksman, J. Huang, J. Lin, Y. Jiang, N. Kobayashi, X. Luo, A. Scott, C. Yang, and J. Yoon at the Massachusetts Institute of Technology for their contributions to the following figures: 6, 7, 9, 10, 11, 12, 13, 14, 15, 16, 17, and 18.

REFERENCES

Allard, F. 1998. *Natural Ventilation in Buildings: A Design Handbook.* James & James Ltd., London.

Allocca, C., Q. Chen, and L. Glicksman. 2003. Design Analysis of Single-Sided Natural Ventilation. *Energy and Buildings*, 35(8): 785-795.

Awbi, H. 1996. *Air Movement in Naturally Ventilated Buildings.* Department of Construction Management & Engineering, University of Reading, U.K.

Bottema, M. 1980. *Wind Climate and Urban Geometry.* Ph.D. thesis. Eindhoven University of Technology, Eindhoven, the Netherlands.

CHAM. 2000. *PHOENICS Version 3.3.* CHAM Ltd., London.

Chandra, S., P. Fairey and M. Houston. 1983. *A Handbook for Designing Ventilated Buildings* FSEC-CR-93-83. Florida Solar Energy Center, Cape Canaveral, FL.

Chen, Q. 1997. Computational Fluid Dynamics for HVAC: Successes and Failures. *American Society of Heating, Refrigerating, and Air-Conditioning Engineers Transactions*, 103 (1): 178-187.

Chen, Q. and L. Glicksman. 2000. Application of Computational Fluid Dynamics for Indoor Air Quality Studies, in *Indoor Air Quality Handbook*, vol. 59. J. Spengler, J. Samet and J. McCarthy, eds. pp.1-22. McGraw-Hill, Inc., New York.

Chen, Q. 2004. Using Computational Tools to Factor Wind into Architectural Environment Design. *Energy and Buildings*, 36: 1197-1209.

CIBSE. 1997. Natural Ventilation in Non-Domestic Buildings. *Applications Manual AM10*. The Chartered Institution of Building Services Engineers, London.

Jiang, Y. 1999. Private communications. Tsinghua University, China

Jiang, Y., H. Xing, C. Straub, Q. Chen, A. Scott, L. Glicksman and L. Norford. 1999. Design Natural Ventilation in Buildings and Outdoor Comfort around Buildings by an Architect and Engineer Team with a CFD Program. *Proceedings of the 3rd International Symposium on HVAC*, 17-19 November 1999, vol. 2: 591-601. Shenzhen.

Linden, P. 1999. The Fluid Mechanics of Natural Ventilation. *Annual Review of Fluid Mechanics*, 31: 201-38.

Lechner, N. 2000. *Heating, Cooling, Lighting: Design Methods for Architects* (2nd edition). John Wiley & Sons, New York.

Marion, W. and K. Urban. 1995. *User's Manual for TMY2s Typical Meteorological Years*. National Renewable Energy Laboratory, Golden, Colorado. Available at <http://rredc.nrel.gov/solar/old_data/nsrdb/tmy2/>.

Moore, F. 1993. *Environment Control Systems: Heating Cooling Lighting*. McGraw-Hill, Inc., New York.

Murakami, S. 1998. Overview of Turbulence Models Applied in CWE-1997. *Journal of Wind Engineering and Industrial Aerodynamics*, 74-76: 1-24.

NOAA. 1983. *Climate Atlas of U.S. National Oceanic and Atmospheric Administration*. NOAA, Washington, D.C.

Olson, D., L. Glicksman, and H. Ferm. 1990. Steady-State Natural Convection in Empty and Partitioned Enclosures at High Rayleigh Numbers. *Transactions. ASME Journal of Heat Transfer*, 112: 640-647.

Phaff, J., W. de Gids, J. Ton, D. van der Ree and L. Schijndel. 1980. The Ventilation of Buildings: Investigation of the Consequences of Opening One Window on the Internal Climate of a Room. Report C 448. TNO Institute for Environmental Hygiene and Health Technology (IMG-TNO), Delft, the Netherlands.

Vickery, B. and C. Karakatsanis. 1987. External Pressure Distributions and Induced Internal Ventilation Flow in Low-Rise Industrial and Domestic Structures. *American Society of Heating, Refrigerating and Air-Conditioning Engineers (ASHRAE) Transactions*, 98(2): 2198-2213.

Visser, G. 1980. Windhindercriteria: een literatuuronderzoek naar en voorstellen voor het hanteren van uniforme TNO-windhindercriteria. Report 80-02746 (in Dutch). IMET-TNO, Apeldoorn, the Netherlands.

Zhai, Z., S. Hamilton, J. Huang, C. Allocca, N. Kobayashi, and Q. Chen. 2000. Integration of indoor and outdoor airflow study for natural ventilation design using CFD. *Proceedings of the 21st AIVC Annual Conference on Innovations in Ventilation Technology*, The Hague, the Netherlands.

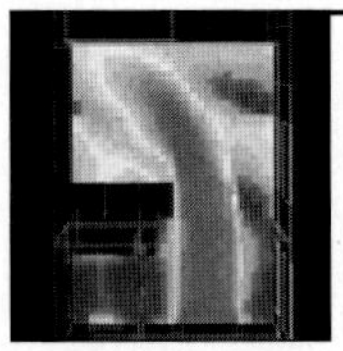

CHAPTER SEVEN

DESIGN OF NATURAL VENTILATION WITH CFD

Qingyan Chen

L. Glicksman and J. Lin (eds), Sustainable Urban Housing in China, 116-123

INTRODUCTION

As the previous chapter demonstrated, if a building is properly designed, natural ventilation can provide both a comfortable and healthy indoor environment as well as energy savings. Most architects and designers know about the principles of passive solar heating and can indicate how they desire small buildings to also take advantage of natural ventilation, as shown in Figure 1. Natural ventilation and thermal comfort, however, are difficult to understand and model, even for simple buildings. It is important that the architects and engineers collaborate early in the design process when key decisions about master planning and building geometry are made.

The present chapter illustrates, using examples from the Sustainable Urban Housing in China case studies, how architects can work with engineers to design a sustainable building. With the help of the computational fluid dynamics (CFD) technique, we suggest a design procedure as follows: First, an architect produces an initial design. Next, an engineer uses the CFD technique to calculate airflows in and around the buildings. Based on the calculated results, the architect modifies the design. Several iterations are often necessary to achieve satisfactory indoor and outdoor environment for the building. Since an architect generally does not have sufficient knowledge of fluid dynamics or numerical skills, the engineer plays an essential role in helping the architect obtain the detailed flow information. The engineer, in turn, relies on the architect to make the building comfortable and healthy.

This chapter illustrates this process through the design of a small, naturally ventilated building in Beijing (Beijing City Garden) and the analysis of a building site in Beijing for outdoor thermal comfort and natural ventilation (Beijing Star Garden).

Natural Ventilation Design

Figure 2 shows the location of the six-story (20 meters) demonstration building and its surroundings in Beijing City Garden, Beijing. There is a long, mid-rise building, 40 meters high, to the north and low-rise six-story buildings to the east and south. A wide street runs along the west of the demonstration building. Our design in this case focused on natural ventilation.

The design of natural ventilation in the demonstration building required data on indoor and outdoor airflow distributions. The wind rose in Beijing, as shown in Figure 3, indicated that in the summer the prevailing wind is from north and south. Thc corrcsponding mean wind speed is 1.9 m/s. Figure 4 breaks the reference weather data in Beijing into different comfort categories. Although heating is an important issue in Beijing, buildings there also require cooling from June through August. However, the figure illustrates that the period of comfort can be increased substantially, and the level of mechanical cooling reduced, if mean air speed in the building is at least 2 m/s through natural ventilation or internal overhead fans. Indeed, air-conditioning may not be necessary in Beijing if natural ventilation is combined with night cooling. Excess wind speed, however, is detrimental, causing discomfort to pedestrians. In Beijing, winter winds around some buildings can reach speeds of 14 m/s. Considering the low air temperature in the

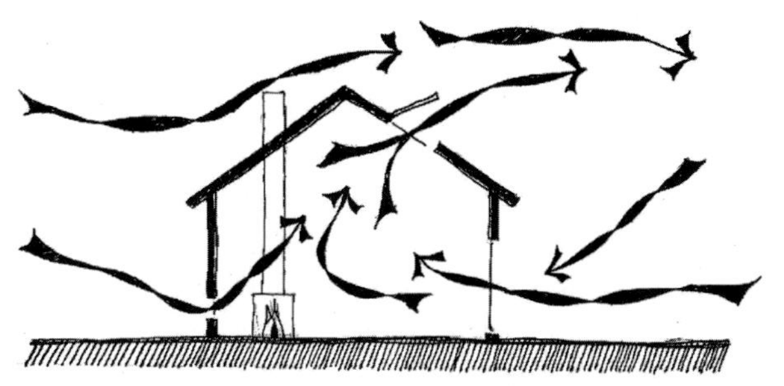

Figure 1 Smart arrows used by architects (Source: adapted from Moore 1993)

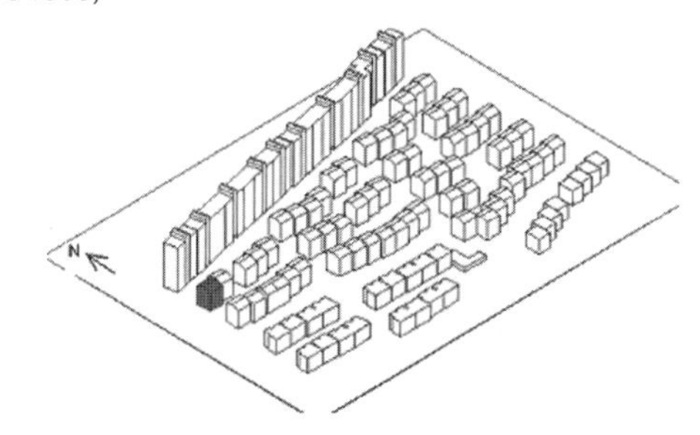

Figure 2 The demonstration building (shaded) and its site (Beijing City Garden)

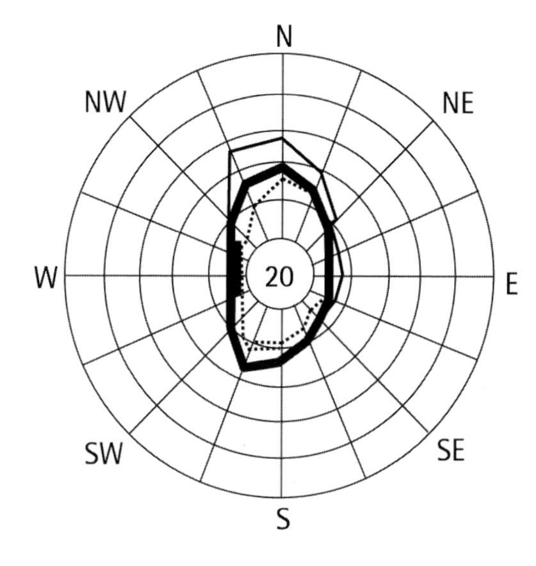

Figure 3 Wind rose for Beijing. Thick solid line: whole year; thin solid line: winter (December, January, and February); dashed line: summer (June, July, and August).

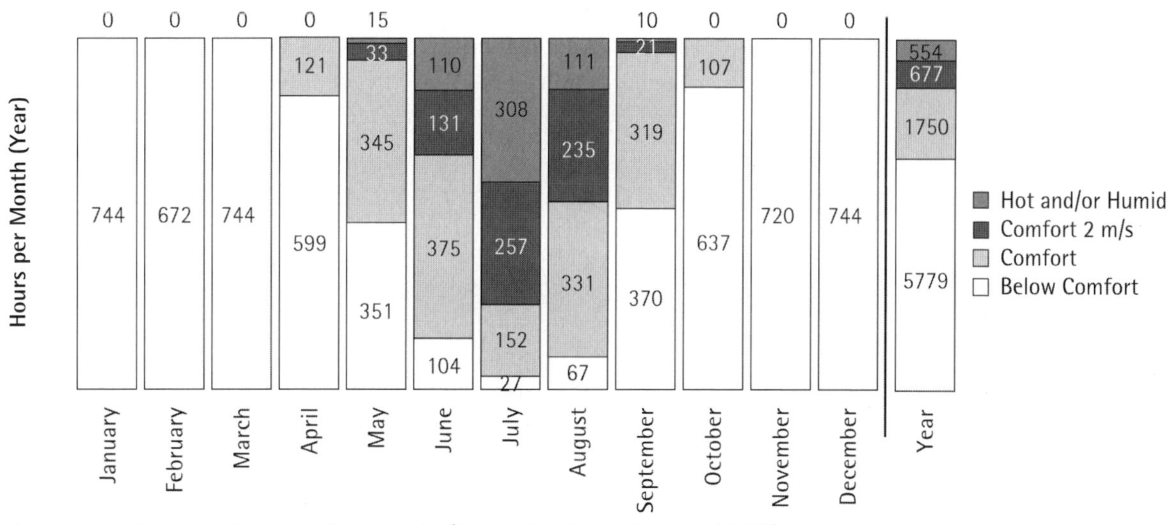

Figure 4 Comfort zones for developing countries (Source: Carrilho da Graça et al 2002)

winter, the chilling effect of this wind prevents pedestrians from walking comfortably and safely.

Using the wind information and building site, a CFD program calculated the airflow around the demonstration building, as shown in Figure 5. When wind was from the north, the air speed around the building was about 0.1 m/s ~ 0.4 m/s, which is too low to support natural ventilation of units. This reduction in air velocity occurs because the building site is in a recirculation zone, created by the tall building to the north. When the wind was from the south, the air speed around the demonstration building was higher but still low, between 0.3 and 0.8 m/s. The small distance between the demonstration building site and adjacent buildings to the south deflects wind before it reaches the building.

Despite these conditions, the design team chose to use the available, albeit low-velocity, wind to provide natural ventilation to the building. Since the design of natural ventilation does not usually increase building construction and operation costs, it is always worthwhile to try to incorporate such systems, when basic climatic conditions will permit thermally comfortable through-ventilation. Figure 6 shows two design schemes for a typical floor of the demonstration building, both designed to allow free passage of wind. This chapter demonstrates how the CFD technique is used to determine the necessary size of the court to the south. The project team expects that the court will increase natural light in the building, create a social space for the building residents, and channel wind from the south for ventilation.

Figure 7 shows the airflow distribution in and around the demonstration building when the wind comes from the south. The separation of detailed airflow information within the demonstration building from the airflow around Beijing City Garden (Figure 5b) reduced computing time significantly. Because wind speed from the south around the demonstration building is 0.3 ~ 0.8 m/s, CFD models of airflow within the building assumed wind from the south at a uniform speed of 0.5 m/s. The model shows that the building layout permits free, easy ventilation of the units. However, court size has no significant impact on the airflow pattern and flow rate. Furthermore, the results suggest that mechanical ventilation or stack ventilation might be needed in order to enhance the ventilation in the building. Chapter 5 documents the impact on thermal comfort of specified air change rates assisted by mechanical ventilation. With an open interior design for natural ventilation, fan power for adequate stack ventilation will be kept to a minimum. These results were important in refining the final ventilation design. See *Chapter 10, Case Study One – Beijing Prototype Housing* for more information.

Outdoor Comfort and Site Planning

A second example demonstrates how architects and engineers can work together to design a comfortable outdoor and indoor environment. Acceleration of airflow among high-rise buildings may create outdoor comfort problems. This study uses the Beijing Star Garden Project as an example of how to design a comfortable outdoor environment.

Because it is mainly the chilling effect of the wind in the winter that causes the outdoor discomfort problem, the design analyzed the airflow distribution on a winter day. The wind rose for Beijing (Figure 3) illustrates that in the winter the prevailing wind is from the north (5° inclined to the west). In a typical year, there are nine days during which the wind speed is higher than 7.6 m/s in Beijing (ASHRAE 1997), and high wind days generally occur in the winter. The present investigation studied a scenario with a north wind of 7.6 m/s for outdoor thermal comfort consideration.

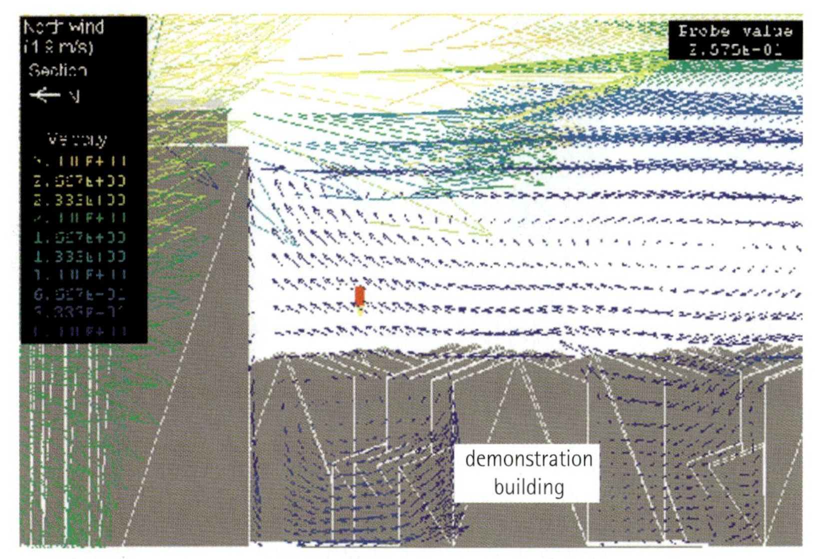

Figure 5a The airflow distribution around the demonstration building section with a north wind; existing tall buildings shown on the left-hand side of the figure

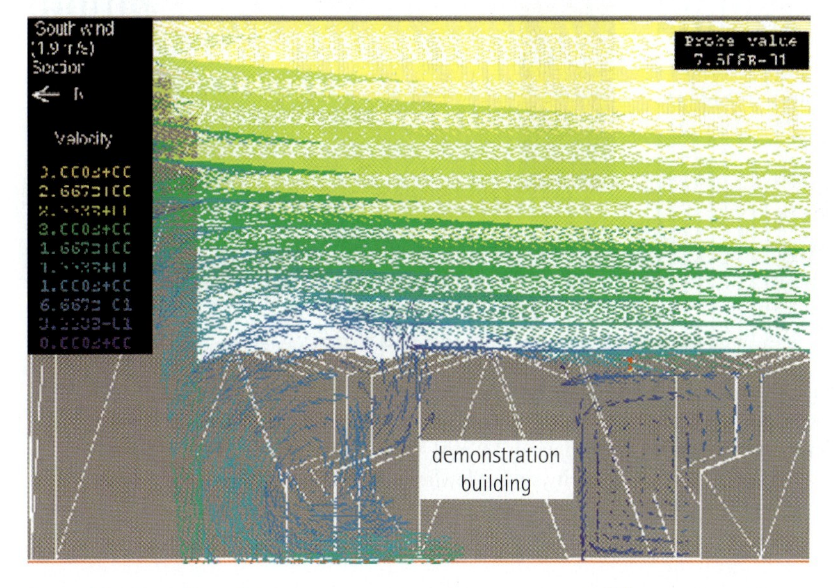

Figure 5b The airflow distribution around the demonstration building section with a south wind; existing tall buildings are shown on the left-hand side of the figure

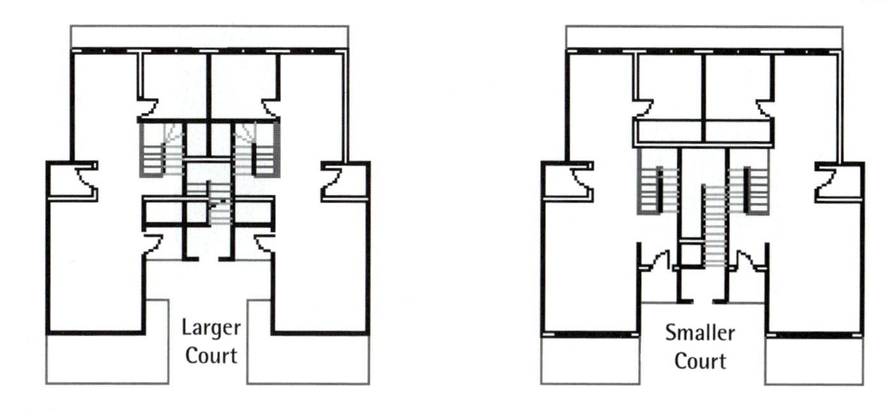

Figure 6 A typical floor plan of the demonstration building with different court sizes

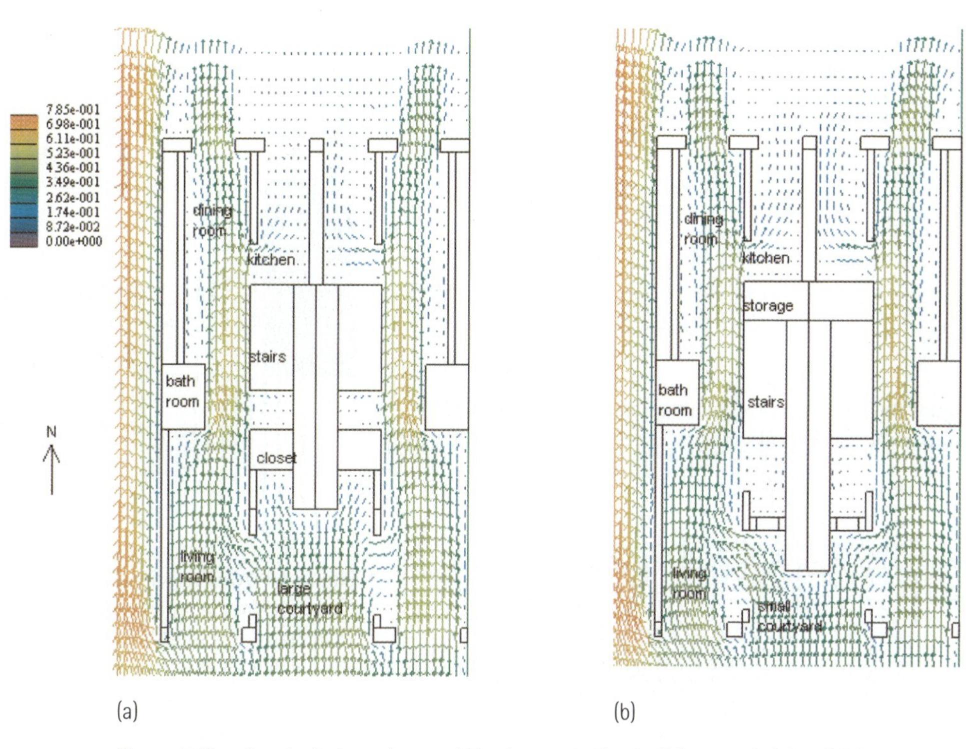

Figure 7 The air velocity in and around the demonstration building at a height of 1.2 meters above the floor: (a) shows air velocity in the plan with the large court, and (b) shows a similar plan with a smaller court

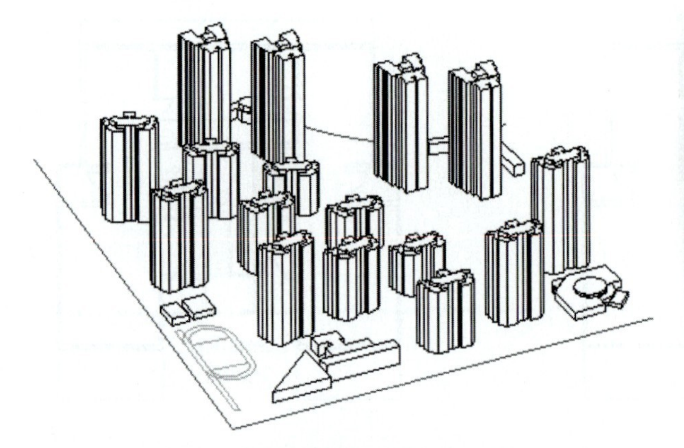

Figure 8a Original design for Beijing Star Garden by an architectural firm (scheme I)

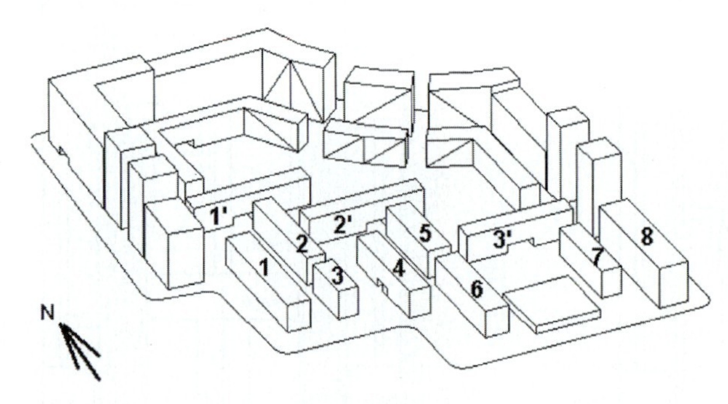

Figure 8b Preliminary redesign for Beijing Star Garden by MIT project team (scheme II)

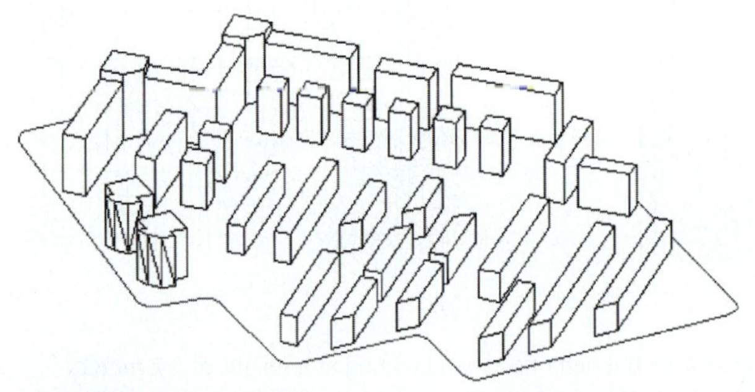

Figure 8c Second redesign for Beijing Star Garden by MIT project team (scheme III)

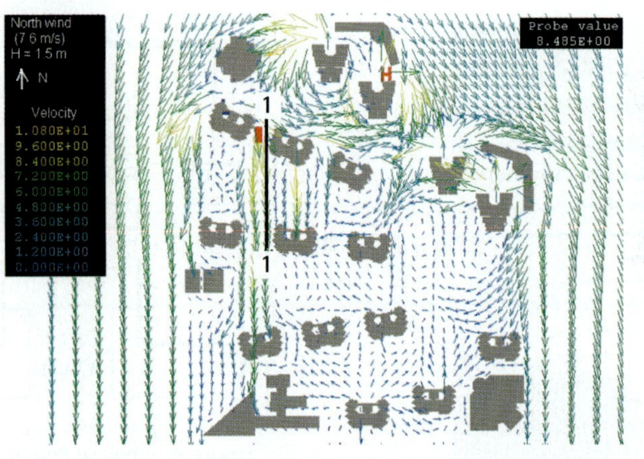

Figure 9 Wind velocity distribution at the height of 1.5 m above the ground around the buildings for scheme I with a north wind (blue - low velocity, green - moderate velocity, and yellow/red - high velocity)

Figure 11 Wind velocity distribution at the height of 1.5 m above the ground around the buildings for scheme III with a north wind (7.6 m/s) (blue - low velocity, green - moderate velocity, and yellow/red - high velocity)

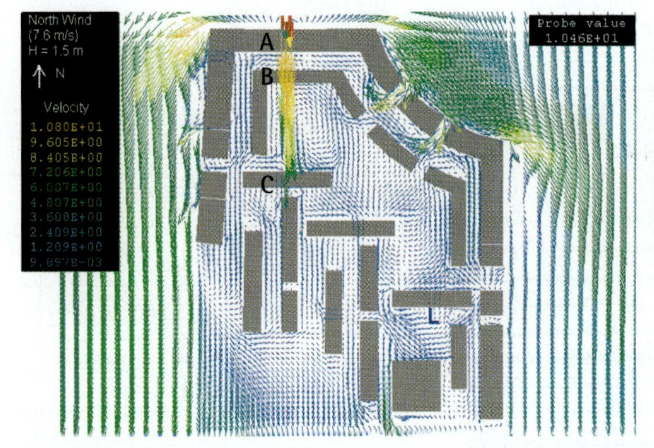

Figure 10 Wind velocity distribution at the height of 1.5 m above the ground around the buildings for scheme II with a north wind (blue - low velocity, green - moderate velocity, and yellow/red - high velocity)

Figure 12 Wind velocity distribution at the height of 1.5 m above the ground around the buildings for scheme III with a south wind (1.9 m/s) (blue - low velocity, green - moderate velocity, and yellow/red - high velocity)

Figure 8a shows the developer's original site design (scheme I). The design for scheme I consisted of sixteen high-rise buildings ranging from 33 to 90 meters high. This chapter presents the wind speed distribution at the height of 1.5 meters above the ground to evaluate pedestrian comfort (Figure 9). The wind speed at section 1-1 is around 8~9 m/s (grade 5), too high to be accepted even for a short stay in the winter. The linear arrangement of the buildings permits wind to pass freely. The CFD calculation also shows that at a height of 30 meters, the wind speed among most of the buildings is 9~10 m/s, and at a height of 70 meters, the wind speed is above 12 m/s (grade 6). The high wind speed leads to excessive infiltration in the winter and difficulties in using the wind for natural ventilation in the summer. Therefore, the MIT team made recommendations to the developer that the building site should be redesigned, and the height of the buildings should be reduced.

Since the CFD calculations showed that the height of buildings in design I causes a serious discomfort problem, the MIT design team produced schemes II and III (Figures 8b and 8c) with the following features:

- lower overall building height (building heights range from 20 to 60 meters in scheme II, and from 20 to 50 meters in scheme III) to reduce winter infiltration and to provide opportunity for summer natural ventilation, without compromising the population density; and
- protection from the north wind in the winter by using relatively high buildings in the north.

Figure 10 shows that scheme II reduces the discomfort problem greatly, but problems remained. For example, in entrances A, B, and C, the wind speed was very high because of the linear arrangement. Staggering the entrances could easily solve this problem. A number of other issues needed to be carefully examined. For instance, natural ventilation in the summer may not be effective in scheme II. As shown in Figure 8b, more than half of the buildings have a long side facing east or west. Since the prevailing wind in the summer is from the south in this site, the buildings with the long side facing east or west may not take advantage of natural cross ventilation. In addition, the orientation was not good for passive heating design, and it was difficult to shade the building from the strong solar radiation in the summer.

Based on the results for scheme II, the MIT team designed scheme III. The low-rise buildings were now tilted 45°, with the long side facing southeast and northwest. In scheme III, both outdoor thermal comfort and natural ventilation were considered. When studying the outdoor thermal comfort, the incoming wind was set to be 7.6 m/s from the north. Figure 11 shows the wind distribution for scheme III for evaluating pedestrian comfort. The high-rise buildings on the north side can block the high wind from the north. As a result, the wind speed in the site is low. While wind speeds at points A and B were high, these are vehicular entrances that will not affect pedestrian comfort.

Since the high-rise buildings were used to block the wind in the winter, it is hard to use north wind for natural ventilation in the summer. In Beijing, protecting pedestrians from the cold winter wind was more important than utilizing natural ventilation in the summer. However, it was feasible to use south wind for natural ventilation. The mean wind speed in the summer from the south is 1.9 m/s. With such a wind speed, Figure 12 shows that the wind speed around most of the buildings at 1.5 meters above the ground is above 1.0 m/s. This wind speed is sufficiently high for natural ventilation. The tilted building arrangement helped to introduce more wind into the site. Furthermore, the staggered arrangement prevented the front buildings from blocking winds. Therefore, scheme III provided good outdoor thermal comfort and potential to use natural ventilation. Note that scheme III was not the final design. In addition to the ventilation studies shown in this chapter, the design team studied other important issues, such as sun availability, natural lighting, and energy in buildings. See *Chapter 11, Case Study Two – Beijing Star Garden* for more information.

SUMMARY

This chapter shows how engineers can use CFD to help architects design natural ventilation in buildings and zones of thermal comfort around buildings. With the help of the CFD technique, engineers can calculate the airflow distributions in and around buildings, and architects can use the resulting information to modify their designs.

The CFD technique allows engineers to quickly and inexpensively analyze airflow in and around buildings. The information can help architects to design buildings that can be more effectively ventilated in summer and that can avoid strong outdoor wind in winter to achieve better thermal comfort. This chapter describes two examples of building design with CFD, one for natural ventilation and another for outdoor thermal comfort. The step-by-step design procedure shows the usefulness of CFD and how engineers and architects can work together to achieve better building design.

In the natural ventilation design, the results showed that the wind direction and building site information were most crucial for the ventilation rate, while the court size was not as important. In the case of outdoor thermal comfort design and site planning, the results indicate that one can improve the design by building shape and orientation. However, other factors, such as solar availability, should also be considered to obtain a sustainable design.

Several iterations may be necessary to design a building with satisfactory indoor and outdoor comfort environment. This design procedure, undertaken by a team of architects and engineers at MIT, shows that the CFD technique is a very useful tool for building design. This kind of cooperation between architects and engineers is necessary to design good buildings.

Although the CFD technique has great potential for building design, it is time consuming. To mitigate this, the architects and engineers on a project should review initial designs based on their experience and knowledge. This discussion will reduce the required number of iterations significantly and can speed up the design process. The CFD technique should be used only to evaluate a few, final design alternatives because of the difficulty of performing the analysis.

REFERENCES

ASHRAE. 1997. *Handbook of Fundamentals*. American Society of Heating, Refrigerating, and Air-Conditioning Engineers, Atlanta, GA.

Carrilho da Graça, G., Q. Chen, L. Glicksman and L. Norford. 2002. Simulation of Wind-Driven Ventilative Cooling Systems for an Apartment Building in Beijing and Shanghai. *Energy and Buildings*, 34(1): 1-11.

Chen, D. and J. Cai. 1994. *Documents Collection for Building Design*, vol. 1 (in Chinese). Chinese Construction Industry Publication Inc., Beijing.

Chen, Q. 1997. Computational Fluid Dynamics for HVAC: Successes and Failures. *American Society of Heating, Refrigerating and Air-Conditioning Engineers Transactions*, 103 (1): 178-187.

Givoni, B. 1997 *Climate Considerations in Building and Urban Design*. Van Nostrand Reinhold , New York.

Moore, F. 1993 *Environment Control Systems: Heating Cooling Lighting*. McGraw-Hill, Inc., New York.

Murakami, S. 1998. Overview of Turbulence Models Applied in CWE-1997. *Journal of Wind Engineering and Industrial Aerodynamics*, 74-76: 1-24.

Vickery, B. and C. Karakatsanis. 1987. External Pressure Distributions and Induced Internal Ventilation Flow in Low-Rise Industrial and Domestic Structures, *American Society of Heating, Refrigerating, and Air-Conditioning Engineers Transactions*, 98 (2): 2198-2213.

CHAPTER EIGHT

LIGHT AND SHADING

Leslie Norford

DAYLIGHTING

Characteristics

Daylighting is usually considered an attractive option for illuminating interior spaces in buildings because it is to a large extent cost-free. There is no need to generate electricity, with its attendant emissions of carbon dioxide and other pollutants, and no need for the end user to pay for that electricity to power electric lights. The infrastructure for lighting – from the generation station to the lamp – must be purchased for use at night and perhaps during overcast conditions, with the result that capital costs for electric lighting are unchanged. Daylighting however, can substantially reduce energy consumption for lighting and can also reduce peak electricity demand.

Daylighting, however, should not be considered totally free. The light emitted by the sun has infrared as well as visible energy. The infrared energy – heat – contributes to the air-conditioning load in mechanically cooled buildings and makes it more difficult to maintain comfortable conditions via ventilation (entirely natural or mechanically assisted). If sunlight were relatively inefficient, in terms of heat content, it would lose some of its appeal. Fortunately, light from the sun is characterized by a relatively high number of lumens, a measure of power in the visible part of the spectrum, for

L. Glicksman and J. Lin (eds), Sustainable Urban Housing in China, 124–133

each watt of total power, including heat. Table 1 compares sunlight with common electric lamps.

This table shows that sunlight is more efficient than the most prevalent lamps – incandescent and fluorescent. It therefore generates less heat for a given amount of light and should be strongly considered as a viable light source in all but the most extreme conditions. In very hot climates where buildings are not mechanically cooled, thermal comfort may be more important than indoor illuminance levels, and windows in vernacular buildings may be small.

One argument for caution in admitting daylight, even in relatively mild climates, is that it is more difficult to control than electric lighting. If windows are sized to admit adequate light on an overcast day, they will admit an excessive amount on a clear day if direct radiation from the sun is not properly controlled via some sort of baffles, shades, or diffusers. While occupants may be happy with more than the minimum illuminance, they may not welcome the associated heat. Ten times more light than needed will produce ten times more heat, overwhelming the benefits in efficiency of the light source.

Solar heat gain is one potential cost associated with daylight. There is also a capital cost associated with windows, shading devices, and such controls as illuminance sensors that can dim electric lamps or control shading. In addition, windows have a higher thermal conductivity than a reasonably well-insulated wall or roof, thereby increasing heating loads in cold weather and cooling loads in summer. Of course, the heat associated with sunlight may be used to augment or replace mechanical systems in winter, in which case the cost of windows should be traded against other equipment.

Another characteristic of sunlight is its abundance and intensity. Qualitatively, there is much more light outside than indoors. Quantitatively, an indoor illuminance level suitable for many tasks is about 500 lumens/m^2, or 500 lux. Direct sunlight produces as much as 100,000 lux, and the sky alone produces about 10,000 lux on a horizontal surface. Even on a cloudy day, about five percent of the light from the sky is adequate for reading. Under sunny conditions, the percentage drops to 0.5 percent. This suggests that relatively small, well-placed windows or skylights should be adequate to achieve reasonable illumination levels.

Figures 1 and 2 show examples of daylit buildings in a school and a residential building in Shanghai. The skylights in the atrium in the school admit more light than is needed for the atrium itself but the atrium serves as a light source for the classrooms, through the large windows shown in the figures.

Benefits

Daylighting has the direct benefits of reducing operating costs for electric lighting and may reduce the cost of air-conditioning, due to its efficiency as a light source. There are indirect benefits that may be more economically valuable. Daylighting meets some of the biological needs of building occupants (Lam 1997). These needs include orientation, physical security, definition of personal territory, relaxation of the body and mind, adjustment of the biological clock, and contact with nature and sunlight.

While daylight offers indirect or personal benefits to occupants of residential as well as non-residential buildings, quantified benefits are readily available only for non-residential structures. Benefits in schools have been documented through analysis of 21,000 standardized math and reading tests given to students in grades 2-5 in three states in the United States (Heschong Mahone Group, Inc. 1999a). Classrooms were rated on a 0-5 scale for three characteristics: the size

Light Source	Efficacy (Lumens/Watt)
Incandescent lamp	7-20 [a]
Fluorescent lamp	68-87 [b]
Compact fluorescent lamp	61 [b]
Metal halide lamp	68-87 [b]
High-pressure sodium lamp	58-79 [b]
Sunlight	94 [a]

Table 1 Efficacy of light sources: values for fluorescent, metal-halide and sodium lamps including ballasts (Sources: a: Murdoch 2003, b: - IESNA 2000)

Figure 1 Daylighting in a school in the Pudong region of Shanghai

Figure 2 Daylighting in a multifamily residential building in the Pudong region of Shanghai

Figure 3 Overhanging roofs on buildings in the Imperial Palace in Beijing protect indoor spaces from direct sunlight in summer

Figure 4 Houses on narrow Beijing streets are shaded by each other and by trees

and tint of windows, the presence and type of skylighting, and the overall amount of expected daylight. The result: children did better with natural light. Correlations between daylight and performance in California schools, significant at the 99 percent level, showed that students in classrooms with the most daylight progressed faster over a school year than did students with the least daylight. There was 20 percent more improvement in math test scores and 26 percent more improvement in reading tests.

Similar correlations showed that students in rooms with the largest windows progressed faster than those with the smallest windows: 15 percent on math tests and 23 percent on reading tests. In classrooms with well-designed skylights, controllable and equipped with a diffuser, students progressed 19 percent faster than those in classrooms with no skylight. Natural ventilation was found to be important, too. Students in classrooms with operable windows progressed 7 percent faster than did those in classrooms with fixed windows.

Studies in Seattle, Washington, and Fort Collins, Colorado, assessed the performance of students on end-of-year tests. Students in classrooms with the most daylight earned scores that were 7 to 18 percent higher than students in classrooms with the least amount of daylighting.

Documented daylighting benefits are not confined to schools. Merchandise sales in a set of 108 nearly identical stores operated by a single company have been analyzed and related to daylighting (Heschong Mahone Group, Inc. 1999b). All stores were in the same geographic region of the United States, with a relatively sunny climate. All were single story. Two-thirds of the stores featured skylights and one-third did not. Daylighting in those stores with skylights often provided two to three times the target illuminance. Average monthly sales per store were correlated with the presence of skylights, the size and age of the store, hours of operation, and economic factors, as indicated by postal code.

The results showed that skylights boosted sales. When all other factors were equal, an average store without skylights would have boosted its sales 40 percent with the addition of skylights. For the set of stores as a whole, adding skylights to the one-third that lacked them would have boosted sales by an estimated 11 percent.

Shading

It is important to account for the thermal impact of solar radiation on buildings. The sun and sky provide welcome daylight but the associated heat may be a source of discomfort in summer. Shading strategies, designed to admit winter sunlight and block direct radiation in summer, should be considered in most regions in China.

Generous roof overhangs were incorporated into traditional Chinese courtyard houses. Overhanging roofs worked well seasonally, screening summer sun and admitting sunlight in winter. Buildings sited along narrow urban lanes provided shading for neighboring buildings but also blocked much-needed breezes and blocked winter daylight even more effectively than summer daylight. Figures 3 and 4 show an overhanging roof from an imperial palace and a narrow Beijing street.

Modern architecture in Beijing has diverged in its approach to controlling solar radiation. A very successful courtyard housing project made good use of shading, as is shown in Figure 5. High-rise towers offer no shading and are overexposed to the sun, as shown in Figure 6. Inspiration for effective shading of schools, office buildings, and multi-story residential buildings can be derived from experiences in other countries, including India (Figure 7) and Zimbabwe (Figure 8).

Figure 5 Vegetation and shading from overhanging surfaces make this house appear cool and comfortable in summer

Figure 6 Which way is south? This high-rise residential building in Beijing lacks any reasonable relationship to the sun.

Figure 7 Shading is used successfully in this school near Ahmedabad, India

Figure 8 Shading can be incorporated into mid-rise and high-rise buildings. This office building in Harare, Zimbabwe, uses shading to reduce heat gains and heavy construction to moderate indoor temperature swings. Ventilation is provided by fans, but there are no chillers.

Figure 9 An image of a design for a Shanghai residential unit. The relatively coarse level of detail is appropriate for design as distinguished from presentation-quality images. Note the existence of direct daylight that produces desired passive-solar heating in winter, but may be a source of unwelcome glare.

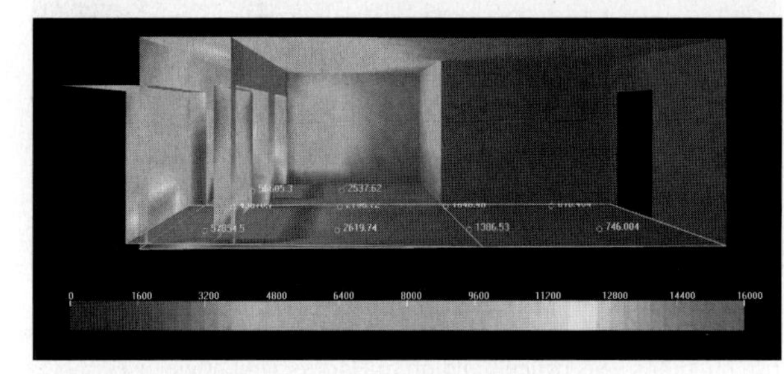

Figure 10 A false-color map of illuminance, for the same residential unit shown in Figure 9. This image gives designers quantitative information about lighting levels. (See color versions of related images in chapter 12, Figures 15-19)

DESIGN TOOLS

Design tools for daylighting and electric lighting include simple methods suitable for hand calculation or computer spreadsheet as well as more complex and powerful ray-tracing and radiant-energy-balance methods that can produce photo-realistic renderings.

Simple Calculations

The simple calculation method is known as the zonal-cavity or lumen method (Murdoch 2003, IESNA 2000). As used for electric lights or skylights, it calculates a single value of illuminance for the entire room. In that sense, it is similar to airflow calculations that assume the air in a room is well mixed. For windows, it calculates illuminance at three fixed points on a line orthogonal to the wall with the window.

The basic idea, applied to electric lighting, is expressed in the following equation:

$$E = \frac{n \cdot \Phi}{A} CU \cdot LLF$$

where

E	=	desired illuminance level
n	=	number of lamps
Φ	=	output of lamp in lumens
A	=	floor area
CU	=	coefficient of utilization
LLF	=	light-loss factor

The fundamental equation can be used to determine the number of lamps in a specific type of fixture that are required to provide the desired illuminance. The coefficient of utilization, CU, accounts for the shape of the room, the reflectances of the surfaces (walls, ceiling, and floor), and the type of light fixture. CU ranges from less than 0.1 to more than 0.9; fixtures that reduce glare tend to have lower values. The light-loss factor, LLF, is typically about 0.7, which indicates that the illuminance level in a room drops about 30 percent over the life of the lamps, due to lamp degradation and dirt in fixtures and on room surfaces.

The same equation can be readily adapted to apply to skylights, as follows:

$$E = E_{outside} \frac{A_{skylight}}{A} CU \cdot LLF$$

where

$E_{outside}$	=	horizontal outdoor illuminance due to the sun and the sky
$A_{skylight}$	=	gross projected horizontal area of the skylights

Here, the coefficient of utilization, CU, accounts for losses in the room and in the skylight, including the glazing material, the framing, and the skylight well that extends above the interior ceiling. For a given outdoor condition with an associated illuminance, the fundamental equation can be solved for the skylight area required to produce the desired indoor illuminance. Outdoor illuminance varies with the position of the sun in the sky as well as sky conditions.

Skylights are very useful in commercial buildings. In the United States, 70 percent of commercial buildings are single story, 62 percent of commercial-building floor area is directly under a roof, and only 25 percent is easily lit with windows (Murdoch 2003). Skylights are also appropriate for the top stories of residential buildings and for common spaces. In houses, windows are of primary importance but pose a more difficult design problem because the illuminance decreases strongly as a function of distance from the window. The lumen method can be adapted to estimate illuminance at three points in a room: as a function of room characteristics, window size, and available light from the sun and sky. For a daylighting design, window size can be determined on the basis of desired illuminance at any of the three points in the room.

The daylight factor method is popular in European countries, where cloudy conditions are so common that daylight design typically ignores direct solar radiation. A daylight factor is the ratio of indoor illuminance at a user-selected point and outdoor illuminance, both on a horizontal surface. This ratio depends on the selected indoor location and the size of the window. The overall daylight factor is the sum of individual factors for light from the sky, light externally reflected off neighboring buildings, and light from the ground internally reflected off the ceiling.

Consider, for example, a point in a room, as described in Murdoch (2003), that has daylight factors of 0.02, 0.002, and 0.012 for direct, externally reflected and internally reflected light, respectively. The total daylight factor is 0.034. If the horizontal illuminance is 8,000 lux, the illuminance indoors would be 272 lux, adequate for many reading tasks. The daylight-factor method is best applied when combined with long-term average data for outdoor horizontal illuminance, making it possible to readily estimate the percentage of hours that an indoor space with a specified daylight factor will have adequate illuminance.

Lighting Simulation with Radiosity and Ray-Tracing Methods

The lumen method assumes that indoor illuminance is uniform wall-to-wall for electric lights and for skylights and provides only three values for windows. It also is based on simple rectilinear spaces. For more detailed illuminance information, it is necessary to consider the location of light sources, either daylight through windows and skylights or electric lights, and inter-reflection of light among surfaces. This can be achieved with either a radiosity method or a ray-tracing method that is modified to account for reflected luminous energy (Cohen and Wallace 1993, Ward Larson and Shakespeare 1998).

Computer codes that embody both methods are available for use by designers. These methods start with a CAD model of a space under consideration. Developing a model with the appropriate level of detail is typically the most challenging step in the lighting analysis. While a highly detailed model is appropriate for photo-realistic images, too much detail during the early stages of design unduly slows the lighting simulations. The CAD model can be organized in ways that make it relatively easy to assign optical properties to surfaces after the model is imported into the lighting-simulation package. Light sources, windows, the position of the sun and sky conditions must also be described. The output for a radiosity-based program includes a view-independent calculation of the light emitted from all surfaces, suitable for a virtual walk-through and for rendering images of specified views. Maps of illuminance and luminance can also be displayed. The radiosity approach, with ray tracing for specular surfaces, was used to generate the images shown in Figures 9 through 11.

ENERGY USED FOR LIGHTING AND HVAC

The design of electric lighting systems makes use of the same tools as those presented for daylighting: the lumen-cavity method or more flexible, powerful, and complex radiosity or ray-tracing methods. Energy consumption

(a)

(b)

Figure 11 Near-photo-realistic images of (a) the design for a school in Bozeman, Montana and (b) analysis of portico design for a Palladian villa near Vicenza, Italy

by electric lights is straightforward to calculate in the absence of occupancy sensors or dimming controls. Of more interest is the potential for daylighting to reduce the energy required for electric lighting. Related to this is the impact of both daylighting and electric lighting on heating and cooling loads. Daylight has a higher efficacy than almost all light sources, as noted in the *Characteristics* section of this chapter and in Table 1, but windows increase heating loads compared to well-insulated walls, and daylight in excess of what is required may bring a cooling penalty as well. Of most use to designers is a calculation that accounts for all energy use. This section describes two methods for performing these calculations. The first is easy to use but is limited to skylights. As such, it is appropriate for schools and low-rise commercial buildings, which are often associated with residential developments. The second makes use of an optimization routine coupled with a more complex calculation program and can be applied to all building types.

Lighting, Heating, and Cooling Energy as Influenced by Skylights

The fundamental equation for illuminance level in a luminous cavity has been implemented in spreadsheet form in the SkyCalc program (Heschong Mahone Group, Inc. 2003), available without charge over the Internet (www.h-m-g.com). This program performs lighting calculations on an hourly basis for a typical day each month, using appropriate outdoor illuminance values. It accounts for a variety of skylight materials and shading devices, used to control glare. It also accounts for heating and cooling energy and electric lighting, in a range of user-selected building functions and a wide range of climates. SkyCalc permits the designer to assess the overall impact of skylights on electric lighting loads as

SkyCalc: Skylight Design Assistant - Basic Inputs

Company Name: Middle School for New Housing Development
Project Description: Skylight Design and Analysis for Classroom

Select Location Boston, MA
Climate data loaded = Boston, MA
Climate data for location is already loaded

Building
Building type: Classroom, K-12
Bldg area: 1,000 ft²
Ceiling height: 15 ft
Wall color: Off-white paint

Shelving/Racks or Partitions?
○ Partitions, ○ Shelves/Racks, ● None/Open
No data required ft
No data required ft
No data required ft
No data required ft

Electric Lighting
Lighting system: Open cell fluorescent
Fixture height: 12 ft
Lighting control: Dimming min 20% light

Skylights:
Number of skylights: 3
Skylight width: 4 ft
Skylight length: 4 ft
Current Skylight to Floor Ratio = 4.8%

Skylight Description
Glazing type: Acrylic
Glazing layers: Double glazed
Glazing color: White 2447

Skylight Well
Light well height: 1 feet
Well color: Off-white paint
Safety grate or screen: ○ Yes, ● No

Heating and Air Conditioning Systems
Air Conditioning: Mechanical A/C
Heating System: Gas/Oil Furnace

Utilities
Average Elec Cost: $0.120 kWh
Heating Fuel Units: Therm
Heating Fuel Cost: $1.000 /Therm

Figure 12 Main input screen for the SkyCalc program (Source: Heshong Mahone Group, Inc. 2003)

SkyCalc: Skylight Design Assistant - Graphic Results

Company Name: Middle School for New Housing Development
Project Description: Skylight Design and Analysis for Classroom

Effective Aperture = 1.10%, Skylight to Floor Ratio (SFR) = 4.80%

Average daylight footcandles (fc)

	1	2	3	4	5	6	7	8	9	10	11	12	13	14	15	16	17	18	19	20	21	22	23	24
Jan	0	0	0	0	0	0	0	2	7	15	22	26	25	20	13	6	1	0	0	0	0	0	0	0
Feb	0	0	0	0	0	0	0	5	13	24	33	38	37	32	22	12	4	0	0	0	0	0	0	0
Mar	0	0	0	0	0	0	3	10	23	34	44	47	46	41	33	19	8	2	0	0	0	0	0	0
Apr	0	0	0	0	0	2	8	19	33	43	50	54	54	46	37	27	13	5	1	0	0	0	0	0
May	0	0	0	0	1	5	13	28	41	53	58	62	61	55	46	32	19	8	2	0	0	0	0	0
Jun	0	0	0	0	2	7	17	31	43	53	62	64	64	57	50	37	25	12	4	0	0	0	0	0
Jul	0	0	0	0	1	5	15	30	45	58	64	68	65	59	48	38	24	11	3	0	0	0	0	0
Aug	0	0	0	0	0	3	10	24	39	54	61	64	61	58	46	33	19	7	2	0	0	0	0	0
Sep	0	0	0	0	0	1	7	18	33	45	50	53	51	46	34	21	9	2	0	0	0	0	0	0
Oct	0	0	0	0	0	0	3	10	22	34	41	42	40	34	22	11	3	0	0	0	0	0	0	0
Nov	0	0	0	0	0	0	1	5	11	20	25	29	25	20	12	4	1	0	0	0	0	0	0	0
Dec	0	0	0	0	0	0	0	2	7	14	20	23	22	17	10	4	0	0	0	0	0	0	0	0

Design Illuminance = 50 fc

< 1 fc; < 25 fc; < 50 fc; > 50 fc;

Figure 13 Indoor illuminance levels from daylight as a function of month and hour; data are in lumens/ft², or footcandles, where 10.76 lux = 1 footcandle (Source: Heshong Mahone Group, Inc. 2003)

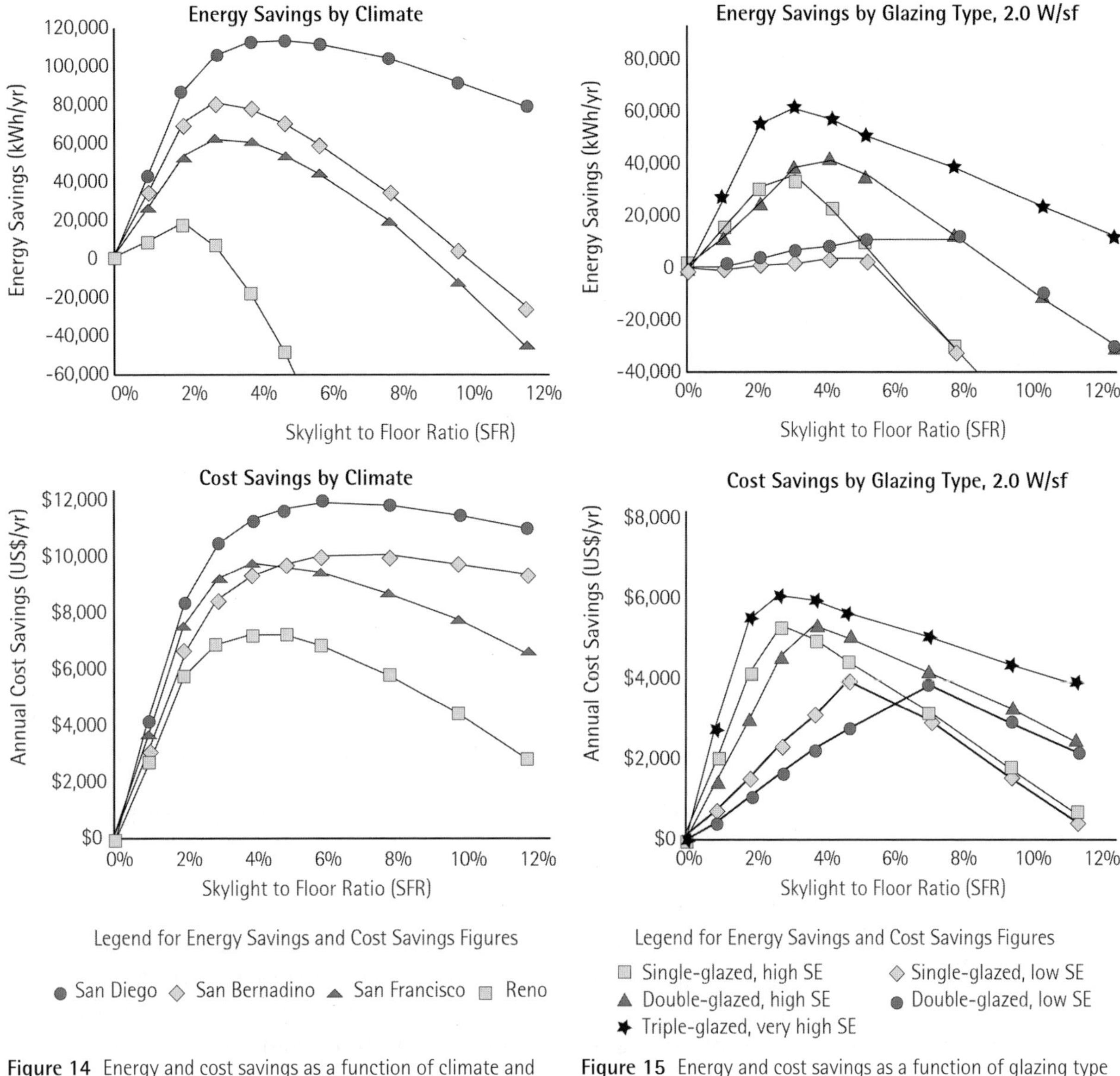

Figure 14 Energy and cost savings as a function of climate and skylight to floor ratio, for an office building (Source: Heshong Mahone Group, Inc. 1999b)

Figure 15 Energy and cost savings as a function of glazing type (with a lighting power density of 21.5 W/m^2) and skylight to floor ratio. A high value of the skylight efficacy, SE, indicates that the skylight transmits a high percentage of visible light but blocks

well as heating and cooling. It is a model of what would be developed for use by Chinese architects and engineers.

Figure 12 shows the SkyCalc input screen. Figure 13 shows the variation of indoor illuminance from daylight, as a function of time and month. Figures 14 and 15 show the energy and cost savings as a function of skylight area and the illuminance produced by daylight alone for a given configuration. Of particular interest is the difference between the savings for energy, as measured at the building, and the cost for the energy. For instance, in a city such as Reno, Nevada, the energy savings quickly become negative as the skylight area increases, due to additional heat losses over a relatively cold winter. The cost savings however, are positive, because the natural gas used for heating is cheaper than the electricity used for lighting and air-conditioning.

Automated Search for Optimal Daylighting

For daylighting designs that rely on windows and not skylights, a different approach is needed to evaluate the impact of daylight on building energy consumption. Such simulation programs as DOE-2 (Hirsch 2006, AEC 2006) and EnergyPlus (2005) are suitable. Another is the web-based simulation tool at http://designadvisor.mit.edu.

One difficulty with relatively complex simulations is the need to run a number of cases in order to identify one or more designs that offer better energy performance. An automated search process can be used, by combining an optimization engine with the simulation program. One type of optimization shown to be particularly suitable for buildings applications is genetic algorithms (Caldas 2001, Caldas and Norford 2001, Caldas and Norford 2002).

In one application, the search procedure varied window size and placement, subject to size constraints that accounted for the design aesthetic of the architect.

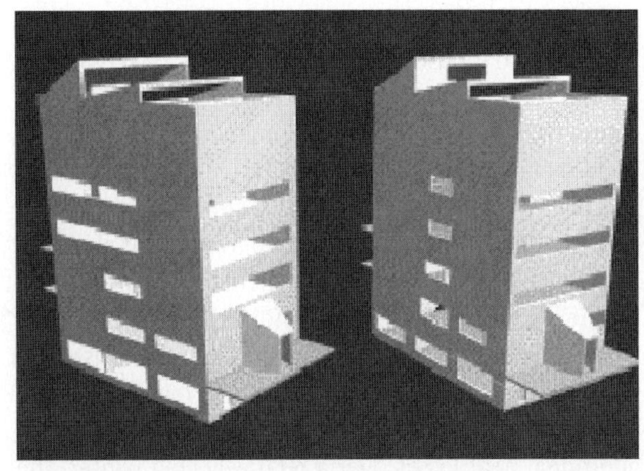

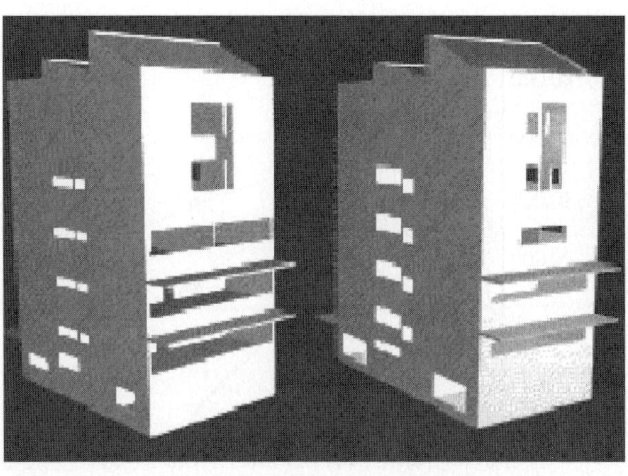

Figure 16 As-constructed and as-optimized academic building in Portugal. The two views on the left are of the east and north façades. The as-constructed building, with larger east-facing windows, is on the far left. The two views on the right are of the south and west façades. Here, the as-constructed building is on the far right. The larger west-facing windows for the as-constructed building led to unwanted heat gain. (Source: Caldas 2001 and Caldas and Norford 2001)

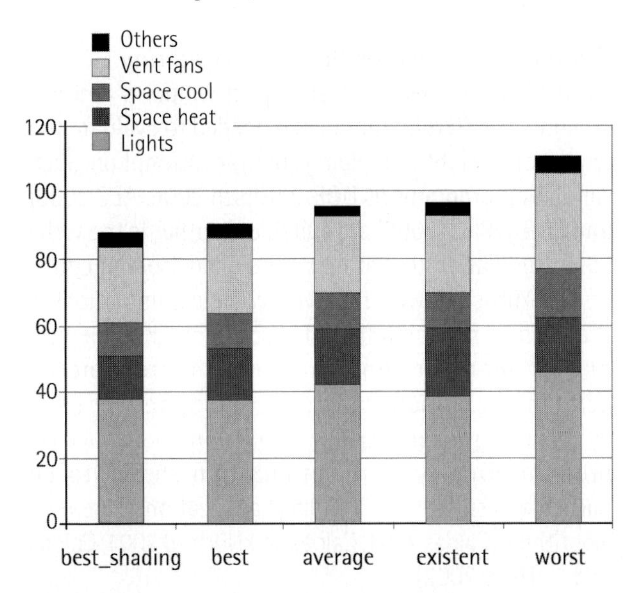

Figure 17 Annual energy consumption for different variations of the building shown in Figure 16. The best_shading design includes variations in window overhangs while the others fix the overhangs at the as-constructed dimensions. (Source: Caldas 2001 and Caldas and Norford 2001)

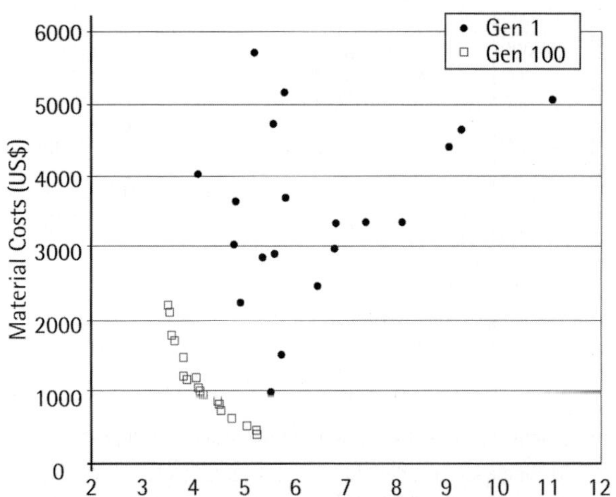

Figure 18 Optimal trade-off of material costs and operating energy for a Beijing residential unit. Gen 1 refers to the first generation of simulations. Gen 100 refers to the 100th generation of simulations. The genetic algorithm identifies simulations that offer the lowest material costs for a given operated energy, and vice versa.

The test building was a school of architecture in Portugal. Figure 16 shows the as-constructed and as-optimized building, and Figure 17 summarizes the energy consumption. The best solution, with an annual energy consumption of 87.6 MWh, was estimated to save about 10 percent relative to the as-built solution, at 96.4 MWh, which in turn was about 10 percent better than the worst solution, at 110.6 MWh.

The design of a Beijing residential unit was optimized using the same search engine and energy-simulation software that was applied to the school in Portugal. The unit had northern and southern exposures; all other walls were shared with other units. The optimization program was set up to vary window size, glazing type, and wall construction in each of the two exposed walls. The optimization accounted for both construction costs and annual operating energy. It is relatively easy to combine the two and optimize life-cycle cost over a specified lifetime. However, it may be more informative to consider the two separately, particularly in projects where capital and operating budgets are separate. If the two criteria are separate, as was the case for this example, the goal of the multi-criteria optimization is to minimize one criterion for any given value of the other. Figure 18 shows the form of the output. The genetic algorithm modified the initial population to eventually produce a set of optimal solutions. Qualitatively, the results are reasonable: a larger investment in materials yields a lower operating energy.

The populations in the first few generations of the search procedure included random values for the variable parameters. In contrast, the optimal solutions had much in common. All windows were double-glazed. The dimensions of northern windows were generally very small. South-facing windows were of average size for the solutions that had the lowest energy consumption, but considerably smaller than the constrained maximum size. Smaller south-facing windows reduced construction cost but also reduced daylight and useful solar energy. The south-facing wall had a high solar absorptivity and less insulation than the north-facing wall.

SUMMARY

This chapter has identified the benefits of using daylight in buildings. Light from the sun is associated with less heat than light from most electric lamps, reducing cooling loads. It has been shown to improve the performance of school children and to promote more activity in retail stores. Designers should carefully design the size of windows and skylights to obtain the optimum conditions, assisted by tools identified in this chapter.

ACKNOWLEDGMENTS

The author would like to thank Heschong Mahone Group, Inc. for permission to include SkyCalc program images and examples of SkyCalc output.

REFERENCES

AEC. 2006. *VisualDOE 4.0.* Architectural Energy Corporation. <http://www.archenergy.com/products/visualdoe>

Caldas, L. 2001. *An Evolution-Based Generative Design System: Improving the Environmental Performance of Buildings*. Ph.D. thesis. MIT, Cambridge, MA.

Caldas, L. and L. Norford. 2001. Architectural Constraints in a Generative Design System: Interpreting Energy Consumption Levels. *Proceedings of Building Simulation 2001, International Building Performance Simulation Association*, pp. 1397-1404. 13–15 August 2001, Rio de Janeiro.

Caldas, L. and L. Norford. 2002. A Design Optimization Tool Based on a Genetic Algorithm. *Automation in Construction*, 11(2): 173-184.

Cohen, M. and J. Wallace. 1993. *Radiosity and Realistic Image Synthesis*. Morgan Kaufman, San Francisco.

EnergyPlus. 2005. <http://www.eere.energy.gov/buildings/energyplus>

Heschong Mahone Group, Inc. 1999a. *Daylighting in Schools: An Investigation into the Relationship Between Daylighting and Human Performance*. Report to Pacific Gas & Electric Co. by the Heschong Mahone Group, Fair Oaks, CA.

Heschong Mahone Group Inc.,1999b. *Skylighting and Retail Sales: An Investigation into the Relationship Between Daylighting and Human Performance*. Report to Pacific Gas & Electric Co. by the Heschong Mahone Group, Fair Oaks, CA.

Heschong Mahone Group, Inc. 2003. *SkyCalc V. 2.0 and User's Manual*. <http://www.h-m-g.com>

Hirsch, J. 2006. *eQUEST*. James J. Hirsch and Associates. <http://www.doe2.com/equest>

IESNA. 2000. *Lighting Handbook Reference & Applications*. Illuminating Engineering Society of North America, New York.

Lam, W. 1977. *Perception and Lighting as Formgiver for Architecture*. McGraw-Hill, New York.

Murdoch, J. 2003. *Illuminating Engineering from Edison's Lamp to the LED*. Visions Communications, New York.

Ward Larson, G. and R. Shakespeare. 1998. *Rendering with Radiance*. Morgan Kaufman, San Francisco.

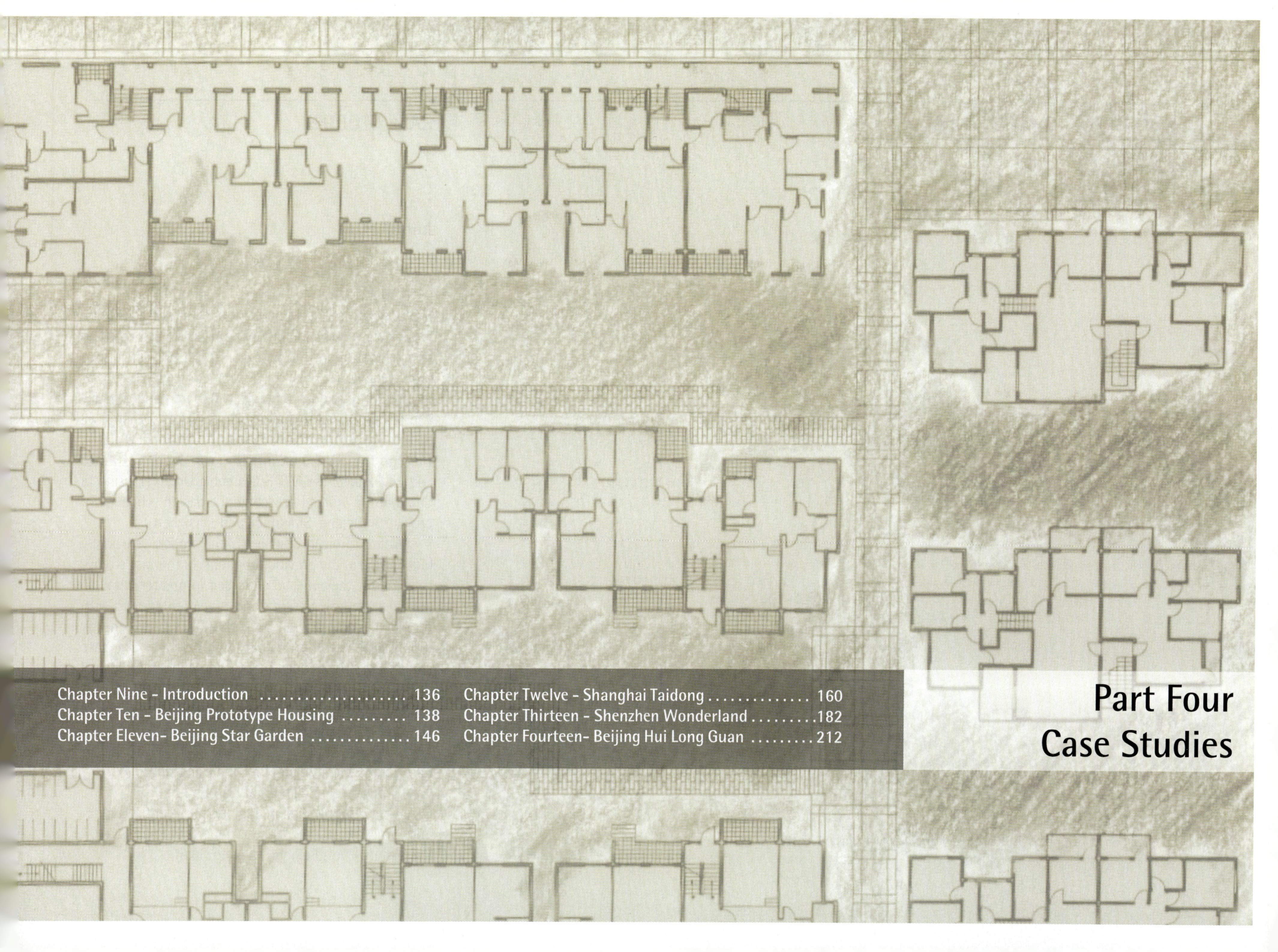

Part Four
Case Studies

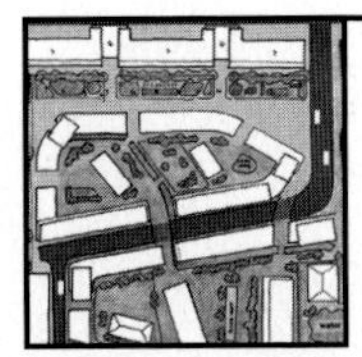

CHAPTER NINE

CASE STUDIES

Juintow Lin

INTRODUCTION

The goal of the Sustainable Urban Housing in China Project was to develop research on sustainable building principles as applied to housing projects within varying urban regions in China. The lessons learned are the result of a direct relationship between research and practice and are grounded in the diverse range of sociological, political, and economic factors of professional practice. Through the collaboration with local design institutes and developers, the MIT Sustainable Urban Housing in China group worked on schematic design schemes for projects in Beijing, Shanghai, and Shenzhen.

All of the projects described in this section were briefly highlighted in the previous chapters. In the case studies that follow, they will be described in greater detail to demonstrate how design options were integrated with technical studies such as climate research, energy simulations, shading and ventilation studies. The case studies listed below will be presented in chronological order. Several other projects, not included in this publication, can be further researched at the project website, http://chinahousing.mit.edu.

Case Study One – Beijing Prototype Housing
Case Study Two - Beijing Star Garden
Case Study Three - Shanghai Taidong Residential Quarter
Case Study Four - Shenzhen Wonderland Phase IV
Case Study Five - Beijing Hui Long Guan

L. Glicksman and J. Lin (eds), Sustainable Urban Housing in China, 136-137

Design Process

The aim of these case studies was to develop designs and technologies at a sufficiently detailed level to be integrated into a wide range of high-density low-rise and high-rise residential projects. The primary goal was to develop sustainable designs that would reduce energy consumption. Passive-based environmental principles were deployed that offer the greatest economic potential based upon an understanding of the prevailing climate. In addition to addressing the ecological and environmental concerns associated with energy-efficient housing, the case studies sought to provide for rich and diverse spaces that both catered to the residents' needs, and encouraged the development of community interaction. A general outline of the typical design process is as follows:

- research climate, building codes, cultural issues, and precedent studies;
- perform energy simulations as well as technical studies such as computational fluid dynamics (CFD) and shading analyses to identify appropriate technologies;
- produce design schemes that incorporate applicable design principles and technologies through the integration of design, sociological, and technological considerations in a multi-step process; and
- incorporate feedback from the client and local design institutes and revise the design scheme as needed.

Collaboration

The case studies represent a collaborative effort between MIT's Sustainable Urban Housing in China Group and researchers at Tsinghua University, as well as various developers and design institutes from China. The projects are the result of several workshops, research fellowships, and independent study projects carried out from 1998 to 2001 by the Building Technology Group of MIT's Department of Architecture. The project in general sought to promote and develop cross-disciplinary research focused on "sustainable" building principles within the design process. Students and faculty from varying disciplines including architecture, landscape, urban design, building technology, and policy worked together on design solutions and technologies that formed the foundation of these case studies.

ACKNOWLEDGMENTS

The work discussed in the following case studies was made possible by the concerted efforts of students, faculty, research fellows, and visiting scholars of MIT's Building Technology Group of the Department of Architecture. Students and visiting scholars include: Winnie Alamsjah, Ozgur Basak Alkan, Camille Allocca, Meredith Atkinson, Becca Brezeale, Xantha Bruso, Hongyu Cai, Erica Chan, Henry Chang, Catherine Chen, Eva Chiu, Brian Dean, Guilherme Carrilho da Graça, Rocelyn Dee, Shaohua Di, Stephen Duck, Janet Fan, Mingzheng Gao, Lara Greden, Sephir Hamilton, Joy Hu, Elsie Huang, Perry Ip, Yi Jiang, Andy Jonic, Myeoung Kim, Yongjoo Kim, Nobukazu Kobayashi, Sean Kwok, Junjie Liu, Xiaofang Luo, Xiaoyi Ma, Karl Munkelwitz, Eric Olsen, Christoph Ospelt, Sam Potter, Paul Rafiuly, Daniel Steger, Kavita Srinivasan, Carolyn Straub, Pearl Tang, Joli Thomas, Joy Wang, Jesse Williamson, Helen Xing, Jae-ock Yoon, and John Zhai. Faculty members included Leon Glicksman, Leslie Norford, Andrew Scott, John Fernandez, and Qingyan Chen. Research fellows included Juintow Lin and Zachary Kron.

The editors would also like to thank Michael Fox, for editing all the case studies several times over. Leslie Norford, Stephanie Harmon, Zachary Kron, Sephir Hamilton, Brian Dean, Ozgur Basak Alkan, Lara Greden, Camille Alloca, and Guilherme Carrilho da Graça all provided additional assistance in the writing or editing of the case studies chapters. In addition, the editors would also like to thank the Beijing Co. Ltd., Vanke Architecture Technology Research Center in Shenzhen and Tian Hong, in addition to Tsinghua University's and Tongji University's design institutes for making these projects possible.

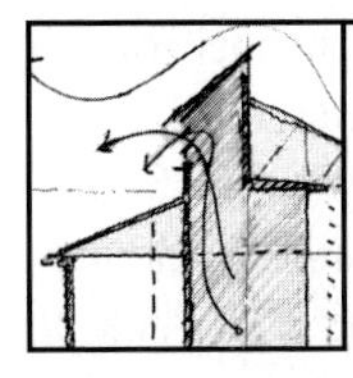

CHAPTER TEN

CASE STUDY ONE – BEIJING PROTOTYPE HOUSING

Andrew Scott and Juintow Lin

PROJECT DESCRIPTION

The Beijing Prototype Housing Projects, including the 12 x 12 house and the 12 x 24 prototypes, were designed in 1998 in collaboration with Tsinghua University for the Beijing Vanke Co. Ltd. The goal was to design a modular prototype for sustainable urban housing that could be applicable to various medium-density sites. These prototypical designs were the first projects explored by the Sustainable Urban Housing in China Group.

The demonstration building for the 12 x 12 residence was located in Beijing in a development called Beijing City Garden. The building and location of the prototype design is shown in Figure 1, and the size of the prototype was 12 meters by 12 meters. To the west of the demonstration building was a wide street. To the north, there was a long, mid-rise building, and several lower six-story buildings to the south and east.

DESIGN APPROACH

A major design intention of the housing prototype modules was to see how the tradition of the courtyard house could be reused and reinterpreted in new urban housing forms. The traditional form of the Chinese courtyard housing was admired for its ability to offer

L. Glicksman and J. Lin (eds), Sustainable Urban Housing in China, 138-145

medium density low-rise solutions while also offering privacy, community, and integration with landscape and climate. The 12 x 12 house attempted to create a large house for four families, where the courtyard acted as a communal semi-private domain, an outside shared room and landscape space, and also functioned as an entrance or gateway off of the street. The module also created an urban spatial hierarchy and provided for variety in the urban scale and street elevation.

The goal was to make the house 'bioclimatic,' which would enable it to be a passive-based environmental design that was responsive to climatic forces. This would allow the housing block to use less energy through relying more on passive-based sources of environmental control and less on mechanical air-conditioning. Several principles were adopted and are listed as follows:

- the landscaped courtyard was planted with trees to create shade in summer, enabling it to function as the communal outdoor room;
- the courtyard plan enabled the perimeter wall area of the living spaces to be maximized, thereby creating more natural daylight;
- the screen to the south-facing street elevation was constructed as a layer away from the building façade and was covered in planting in the summer. This was intended to significantly reduce solar heat gain in the façade. Various ideas for shading were explored and tested. The screens created shade to the houses in summer to counteract the summer heat and sun intensity, and also provided some privacy for a terrace or glasshouse in the summer (Figures 2-4, 7);
- the courtyard was a location for parking bikes and could be a secure communal play space for the families living in the houses; and
- the courtyard enabled natural ventilation to penetrate deeper in the plan of the houses through stack ventilation integrated with the stair and access system. The effectiveness of this system was tested using computational fluid dynamics (CFD) airflow analysis. Each house had three exposed elevations.

The plan was centered on the courtyard with two 2-story units on each side of the entrance. The bottom units were half a floor above the courtyard, while the top units had an extra half 'loft' floor. The bottom units had direct access to the courtyard, while the upper units had access to the roof and a possible roof terrace as a part of the loft space. The central idea of the plan was to use the main shared access stair as a space for stack ventilation (Figures 2-3, 8-10). The intention of the tall stair was that the buoyancy of air in the shaft would induce natural ventilation flow through the units' external windows, into the internal stair, and then into the shared main stair space, exiting through the roof level. The central location of the stair core within the 12 x 12 block, together with the courtyard, cut down on travel distances for the air movement, and the profile of the roof of the stack induced airflow through a negative wind pressure on the leeward side of the chimney. Drawings and sketches for the 12 x 12 scheme are shown in Figures 2–10.

Due to market changes, the developer asked the designers to reduce the unit area. The design team modified the 12 x 12 scheme to create a new 12 x 24 scheme, a 12-meter-deep x 24-meter-wide prototype that incorporated similar design and technical principles (Figures 11 and 12). The units were changed to single-story units accessed by a common public stair, with the exception of the penthouse units, which encompassed a larger area among two stories.

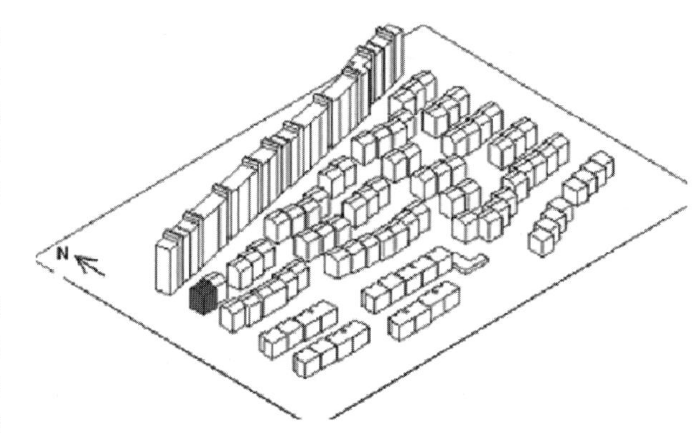

Figure 1 The demonstration building (shaded) and the site, Beijing City Garden

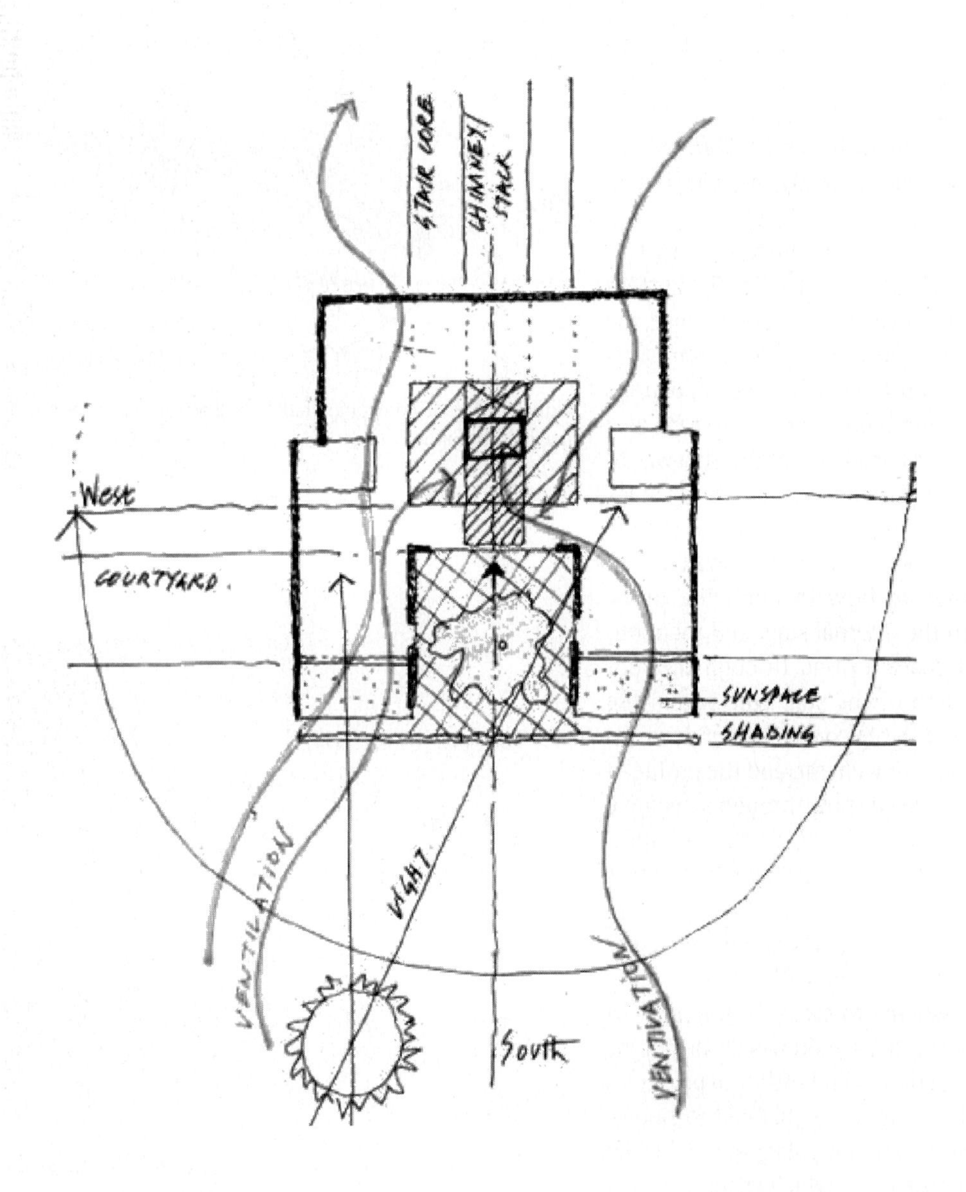

Figure 2 Beijing 12 x 12 prototype house plan diagram showing ventilation and daylighting concepts (north is up)

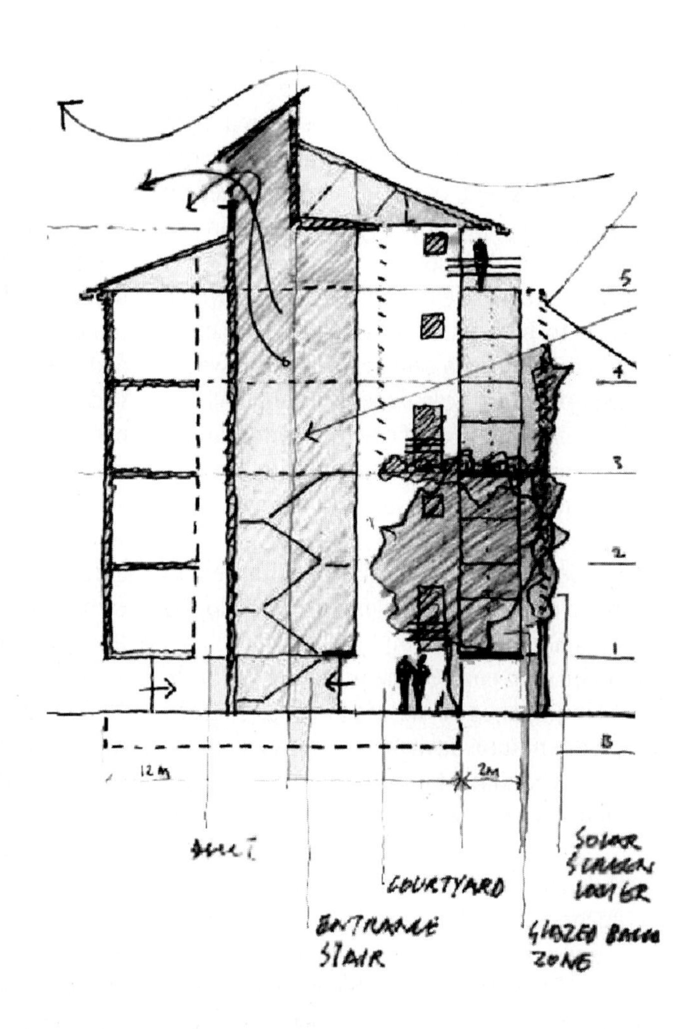

Figure 3 Beijing 12 x 12 prototype house section diagram (looking east)

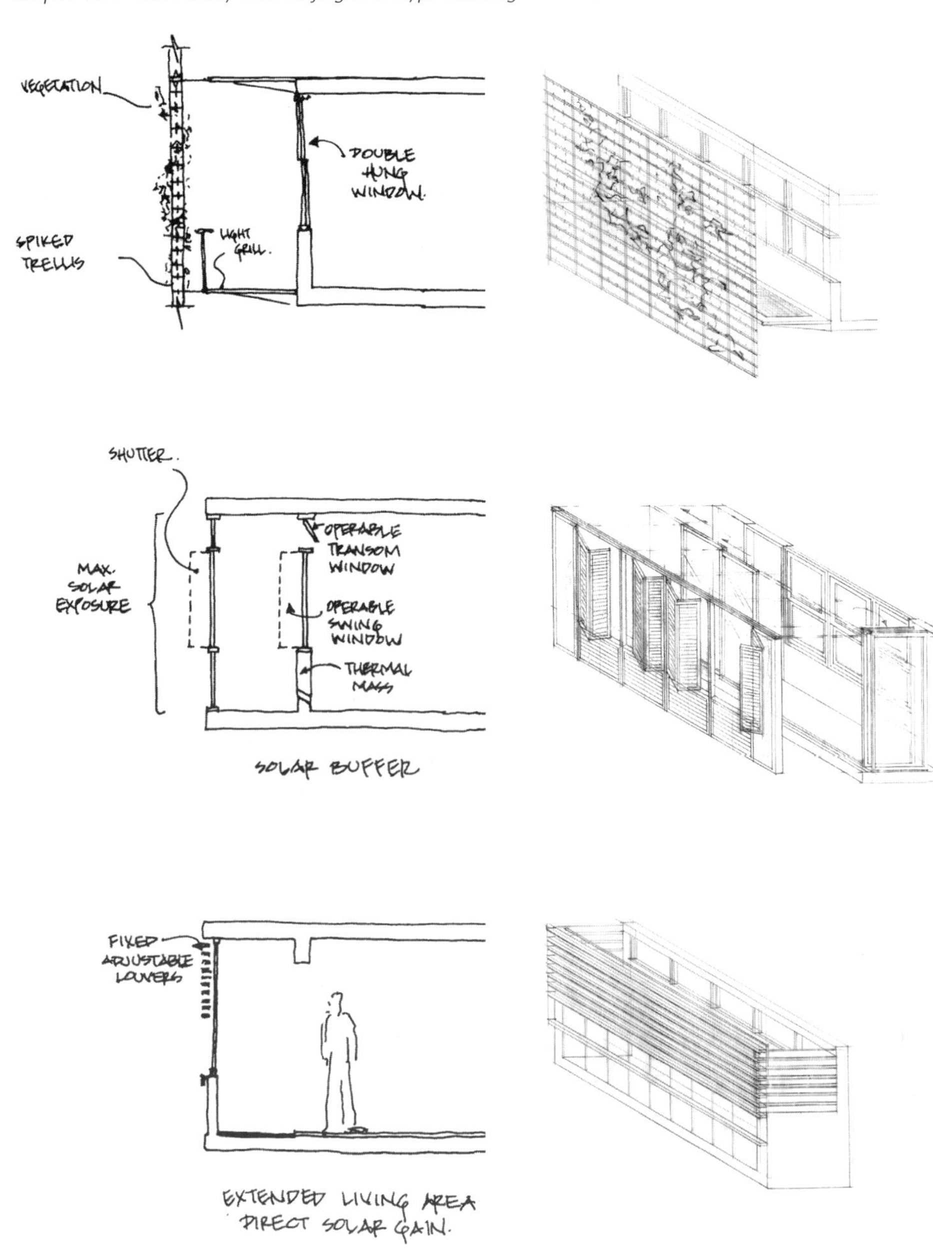

Figure 4 Sketch and axonometric studies showing three shading options

Figure 5 Beijing 12 x 12 prototype house - view of model of an earlier design, showing north and west façades

Figure 7 Beijing 12 x 12 prototype house - elevation study of final design

Figure 6 Beijing 12 x 12 prototype house - view of model of an earlier design, showing south façade

Figure 8 Beijing 12 x 12 prototype house - section model of final design, showing ventilated stair

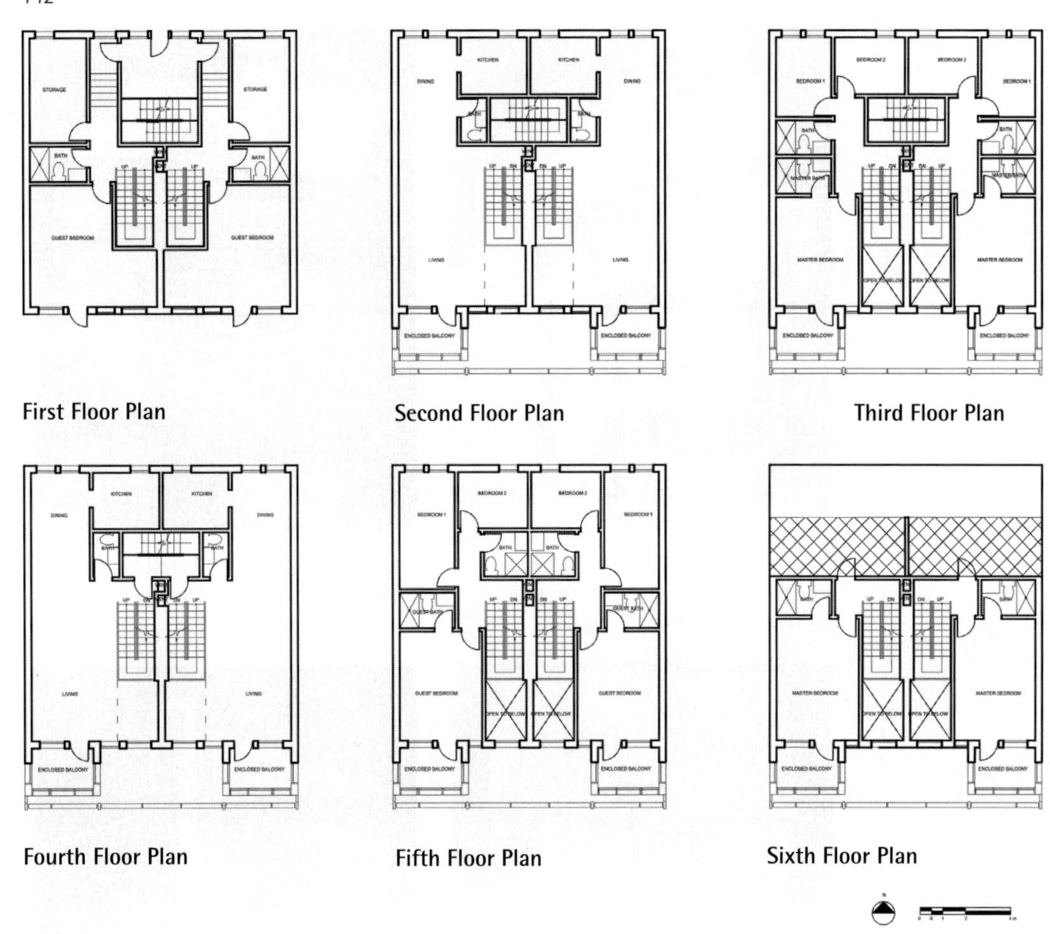

Figure 9 Floor plans of the 12 x 12 prototype house

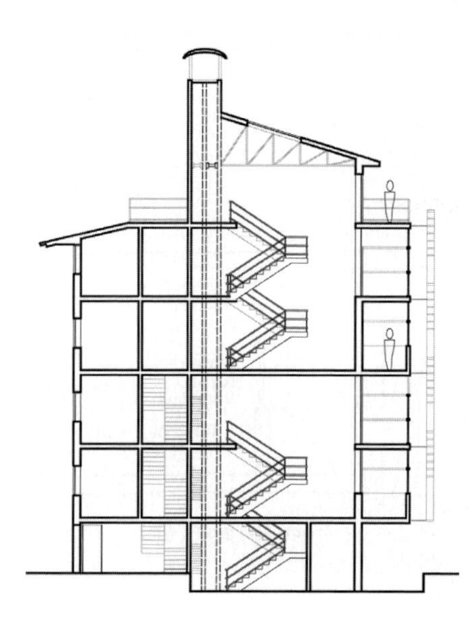

Figure 10 Section through typical 12 x 12 prototypc house

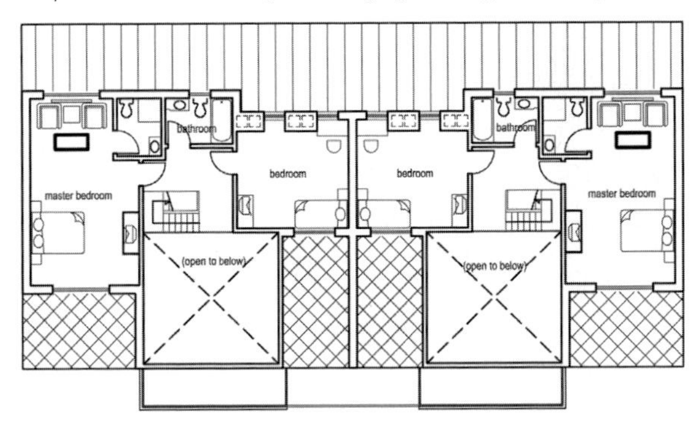

Penthouse Upper Plan

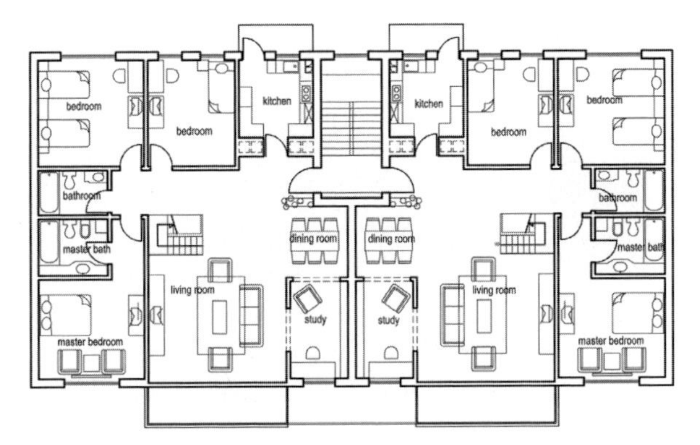

Penthouse Lower Plan

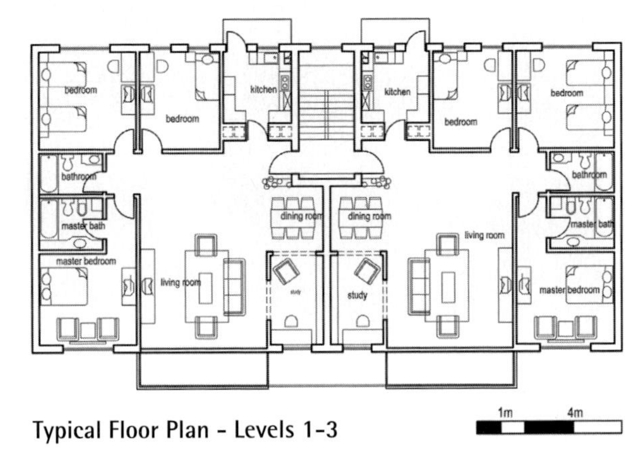

Typical Floor Plan - Levels 1-3

1m 4m

Figure 11 Typical floor plans of the 12 x 24 prototype house

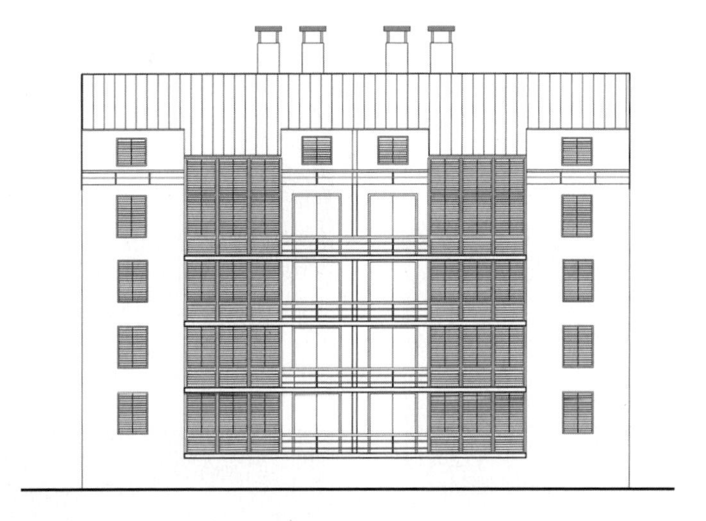

Figure 12 Typical south elevation of the 12 x 24 prototype house

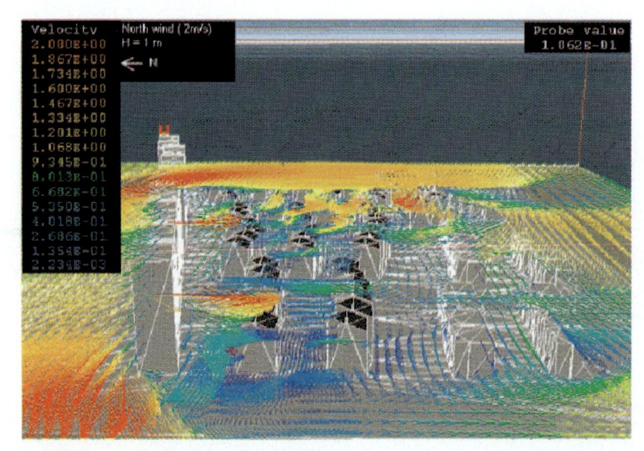

Figure 13 Wind velocity distribution 1 m above the ground when the wind is from the north (3-D view) (north is to the left in Figures 13-17)

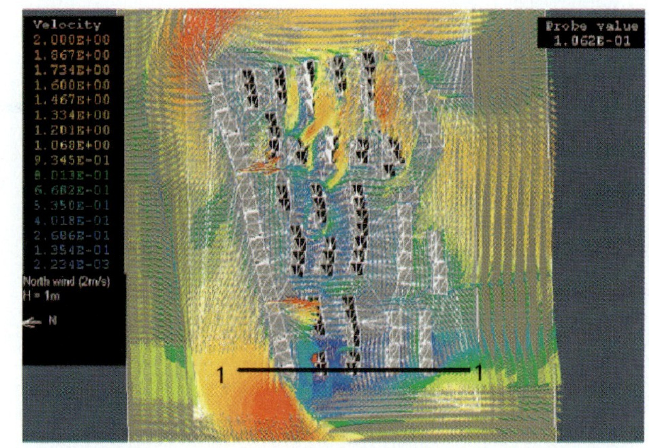

Figure 14 The wind velocity distribution 1 m above the ground when the wind is from the north (2-D view)

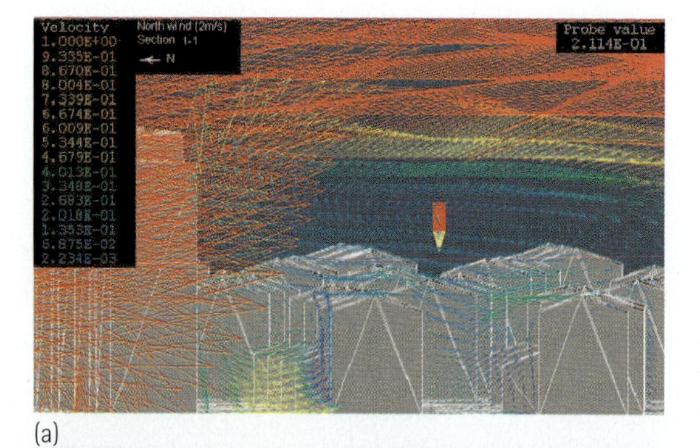

(a)

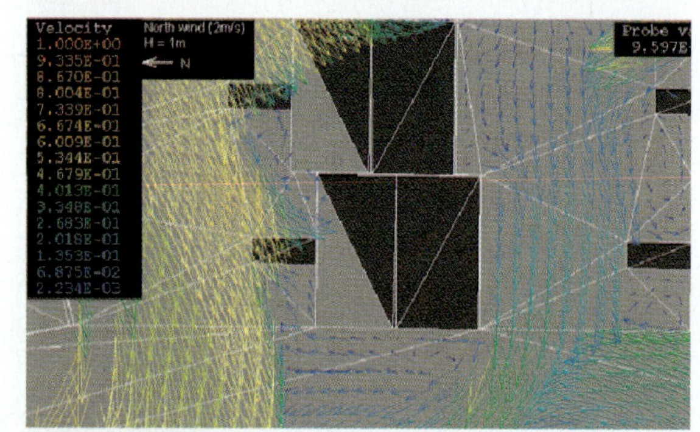

(b)

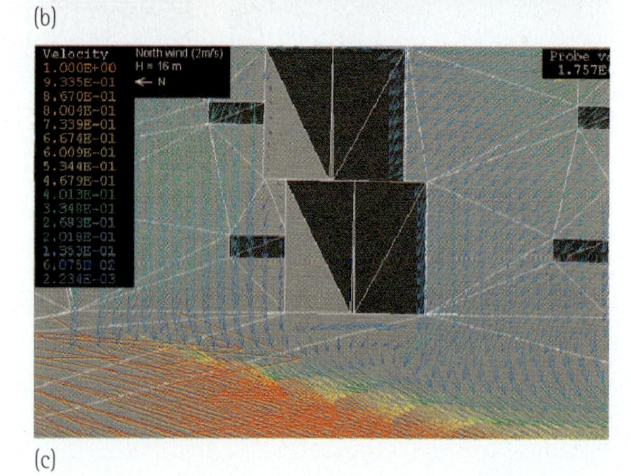

(c)

Figure 15 Wind velocity distributions (m/s) for the prototype building with northern winds shown at (a) section 1-1, (b) 1 m above the ground, and (c) 16 m above the ground

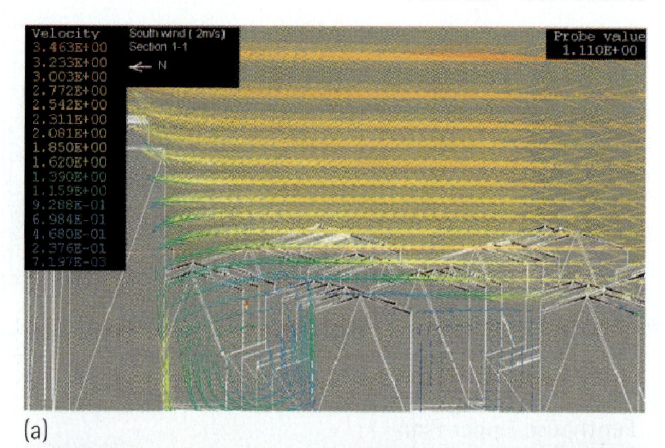

(a)

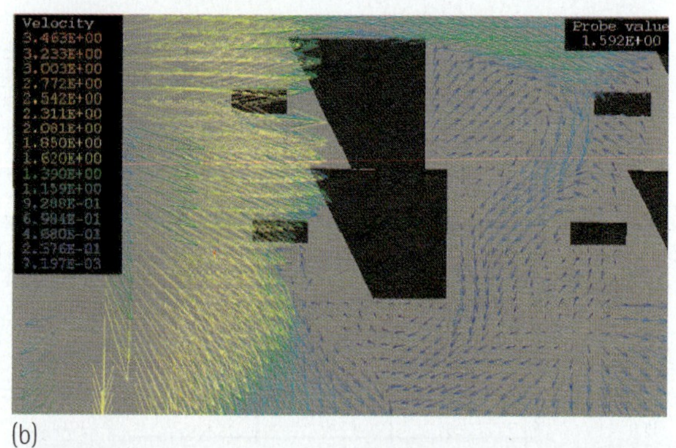

(b)

(c)

Figure 16 Wind velocity distributions (m/s) for the prototype building with southern winds shown at (a) section 1-1, (b) 1 m above the ground, and (c) 16 m above the ground

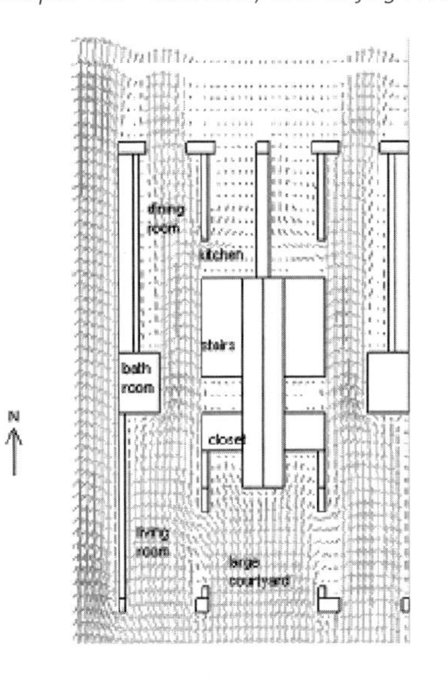

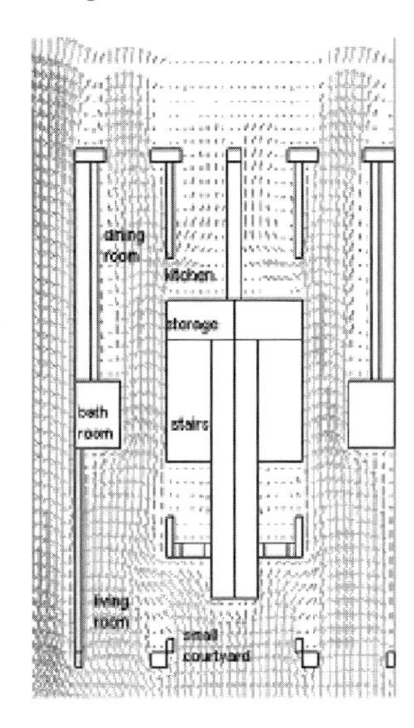

Figure 17 The air velocity of the demonstration building (12 x 12 house) at a height of 1.2 m above the floor assumes winds from the south. The upper plan shows a large court, while the lower plan shows a smaller court. (see color version in chapter 7, Figure 7)

TECHNICAL RECOMMENDATIONS

Beijing weather data was analyzed for comfort categories. As discussed in chapter 7, results indicated that cooling would be required from June to August, and heating is required for colder months. One of the design goals was to promote natural ventilation by increasing the wind speed during summer months, while taking care to reduce wind speed in the winter in order to reduce energy loss through infiltration and increase pedestrian comfort. In Beijing, prevailing summer winds are from the north and the south. Architects used this information to design the floor plan and elevations to maximize natural ventilation.

Exterior studies were performed to understand the site conditions on the 12 x 12 prototype house (Figures 13-16). These studies showed that natural ventilation was not realistic when the winds were from the north, with wind speeds up to 0.4 m/s, due to the tall building located just north of the demonstration building. When winds were from the south, the air speed was increased to up to 0.8 m/s. However, these values were still less than the recommended 2 m/s. The reason for the low values was due to the proximity of other buildings directly to the south, which blocked or deflected the wind. Proper design for natural ventilation could increase comfort levels and reduce mechanical cooling needs in hotter months, but may not be able to fully substitute for mechanical cooling.

Two design schemes were carried out for the 12 x 12 prototype building and were tested with CFD analyses. Both scheme 1, with a larger courtyard, and scheme 2, with a smaller courtyard, were analyzed (Figure 17). The courtyard was enlarged in scheme 1 to test if the larger courtyard could increase wind speed. Results showed that internal natural ventilation was not affected by courtyard size. The architects proceeded to size the courtyard based on other factors.

The interior layout of designs was also tested in CFD. The results showed that the layout was conducive to ventilation in the units. However, the unit tested, with wind values of 0.5 m/s from the south, showed the need for mechanical or stack ventilation. As discussed in the *Design Approach* section of this chapter, the designers focused on the design of the access stair to incorporate principles of stack ventilation.

SUMMARY

The buoyancy-driven ventilation design was used because the site of the 12 x 12 house was directly behind a wide high-rise building complex. Flow predications using CFD revealed that the wind velocity would be substantially reduced around the 12 x 12 house. Therefore, wind-driven ventilation rates would be too low to provide comfort conditions during the cooling season. A good rule of thumb is that the effects of an obstacle such as a tall building on the wind pattern will persist downwind for a distance approximately equal to six times the building height. If the 12 x 12 design was used in an open area where it was surrounded by houses of similar design, wind-driven natural ventilation would be more useful. The design of a satisfactory natural ventilation system must take into account the interaction of exterior conditions as well as interior building geometry. The exterior conditions are not only a function of the individual building design, but must also account for the local built environment surrounding the building.

ACKNOWLEDGMENTS

Tsinghua University provided invaluable assistance with client coordination and was instrumental in the design development of this project.

CHAPTER ELEVEN

CASE STUDY TWO - BEIJING STAR GARDEN

Juintow Lin

PROJECT DESCRIPTION

MIT and Tsinghua were asked by the Design Department of Beijing Vanke Co. Ltd. to develop the site plan and architectural design for a residential development in the area of the former Asian Games site in Beijing (Figure 1). Throughout the process, architects and engineers worked together to design a site plan and building designs that would promote comfortable outdoor and indoor environments.

A number of designs and studies were carried out for this project. The first task was to perform an analysis of the developer's existing designs (Figures 2 and 3). After the initial analysis, the MIT team developed low-rise development options as alternatives to the originally proposed design. The new schemes consisted of various buildings comprising a low-rise high-density development. Because the developer already secured a government permit for the original high-rise master plan, a scheme that sought to redesign the four high-rise towers was developed in greater detail. Throughout the design process, technical studies were carried out for the various design iterations. These studies analyzed building features and specification upgrades, as well as natural ventilation and shading performance.

L. Glicksman and J. Lin (eds), Sustainable Urban Housing in China, 146-159

CLIMATE RESEARCH AND ENERGY STUDIES

The project began with a basic consideration of Beijing's climate data and existing site conditions. At 40° latitude, the climate in Beijing requires heating for a better part of the year. Refer to the comfort zone table in chapter 7, Figure 4 for further information. The demand for cooling is also high, and will increase as the standard of living rises.

A study was performed to see what technologies would help to reduce heating and cooling energy needs in a Beijing building by a target of 30 to 50 percent. A simple steady-state model of a "shoebox"-style unit was developed using monthly average temperatures and solar heat gain factors. The base case consisted of uninsulated concrete walls with gypsum board interior (U=1.73 W/m^{2}K), 3.2-millimeter-thick single-glazed windows (U=7.24 W/m^{2}K), no shading, and infiltration rates of 1.5 and 2 air changes per hour (ACH) in summer and winter, respectively.

The model was generated in Microsoft Excel and served as a simple tool for analyzing trade-offs in technical specifications of various features, including types of walls (encompassing insulation), window area and window type, infiltration rate, and window shading coefficients. This tool was interactive, and engineers and architects were able to make changes to the model to analyze individual and combined upgrades.

Two generic unit designs were considered in the analysis. Both had exposed southern façades. One had an exposed eastern façade, and the other had an exposed western façade. Major findings were as follows:

- decreasing infiltration alone reduced energy consumption by about 20 to 35 percent;

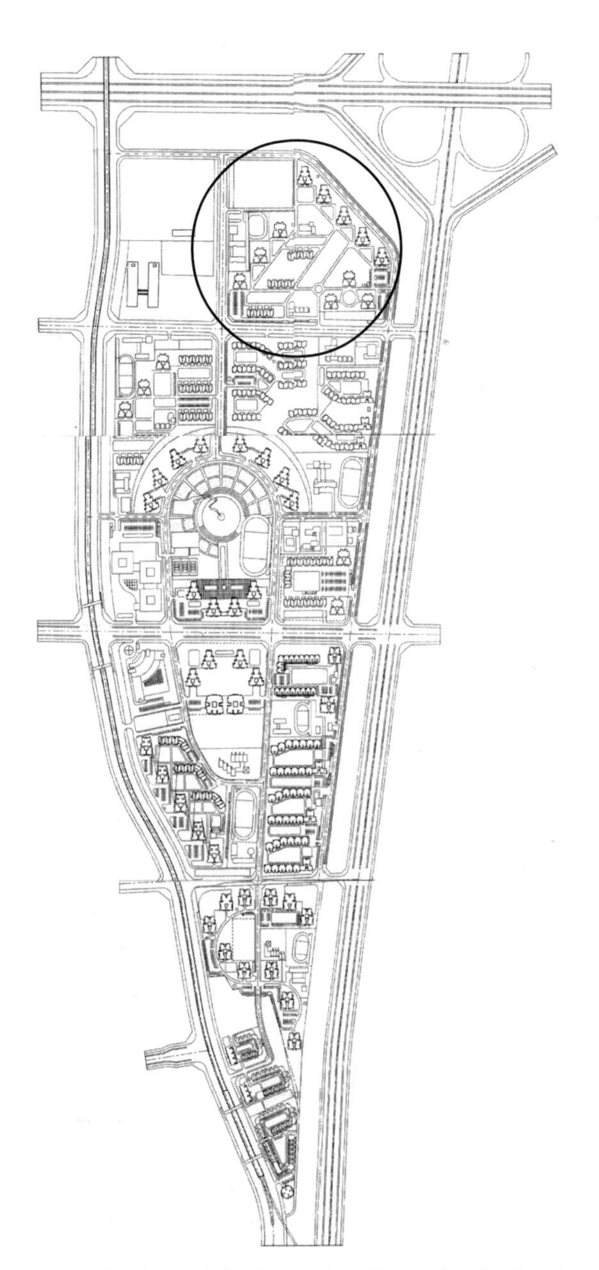

Figure 1 The former Asian Games site - Beijing Star Garden, the project site, is shown circled (north is to the top)

Figure 2 Vanke's original design for Beijing Star Garden master plan (north is to the upper right)

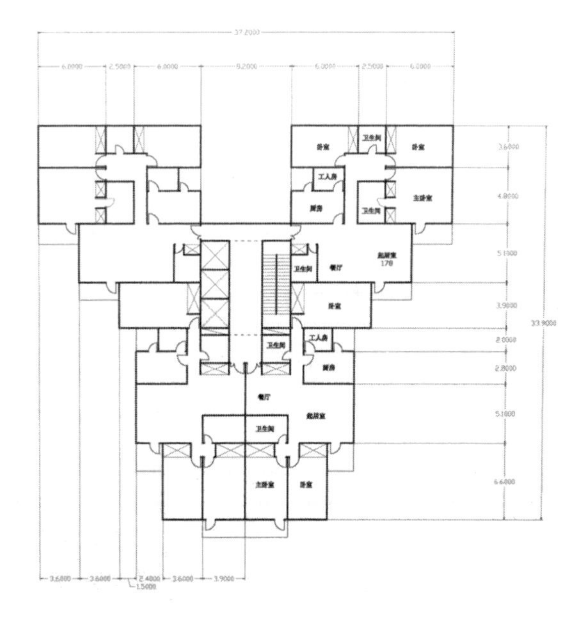

Figure 3 Vanke's original design for the four high-rise towers at the northeast area of the site (north is to the top)

Building Property Descriptors			
Walls			
Case	Description	U (W/m^2 K)	
A	200 mm concrete blocks, inner gypsum plaster board	1.73	*
B	200 mm concrete blocks with UF foam core, inner gypsum plaster board	1.03	**
C	200 mm concrete blocks with perlite-filled core, inner gypsum plaster board	0.85	***
D	200 mm concrete blocks, 50 mm expanded rubber insulation, inner gypsum plaster board	0.47	*****
E	200 mm concrete blocks, 100 mm expanded rubber insulation, inner gypsum plaster board	0.27	******
F	100 mm brick, 50 mm expanded rubber insulation, inner gypsum plaster board	0.54	****
Windows			
Case	Description	U (W/m^2 K)	
A	3.2 mm glass, single-glazed, alum w/out thermal break	7.24	*
B	3.2 mm glass, single-glazed, alum w/ thermal break	6.12	**
C	6.4 mm airspace, double-glazed, alum w/out thermal break	4.93	***
D	6.4 mm airspace, double-glazed, insulated fiberglass vinyl	2.77	*****
E	6.4 mm airspace, double-glazed, alum w/ thermal break	3.7	****
	Winter condition for both double- and single-glazed windows, shades not in place	0.9	
Shading			
Case	Description	SC	
A	No shading	1	*****
B	Single-glazed light venetian blinds	0.67	***
C	Single-glazed medium venetian blinds	0.74	****
D	double-glazed, light venetian blinds	0.33	*
E	double-glazed, medium venetian blinds	0.36	**

Infiltration				
Case	ACH	Winter	Summer	
A	More loose	2	1.5	*
B	Loose	1.15	0.68	**
C	Medium	0.73	0.46	***
D	Tight	0.43	0.33	****

Table 1 Building parameter descriptors for the steady-state model

Results - Change 1 Value Compared to Base Case				
% Surface windows	Heat power	Cooling power	Total (Whr/yr)	% change
20%	1.75E+07	-1.00E+07	2.76E+07	0%
10%	1.90E+07	-6.91E+06	2.60E+07	-6%
30%	1.67E+07	-1.39E+07	3.07E+07	11%
Wall - vary U	Heat power	Cooling power	Total (Whr/yr)	% change
A	1.75E+07	-1.00E+07	2.76E+07	0%
B	1.50E+07	-9.97E+06	2.49E+07	-10%
C	1.43E+07	-9.98E+06	2.43E+07	-12%
D	1.30E+07	-1.00E+07	2.30E+07	-17%
E	1.23E+07	-1.00E+07	2.23E+07	-19%
F	1.32E+07	-1.00E+07	2.32E+07	-16%
Window - vary U	Heat power	Cooling power	Total (Whr/yr)	% change
A	1.75E+07	-1.00E+07	2.76E+07	0%
B	1.65E+07	-9.99E+06	2.64E+07	-4%
C	1.53E+07	-9.96E+06	2.53E+07	-8%
D	1.34E+07	-1.00E+07	2.34E+07	-15%
E	1.43E+07	-9.99E+06	2.42E+07	-12%
Window - vary SC	Heat power	Cooling power	Total (Whr/yr)	% change
A	1.75E+07	-1.00E+07	2.76E+07	0%
B	2.04E+07	-8.18E+06	2.86E+07	4%
C	1.98E+07	-8.58E+06	2.84E+07	3%
D	2.33E+07	-6.26E+06	2.96E+07	7%
E	2.31E+07	-6.43E+06	2.95E+07	7%
F	2.12E+07	-7.67E+06	2.88E+07	5%
G	2.08E+07	-7.90E+06	2.87E+07	4%
H	1.84E+07	-9.48E+06	2.79E+07	1%
Infiltration	Heat power	Cooling power	Total (Whr/yr)	% change
A	1.75E+07	-1.00E+07	2.76E+07	0%
B	1.11E+07	-1.10E+07	2.21E+07	-20%
C	8.03E+06	-1.15E+07	1.95E+07	-29%
D	5.82E+06	-1.18E+07	1.76E+07	-36%

Table 2 Analysis showing effects of individual upgrades compared to the base case

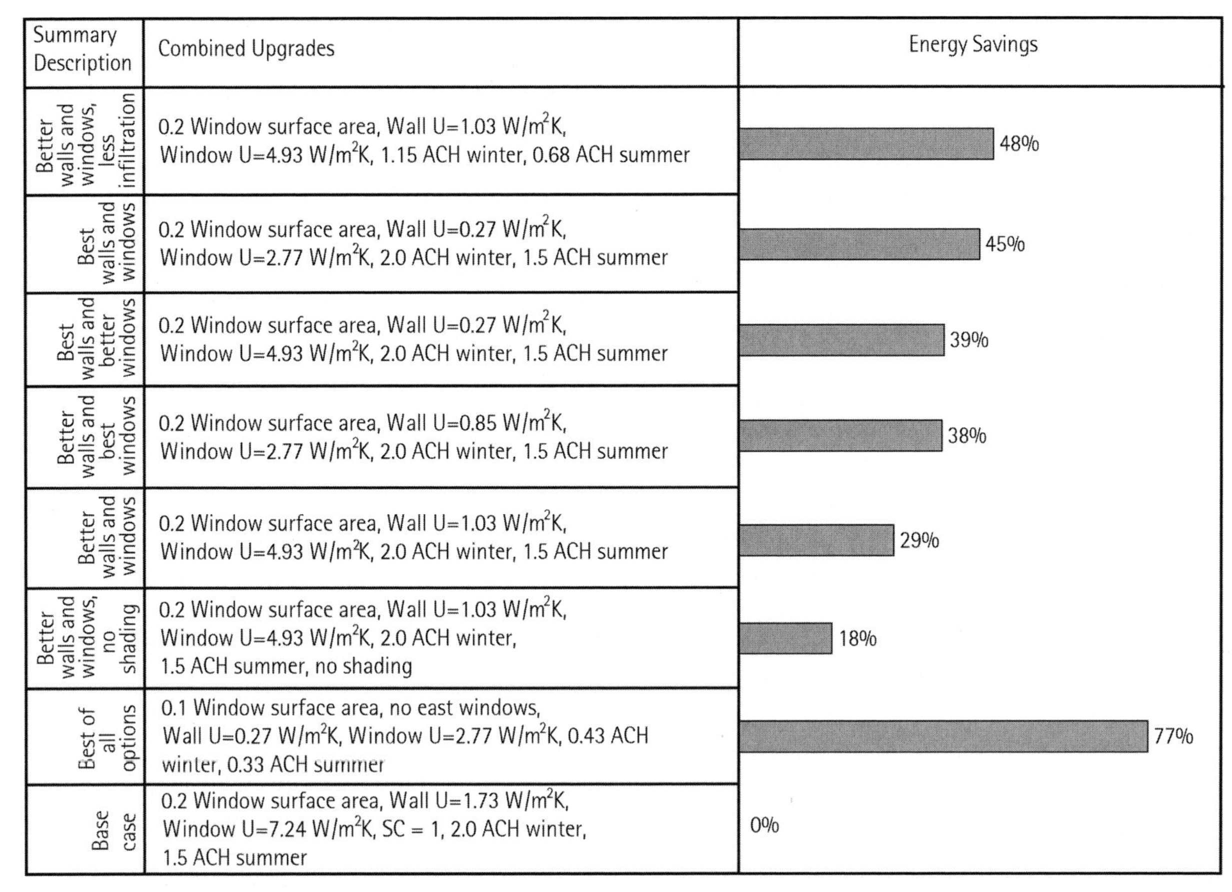

Always shade (except as noted):
SC=0.33 (summer - double-glazed windows with light venetian blinds), and
SC=0.9 (winter - double-glazed windows, blinds drawn)

Figure 4 Combined upgrades and resulting energy savings

- using insulation and higher quality windows decreased energy consumption by 30 to 45 percent depending on the degree of improvement; and
- while shading in the summer is essential, solar heat gain in the winter was similarly beneficial.

Overhangs and fins were not able to be quantified using shading coefficients. With the simple strategy of understanding overhang lengths due to changing sun angles during the seasons, one can provide the benefits of sun in the winter while at the same time reducing the solar gain during the summer.

The results were similar for the both units tested, which differed by having either a west- or an east-facing façade. Results were expressed in both annual heating and cooling energy consumption and as a percent savings in energy compared to a base case, with a possible savings ranging from 17 to 77 percent (Tables 1 and 2, Figure 4).

Recommendations

Based on results from climate findings, site considerations, and energy simulations, the design team developed the following design priorities:

- use natural ventilation at all levels of design, from site planning to unit design;
- protect the site from the northern winter winds;
- design for the macroclimate surrounding each individual building; and
- use careful placement of windows to allow winter sun to penetrate building while keeping the summer sun out.

Natural Ventilation

As discussed in *The Performance of the Passive Cooling Systems in Beijing* section of chapter 5, the need for mechanical cooling can be reduced through the use of natural ventilation and could potentially be eliminated in cases where it is combined with night cooling strategies (Figures 5-7).

Though natural ventilation can be beneficial in the summer, the potentially adverse effects in the winter must also be considered. The wind rose in Beijing (Figure 8) illustrates that in the winter, the prevailing winds are from the north; thus it is important to consider the adverse effects of winter winds on infiltration and pedestrian comfort. However, in the summer, prevailing winds are from the north and south. As a result, the design approach to the site plan was to maximize openings to the south and minimize openings to the north while still allowing for natural ventilation.

Ventilation was also considered at the site level. The existing site design has several high-rise towers, and it was important to consider the potential for an outdoor comfort deficiency at the ground level where the airflow around the buildings could be accelerated.

LOW-RISE DESIGN

The site design was based on unit density studies arranged to optimize the benefits of natural ventilation. As discussed in chapter 7, CFD studies were performed to analyze three schemes, including one by the original architect, and two by MIT's design team. Tests showed that Vanke's original scheme (Figure 9a), which consisted of 16 buildings ranging from 33 to 90 meters high, showed high levels of wind speed at all levels of building height. Such high values made it difficult to use wind as the primary natural ventilation strategy for the hotter summer months while still maintaining comfortable conditions around the buildings during the winter.

Two low-rise high-density schemes, schemes II and III, (Figures 9b and 9c) with building heights ranging from 20 to 60 meters were proposed in order to reduce winter infiltration and to utilize natural ventilation in summer, without compromising density. The low-rise schemes would also promote social interaction. In both schemes, taller buildings were placed to the north to protect the interior of the site from harsh northerly winter winds. These schemes were then analyzed with computational fluid dynamics (CFD) to see if there were any improvements over the original tower design (Figure 10).

As discussed in chapter 7, scheme II did not lend itself well to natural ventilation due to the fact that most of the buildings were oriented with the long side facing east. The layout as originally designed was not conducive to natural cross-ventilation strategies because summer winds from the north and south could not penetrate into a building with mostly eastern and western exposures. In addition, the east/west building orientation was not conducive to shading during the summer months.

In order to solve this problem, buildings were rotated 45° in scheme III, so that the long side of the buildings would be facing southeast and northwest. This solution allowed for southern and northern exposure for each unit. The results show that with respect to both northerly and southerly winds, the site-level ventilation was reduced to such an extent as to improve ground-level pedestrian comfort. In addition, the high-rise buildings on the north side would block the high wind from the north, reducing the overall site wind speed.

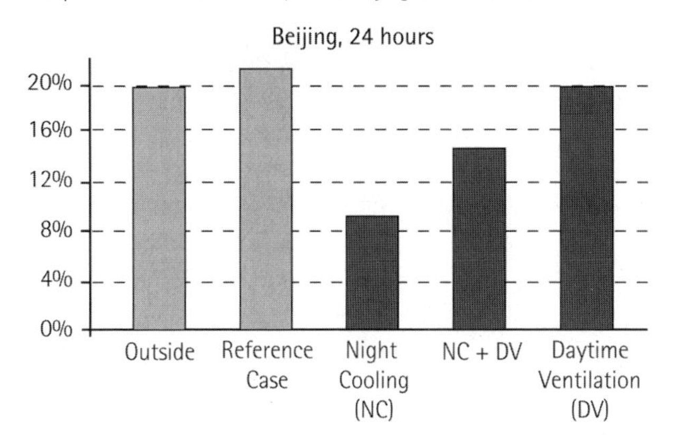

Figure 5 Percentage of discomfort hours using different cooling strategies in Beijing

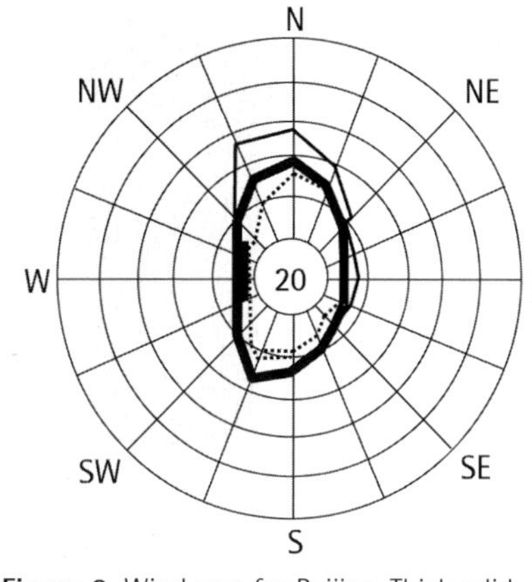

Figure 8 Wind rose for Beijing. Thick solid line: whole year; thin solid line: winter (December, January, and February); dashed line: summer (June, July, and August).

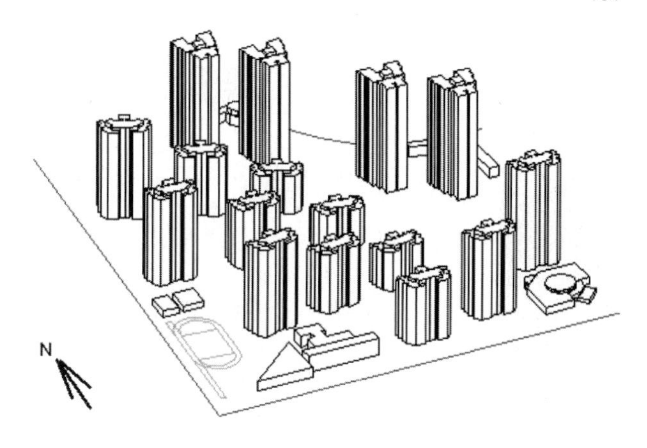

Figure 9a Original design for Beijing Star Garden by another architectural firm (scheme I)

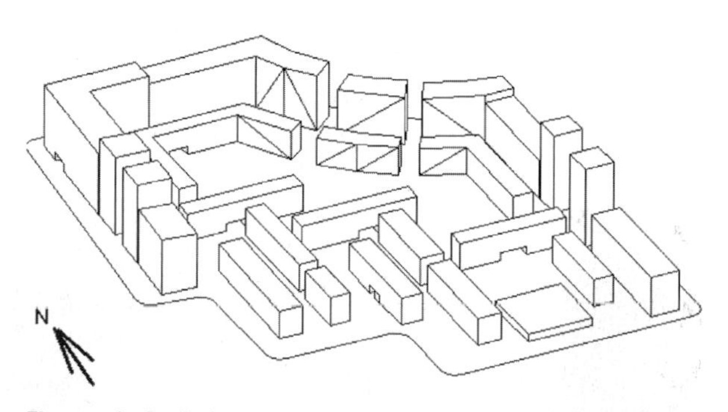

Figure 9b Preliminary redesign for Beijing Star Garden by MIT project team (scheme II)

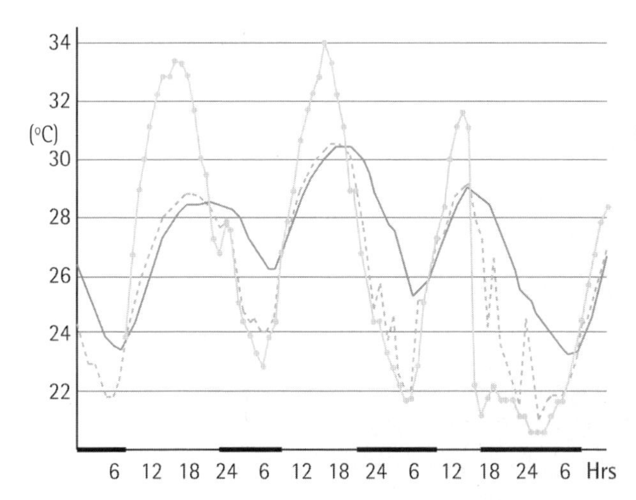

Figure 6 Hourly temperature variation for Beijing night cooling - light gray with circles: outside temperature; medium grey (dashed): wall temperature; dark gray (solid): indoor temperature. The dark line on the horizontal axis indicates periods when the windows were fully opened.

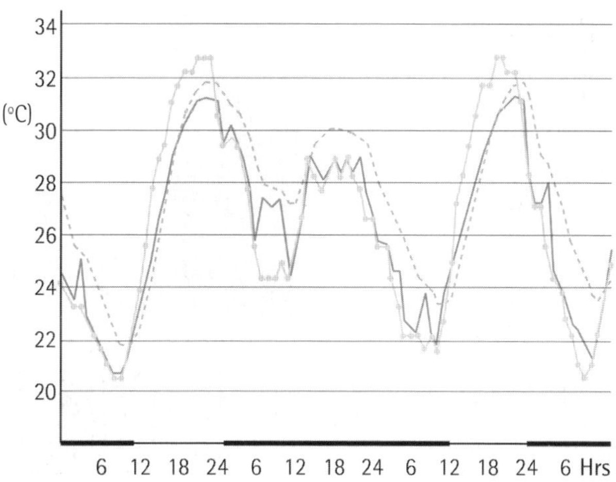

Figure 7 Hourly temperature variation for Beijing daytime ventilation - light gray with circles: outside temperature; medium gray (dashed): wall temperature; dark gray (solid): indoor temperature. The dark line on the horizontal axis indicates periods when the windows were fully opened.

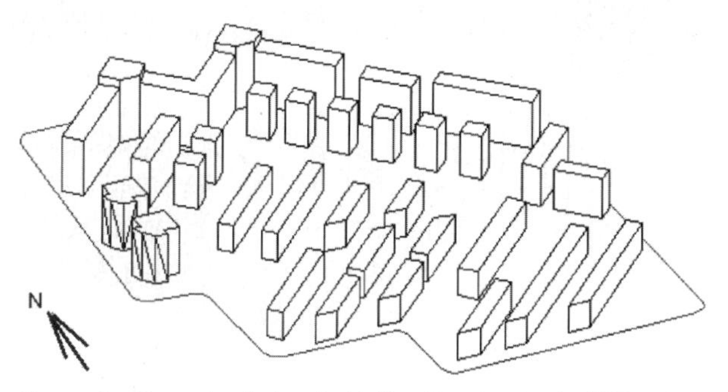

Figure 9c Second redesign for Beijing Star Garden by MIT project team (scheme III)

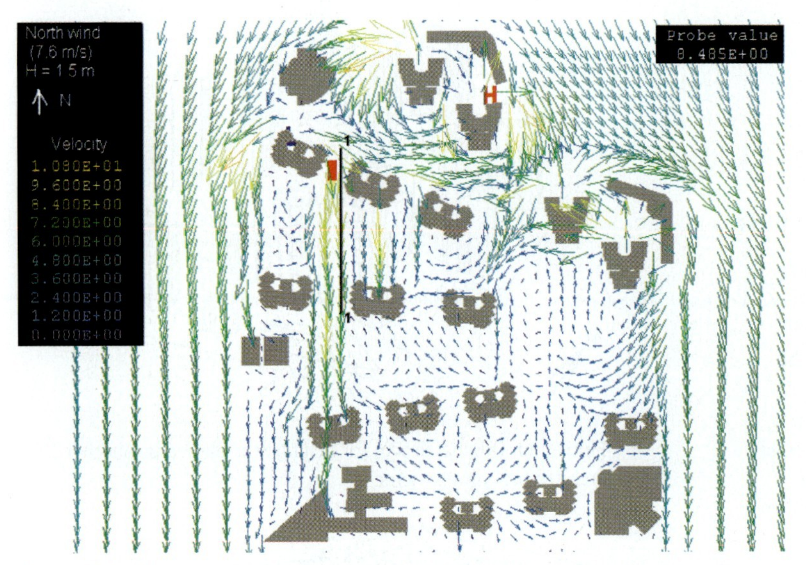

Figure 10a Wind velocity distribution at the height of 1.5 m above the ground around the buildings for scheme I with a north wind (blue - low velocity, green - moderate velocity, and yellow/red - high velocity)

Figure 10b Wind velocity distribution at the height of 1.5 m above the ground around the buildings for scheme II with a north wind (blue - low velocity, green - moderate velocity, and yellow/red - high velocity)

Figure 10c Wind velocity distribution at the height of 1.5 m above the ground around the buildings for scheme III with a north wind (7.6 m/s) (blue - low velocity, green - moderate velocity, and yellow/red - high velocity)

Figure 10d Wind velocity distribution at the height of 1.5 m above the ground around the buildings for scheme III with a south wind (1.9 m/s) (blue - low velocity, green - moderate velocity, and yellow/red - high velocity)

HIGH-RISE SCHEME

The client, however, preferred a high-rise solution similar to its original design illustrated in Figure 2. Thus, MIT and Tsinghua were given the task of designing four 30+-story towers with a total floor area of 85,000 square meters. The total height was not to exceed the planning limit of 83.4 meters. The primary design goals were to reduce operating energy consumption, provide natural ventilation, encourage social interaction in a high-rise setting, and consider site-level issues such as pedestrian comfort and community interaction.

In response to the climatic conditions, a number of recommendations were made, similar in nature to those for the low-rise scheme. In the winter months, it was important to provide protection from the cold winter winds from the north while allowing the warm winter sun to penetrate into the units. In the spring and fall months, it was possible to take advantage of natural ventilation. In the summer months, in order to keep the temperature down, it was essential to protect from the summer sun and minimize solar gains. Due to climate conditions, the siting of the buildings had to simultaneously maximize southern exposure and minimize northern exposure. East and west window areas were also to be minimized.

To promote natural ventilation while minimizing circulation, a vertical circulation scheme was implemented that consisted of a skip-stop elevator service that stopped every three floors, creating a mix of duplex and single-story units. Residents could enter units at the circulation level or take stairs to levels one floor below and above the circulation level (Figure 11). Average circulation area per floor was thereby reduced, which consequently allowed the building footprint to be reduced. Capital and maintenance cost reductions, as well as the creation of shared social spaces, were some other benefits of this three-story cluster system.

Each cluster was comprised of sixteen units, arranged around a core space three stories high (with mezzanines), and shared elevator access and circulation. Functions for the core space included sky-lobbies, play locations for children, laundry facilities, and landscaped terraced gardens. To minimize heat loss in the winter due to northern winds, service functions such as bathrooms, stairs, and elevator cores were placed to the north (Figure 12).

A primary goal in the creation of duplex units was to improve air quality and ventilation at the unit level. Duplex units were designed with double-height spaces as well as northern and southern exposure to promote air movement for potential natural ventilation (Figures 13 and 14). South-facing rooms were designed to have proper shading in the summer yet allow low-angled winter light to pass through.

At an architectural level, the aim was to design a physical form that explicitly expressed a response to the environmental context. The three-story cluster, duplex units, and double-height social spaces were expressed on the façade (Figures 15 and 16). The south side of the building was glazed with overhangs, whereas the north, east, and west sides were much more solid.

CFD STUDIES

To promote site-level interactions, it was imperative to understand the "microclimate" created at the lower zone of a building. Ground-level wind simulations were carried out to study this effect, and the site plan was modified based on this information. As previously discussed, the four towers originally designed for Vanke produced some undesirable high wind speeds at the pedestrian level. The

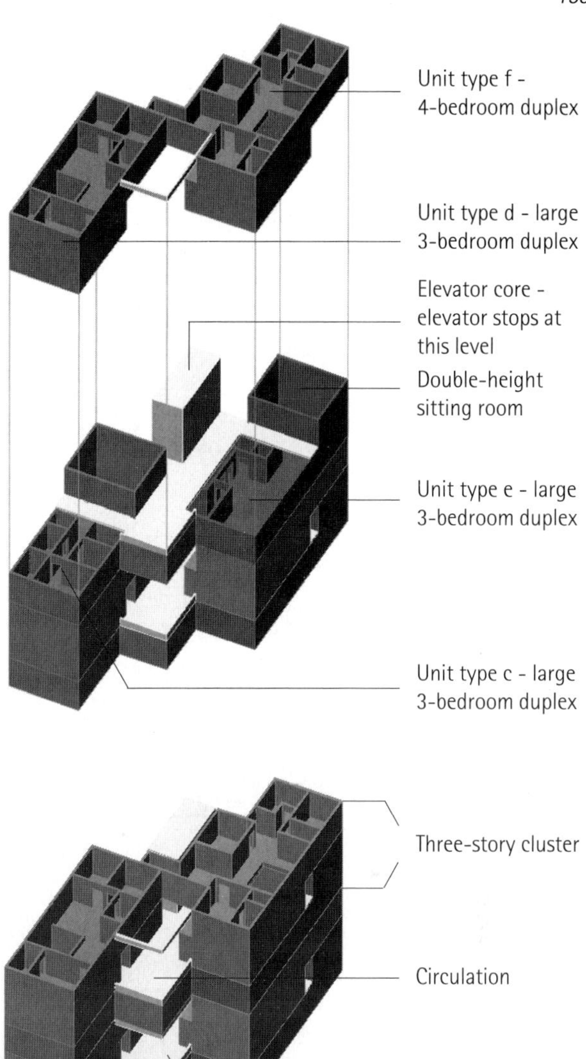

Figure 11 Diagrams of the middle units of a typical tower illustrate skip-stop circulation and duplex unit organization. The units interlock sectionally, allowing a circulation corridor to pass through at the entry level, thus promoting natural ventilation.

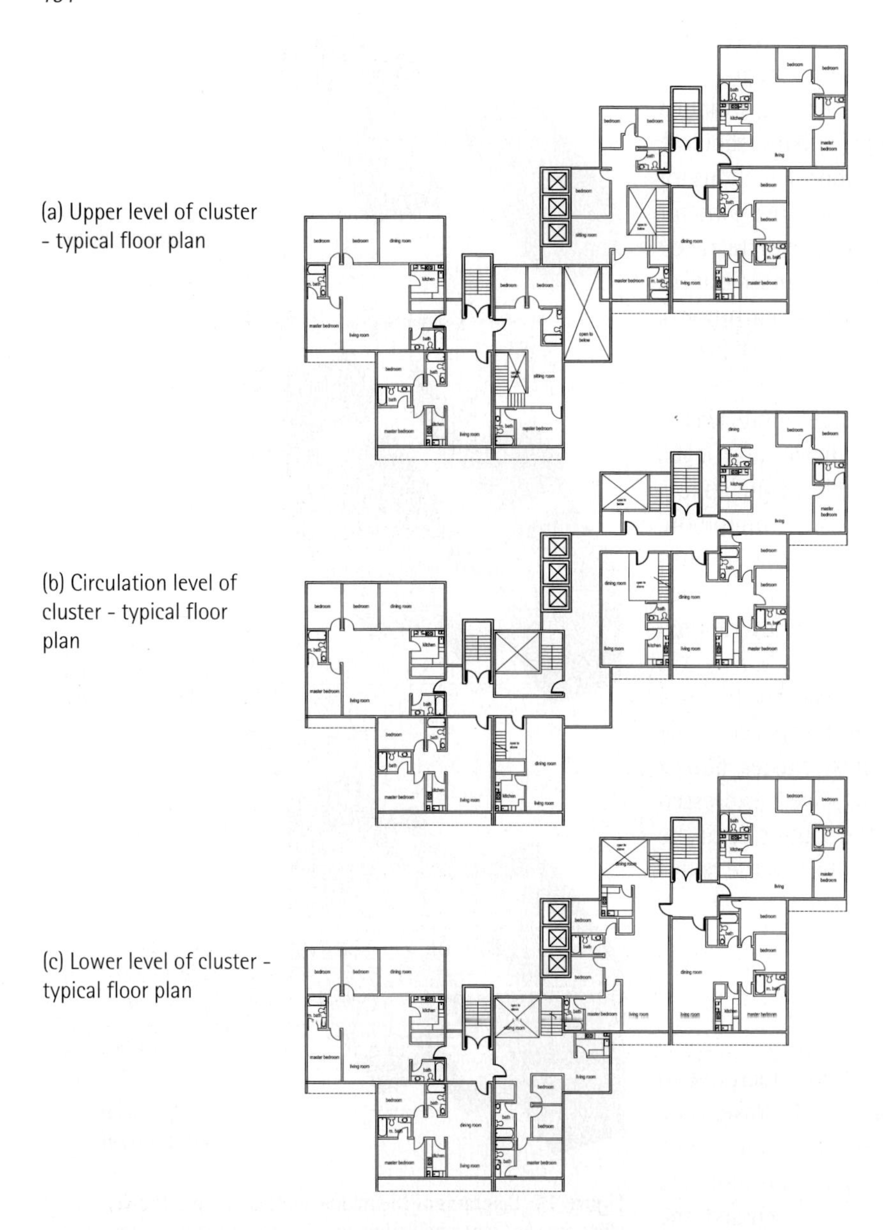

Figure 12 Typical floor plan diagrams at (a) levels 5, 8, 11, 14, 17, 20, 23, and 26; (b) levels 4, 7, 10, 13, 16, 19, 22, and 25; (c) levels 3, 6, 9, 12, 15, 18, 21, and 24

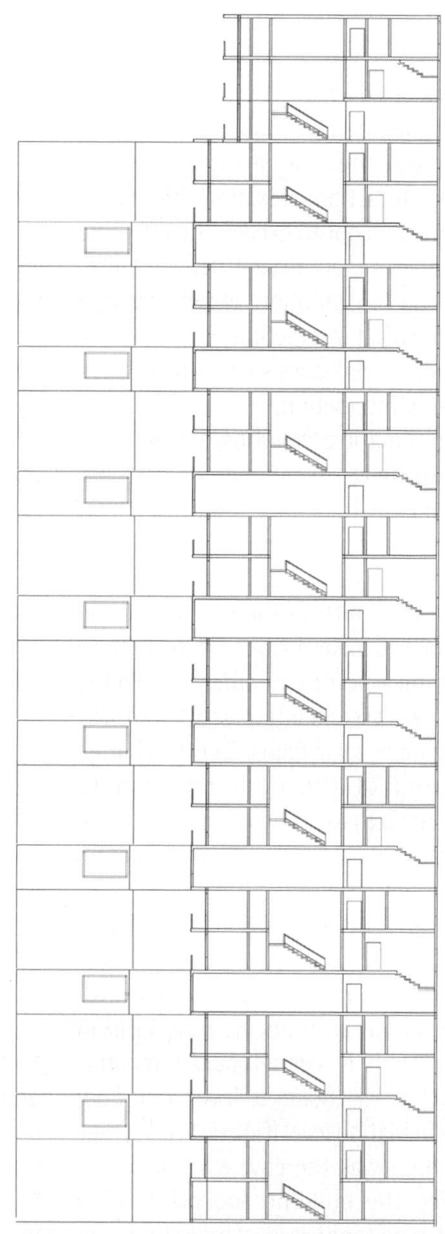

Figure 13 Section through units E and F looking west

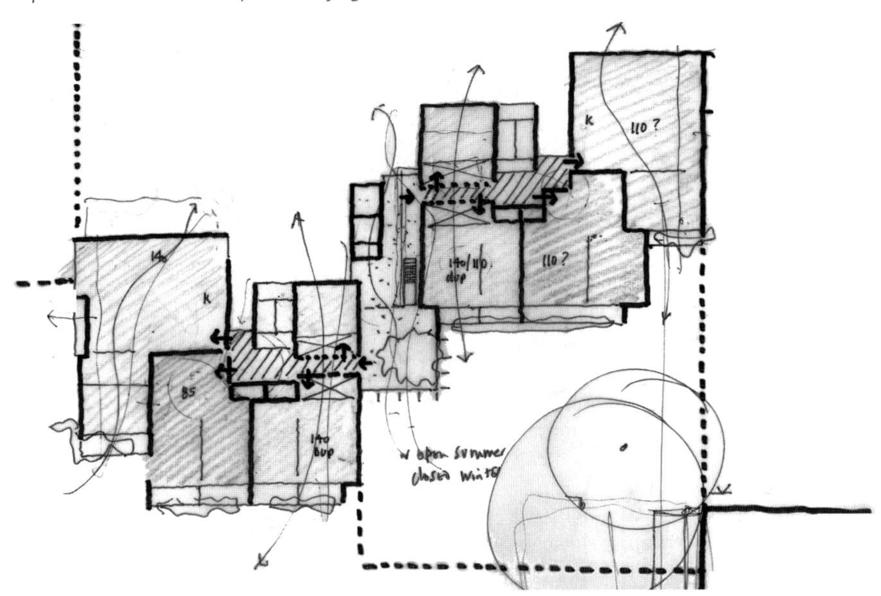

Figure 14 Floor plan diagram of circulation level (see color version in chapter 6, Figure 10)

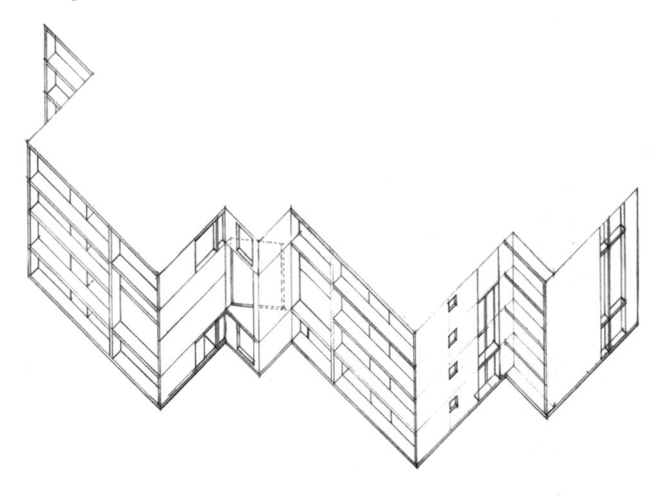

Figure 15 The axonometric shows south- and east-facing elevations and demonstrates one option for façade articulation for the atrium space at the cluster circulation levels

Figure 16 Perspective sketch of the final design illustrates the articulation of duplex units on the façade

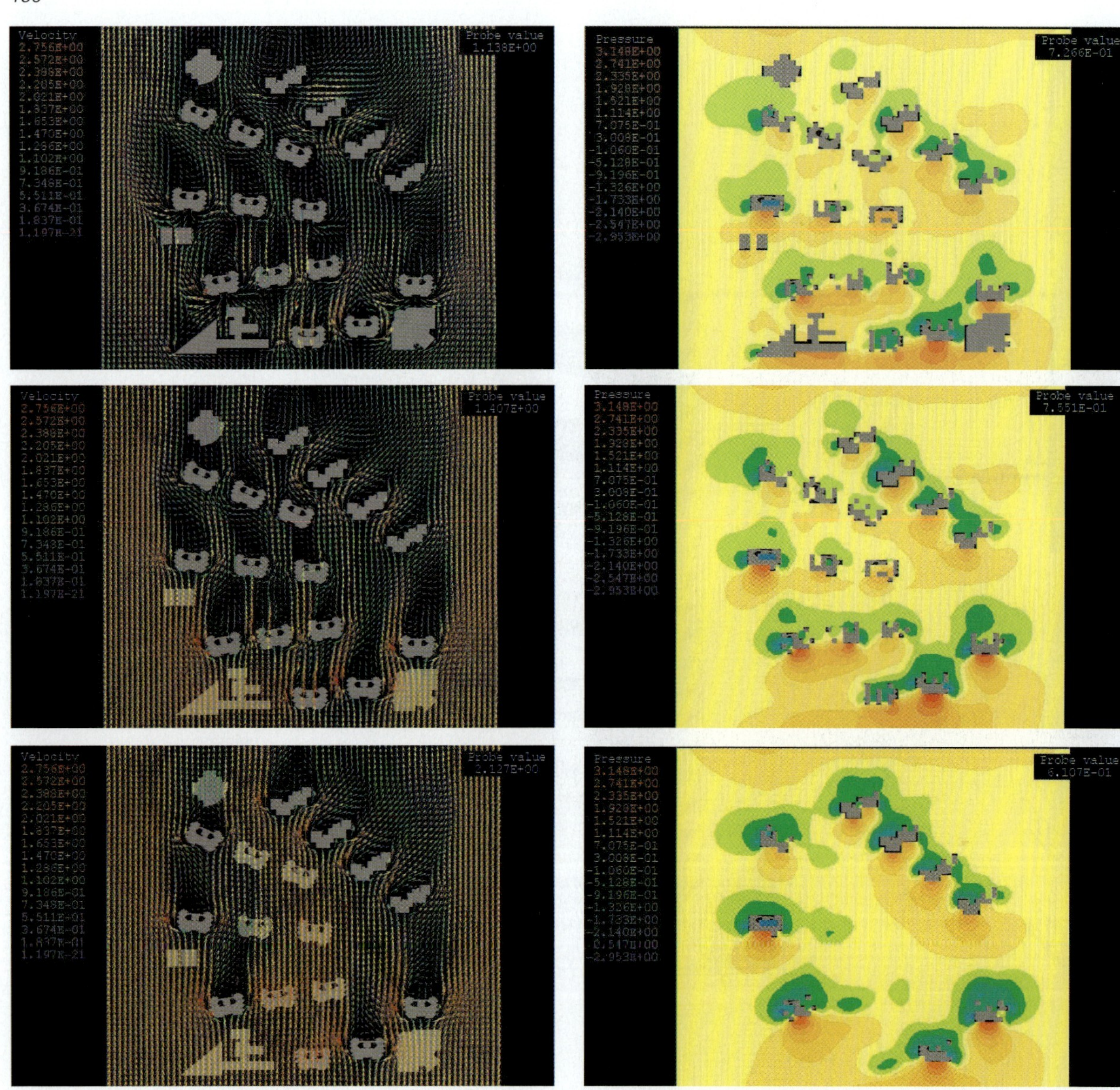

Figure 17 Velocity distribution around four towers in summer months with winds from the south at: (top) +1 m, (middle) +30 m, and (bottom) +60 m (north is to the top in Figures 17-20)

Figure 18 Pressure distribution around four towers in summer months with winds from the south at: (top) +1 m, (middle) +30 m, and (bottom) +60 m

Figure 19 Velocity distribution at +1 m around four towers with winter winds from the north

Figure 20 Pressure distribution at +1 m around four towers with winter winds from the north

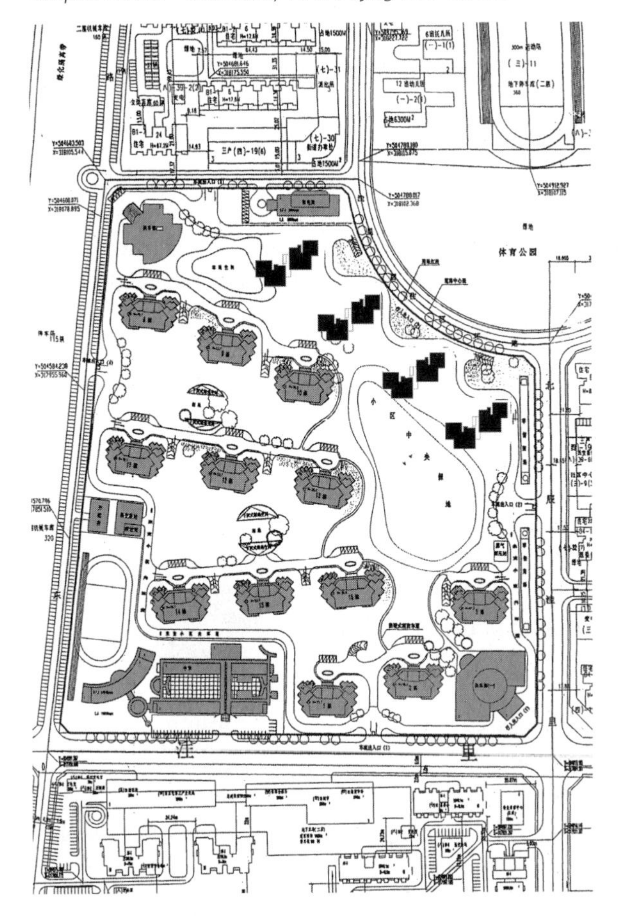

Figure 21 Site plan showing proposed towers in black (north is to the top)

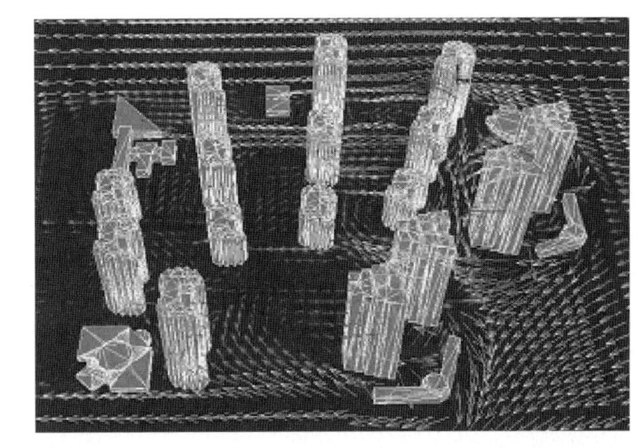

(a)

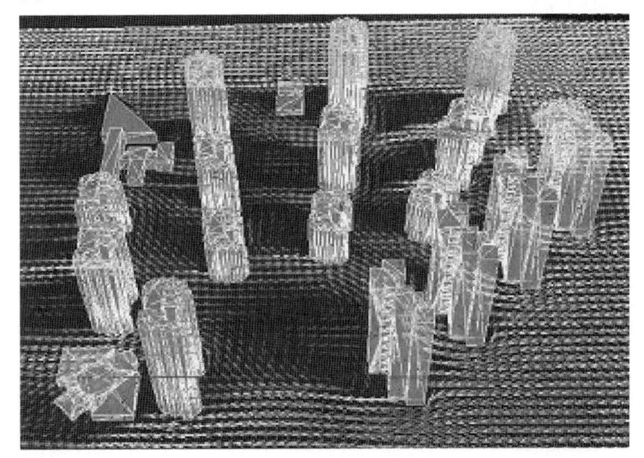

(b)

Figure 22 Wind distribution on the building site: (a) original design and (b) design with four proposed towers to the north (right). Color versions shown in chapter 6, Figure 9 reveal areas of high wind speeds.

new high-rise design sought to change the shape of the buildings in order to deflect wind to the westward direction. This principle was similar to the design of scheme III shown earlier in this chapter.

Exterior CFD studies were performed for the new high-rise scheme. Velocity and pressure levels were measured at different heights of the building, and both summer and winter conditions were examined (Figures 17-20). Due to the layout of the building, individual units were able to take advantage of natural ventilation. The different levels of pressure and velocity are indicated by red color for high levels, yellow color for moderate levels, to blue color for low levels.

As a result of these studies, modifications were made to the design based on the consideration of fluid mechanics. The layout of the buildings was designed so that the four buildings were equally spaced in the east-west direction for more effective and uniform ventilation (Figure 21). Comparisons between the new proposal and the Vanke's original design for the four towers showed reduced high winds that were also better distributed at the site level (Figure 22).

Interior Unit Studies

To effectively test the effect of design decisions on the performance of natural ventilation, interior ventilation was also simulated with CFD. The goal of these simulations was to analyze the floor plan and elevations for optimum window placement, as well as wall and door locations.

One study investigated the typical atrium space that was located on every third level, or at the circulation floor of each three-story cluster. It was found that if the circulation/atrium space was an exterior space, airflow would penetrate the building more thoroughly and enter the corridors. However, positive pressure at the upwind surface would also be decreased, so that airflow was not

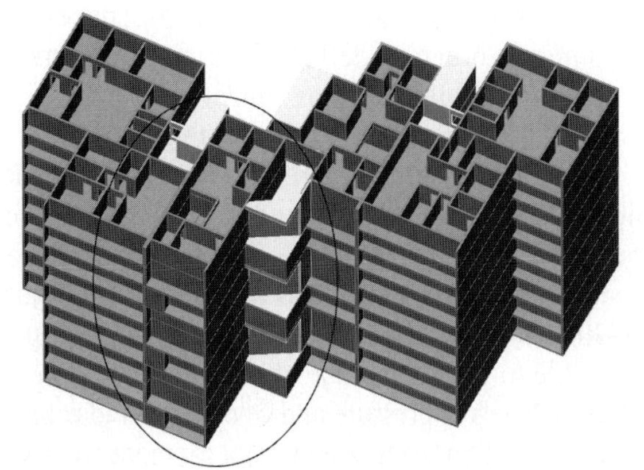

Figure 23 Diagram showing location of unit D (circled), reference for Figures 24-29

Duplex Level	Room	Orien-tation	Window Size (h x w)	Pressure (Pa)	Mean Air Velocity (m/s)
Lower Level	kitchen	south	1 m x 0.5 m	1.93	0.241
	open area	south	2 m x 1.5 m	1.93	0.321
	open area	east	2 m x 1.5 m	1.93	0.336
Upper Level	bathroom	south	1 m x 0.5 m	1.93	0.142
	master bedroom	south	2 m x 1.5 m	1.93	0.236
	northwest bedroom	north	1 m x 1.5 m	-0.92	0.915
	northeast bedroom	north	1 m x 0.5 m	-0.92	0.977

Table 3 Description of window characteristics

Duplex Level	V max	V min
Lower Level (h=1.5 m)	0.43 m/s	0.03 m/s
Upper Level (h=4.3 m)	1.3 m/s	0.01 m/s

Table 4 Velocity patterns

as likely to travel through the units. Thus, it was found that the double-height atrium space should be enclosed. The penetration would not only be detrimental to the ventilation of individual units, but would also bring extra noise and unfavorably strong airflow into the corridors.

Other simulations were carried for each unit design. Values for pressure distribution for each unit were derived from the exterior CFD site analyses. The studies produced a variety of results, including highlighting areas of stagnant air, areas of draft, and areas of uncomfortable flow resulting from flow convergence. After analysis, the placement of windows and doors was revised, and the CFD simulations were run repeatedly until a combination was found that was both aesthetically pleasing and conducive to natural ventilation. In addition to providing information that was useful for wall, door, and window placement, the studies showed that open stairwells with handrails in lieu of tall walls provided for a better airflow between levels of duplex units. The results from the studies also suggested including a mechanical ventilation system such as an exhaust fan in the bath and kitchen areas to prevent smells from entering the rest of the unit. A detailed analysis of unit D is included in this case study (Figures 23-29, Tables 3 and 4).

SUMMARY

The Beijing Star Garden Project was a collaborative team project that provided an inclusively well-informed design solution based on energy simulations, CFD analyses, urban design concepts, developer requirements, and social considerations on the part of the architect. The MIT group began the project by analyzing the original master plan and design scheme. This research provided the architects with informed recommendations, including the primary need to consider ventilation at both the site and unit levels. Based on specific recommendations related to ventilation strategies, the architects were able to design high-rise buildings that were far more efficient in terms of energy use than typical buildings found in Beijing urban developments. Some features included a three-story cluster system and units that had both north- and south-facing windows, promoting natural ventilation. Lastly, the social structure was enhanced to create communities within a building through specific circulation strategies that simultaneously satisfied natural ventilation goals.

ACKNOWLEDGMENTS

Tsinghua University provided invaluable assistance with client coordination and was instrumental in the design development of this project.

REFERENCES

Carrilho da Graça, G., Q. Chen, L. Glicksman and L. Norford. 2002. Simulation of Wind-Driven Ventilative Cooling Systems for an Apartment Building in Beijing and Shanghai. *Energy and Buildings* 34(1): 1-11.

Jiang, Y., H. Xing, C. Straub, Q. Chen, A. Scott, L. Glicksman and L. Norford. 1999. Design of Natural Ventilation and Outdoor Comfort by a Team of Architects and Engineers with the CFD Technique. *Proceedings of the 3rd International Symposium on HVAC*, 17-19 November 1999, vol. 1: 226-36. Shenzhen.

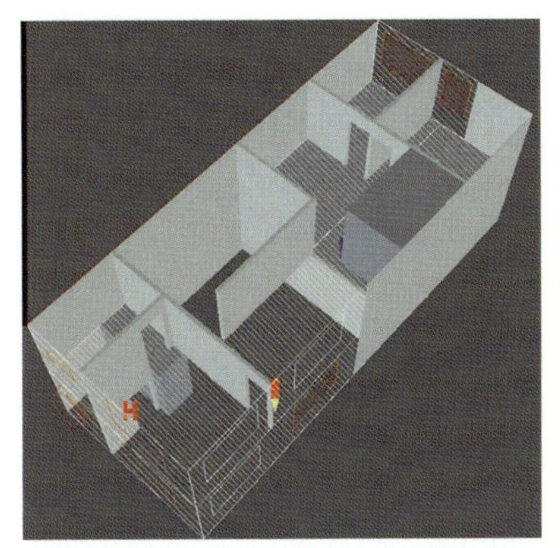

Figure 24a Original layout of unit D

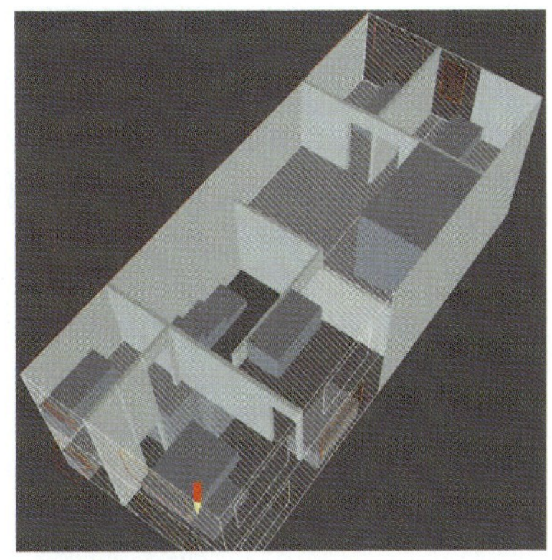

Figure 24b Modified layout (staircase, railing, furniture) of unit D

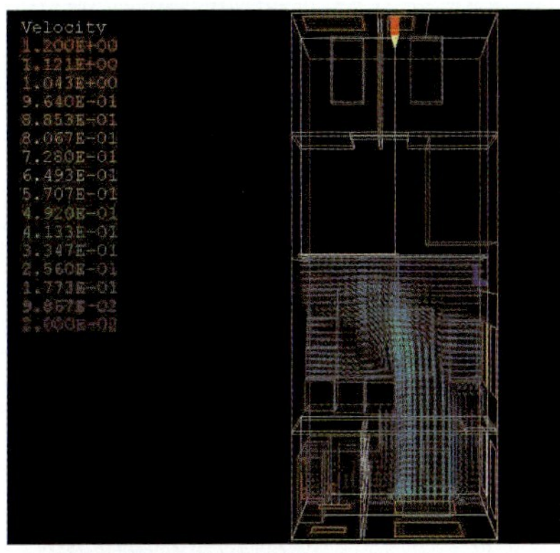

Figure 26a Velocity vectors at bottom level of duplex (height = 1.5 m)

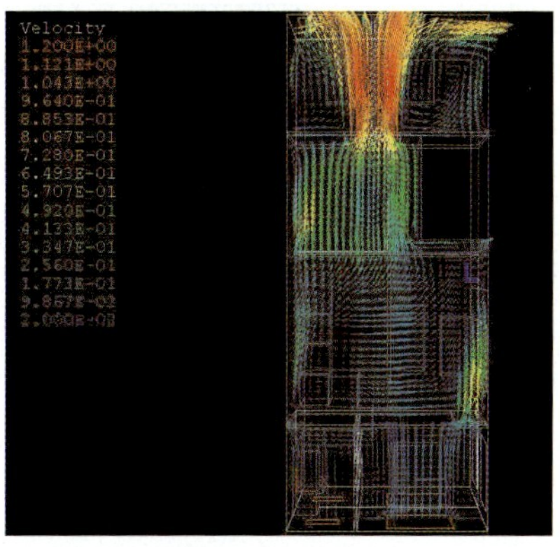

Figure 26b Velocity vectors at upper level of duplex (height = 4.3 m)

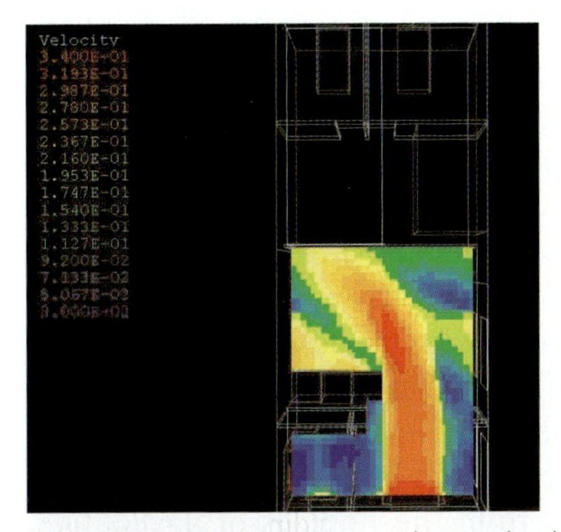

Figure 25a Velocity contours at bottom level of duplex (height = 1.5 m)

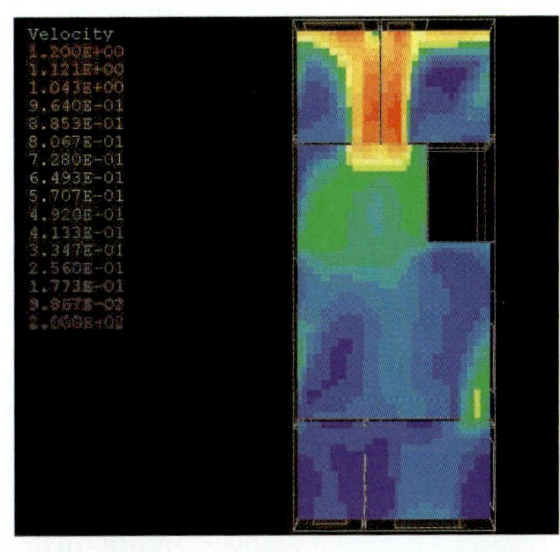

Figure 25b Velocity contours at upper level of duplex (height = 4.3 m)

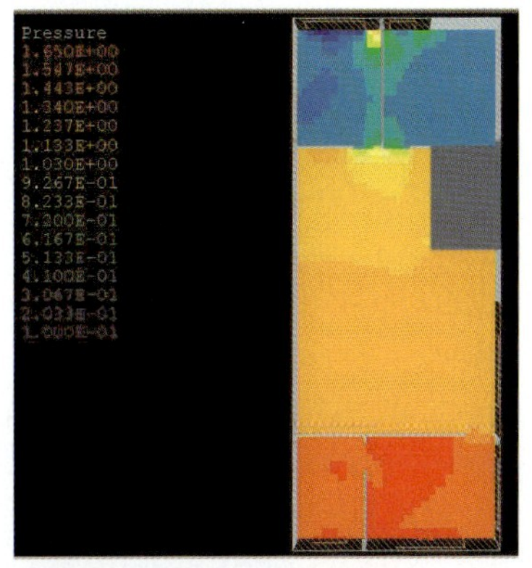

Figure 27 Pressure contours at z = 4.8 m plane

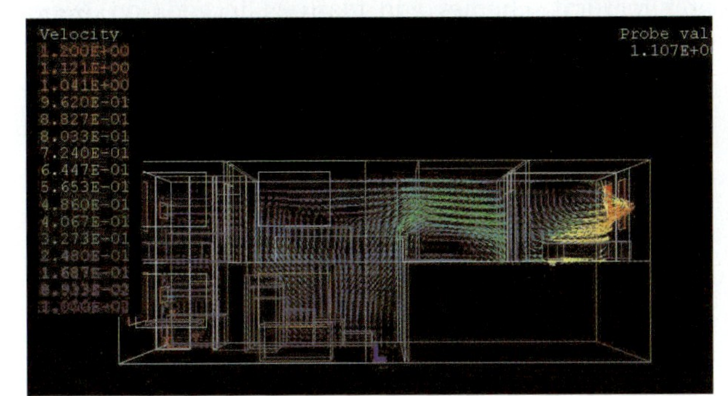

Figure 28a Section showing velocity vectors at x = 1.0 m plane

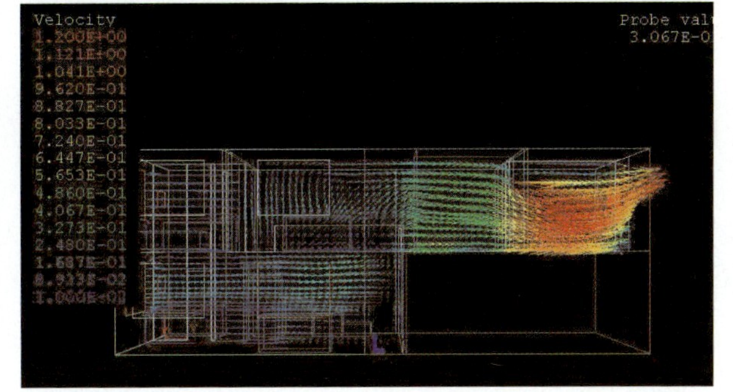

Figure 28b Section showing velocity vectors at x = 3.2 m plane

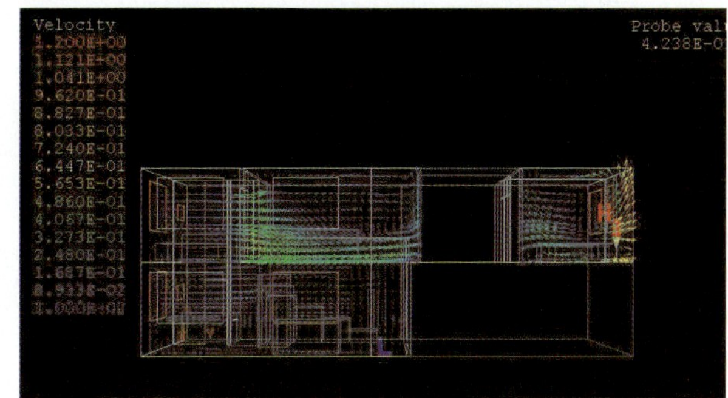

Figure 29 Section showing pressure contours at z = 5.6 m plane

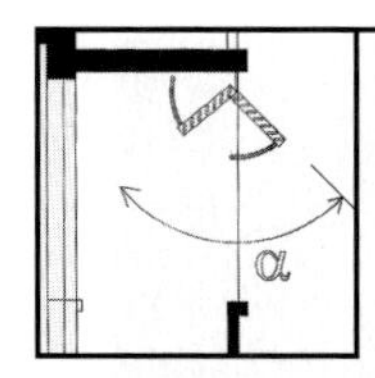

CHAPTER TWELVE

CASE STUDY THREE - SHANGHAI TAIDONG RESIDENTIAL QUARTER

Juintow Lin

PROJECT DESCRIPTION

This project was centered on the design of several mid-rise buildings for the Taidong Residential Quarter in Shanghai. The design was a collaborative effort of MIT's Sustainable Urban Housing in China Group and Tongji University's Construction Engineering Department and Architectural Design and Research Institute. The process was informed by a number of design and technology factors, including the results of energy studies, urban design concepts, as well as daylighting and ventilation studies. The primary goal of the design was to minimize the energy use of mechanical equipment and resultant costs through the use of improved building standards and energy-efficient details. Design decisions were made using guidelines from the findings of daylighting, ventilation, and energy studies that analyzed various architectural schemes and features. In addition, every effort was made to make comfortable and livable spaces that were unique to the typical landscape of Shanghai's urban development.

ENERGY STUDIES

The project began with a series of energy simulation studies conducted to inform architects how to design with an understanding of technologies and architectural

L. Glicksman and J. Lin (eds), Sustainable Urban Housing in China, 160-181

features affecting the energy use of a typical residential unit. These energy simulations were created using the DOE-2 building energy simulation program (AEC 2006). The program performed a dynamic energy balance at each hour of the simulation period, accounting for heat flows across the building's exterior surfaces, heat gains from solar energy, lights and equipment, and heat storage in the building structure. It computed indoor temperatures when the building was not conditioned, and heating or cooling loads when equipment was used to maintain a specified indoor temperature. Simulations took into account the building geometry, equipment, materials, surroundings, occupants, and other details that affect energy usage. These simulations were performed using natural ventilation with airflow of five to ten air changes per hour (ACH); air-conditioning was used to maintain comfort conditions when natural ventilation was not sufficient. In the CFD studies presented later in this case study, the airflow was estimated at 16 to 40 ACH, so the estimated impact on energy should therefore be considered conservative. In addition, the program used hourly weather data for Shanghai (Figure 1) to determine the energy usage for each given hour for the specified condition of the building.

The simulations were performed in two groups. The first set, called "baseline parametric simulations" included only single feature upgrades versus the baseline case described in Table 1. The second set, called "combined feature simulations" combined a number of the best upgrades from the first set to make recommendations for the most appropriate design solution.

Due to the site conditions, including the east-west site shape and prevailing winds from the southeast, it was determined that the units were best suited to be oriented north-south. The baseline case used for all energy studies was therefore a duplex unit with only north- and south-facing exterior walls. East- and west-

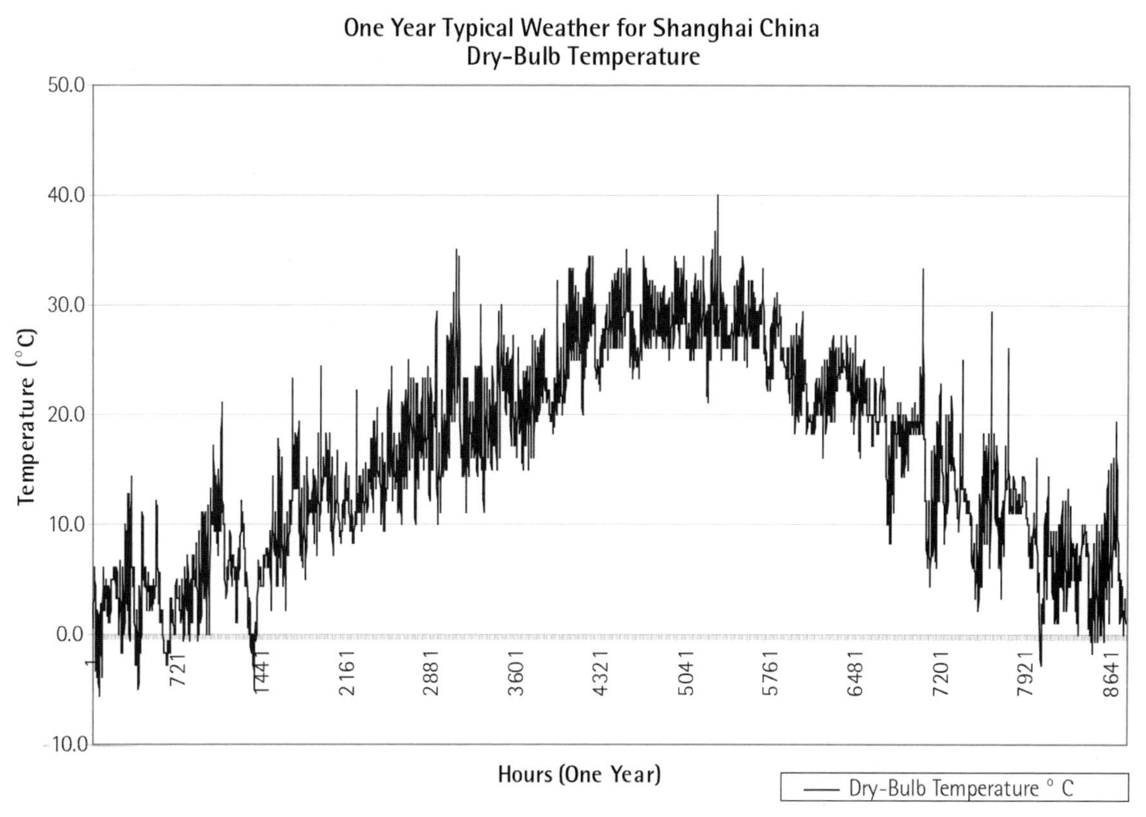

Figure 1 Hourly outdoor dry-bulb temperatures for Shanghai; hours are indicated as hour of the year, from 1 to 8,760 (Source: EnergyPlus Weather Data 2006)

Building Component	Specification
Floor area	2 story, 7 m x 15 m
Wall type	30.4 cm concrete, no insulation
Window type	Single-pane aluminum frame
Window area	33.3% window-to-wall area
Overhang type	No overhang
Electric heat pump	10 SEER / 7.8 HSPF
Heating set point	20° C
Cooling set point	24° C
Ventilation	No ventilation

Table 1 Baseline feature specifications

facing walls, in addition to the ceiling and floor, were modeled as adiabatic. The results corresponded to a typical floor, and were not representative of ground-floor or top-level units. While such analyses were limited in scope, they served as an appropriate design springboard in terms of general building features with respect to the existing climatic conditions. The correct identification of such base issues early on in the design process was critical for many subsequent design development decisions.

Baseline Parametric Simulations

The initial simulation was carried out under the parameters of the baseline case as explained in Table 1. In the first set of simulations, described as "building features," the design of the building was upgraded while the components remain unchanged. In the second set of simulations, described as "detail specifications," the building design was upgraded with architectural components designed within each unit, such as detailed window and HVAC specifications.

In the "building features" set of upgrades, the following upgrade simulations were created, and are shown in Figure 2:

- half-meter overhang on the south façade of the building at each floor;
- one-meter overhang on the south façade of the building at each floor;
- no solar absorption on the exterior walls, such as with ivy or another wall curtain outside the thermal exterior wall;
- 50 percent window-to-wall area ratio (WWA);
- 75 percent WWA;
- concrete wall half the original thickness;
- concrete wall half the original thickness plus one-inch polystyrene foam insulation; and
- concrete wall half the original thickness plus two-inch polystyrene foam insulation.

In the "detail specifications" set, the following upgrade simulations were created, and are shown in Figure 3:

- standard double-pane window with clear glass;
- high-performance double-pane window (low emissivity, vinyl frames, and low infiltration);
- nighttime insulation on windows, such as curtains;
- ventilation at 5 ACH from 1 April through 31 May and from 1 September though 31 October;
- ventilation at 10 ACH during the same moderate seasons;
- heating season nighttime setback on the thermostat from 20°C to 15.6 °C during the night hours from 10 PM to 6 AM;
- cooling season setup to 26.7°C throughout the entire season; and
- cooling season setup to 29.4°C throughout the entire season.

The energy results were converted to cost savings in order to better understand the impact of the simulations. Cost comparisons are shown for the building feature and detail specification set simulations in Figures 4 and 5. The cost was calculated per residential unit based on an electrical rate of US$0.06 per kilowatt-hour (kWh). Through these studies, it was found that as much as one-third of the energy cost could be saved by single upgrades alone, such as changing the cooling set point to 29.4°C.

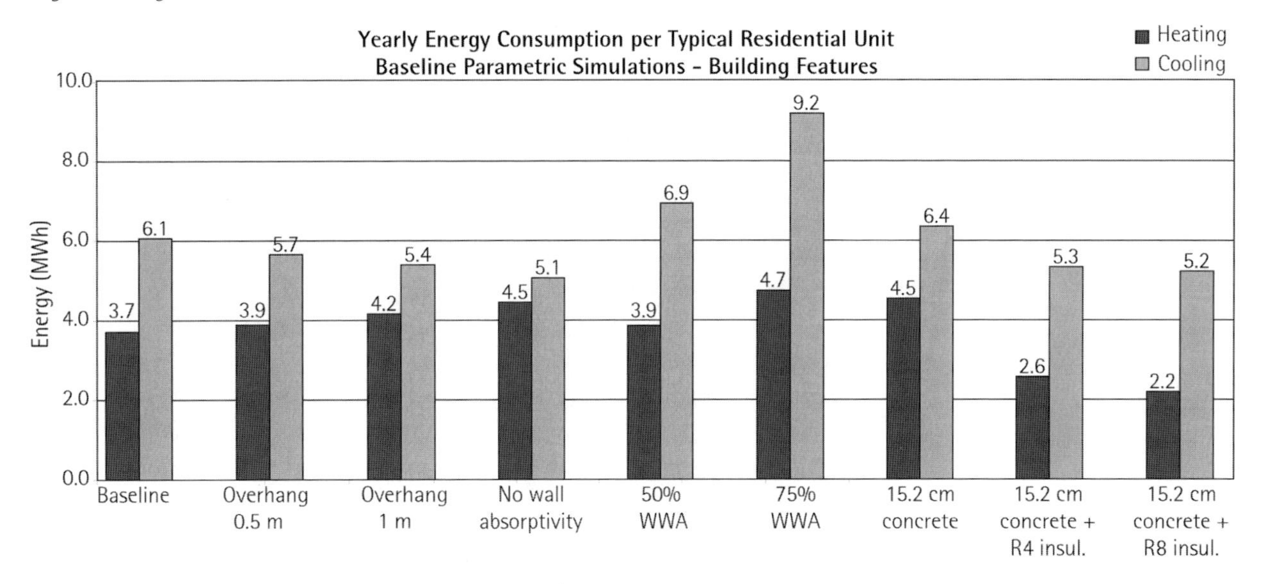

Figure 2 Energy use for building feature upgrades versus the baseline case

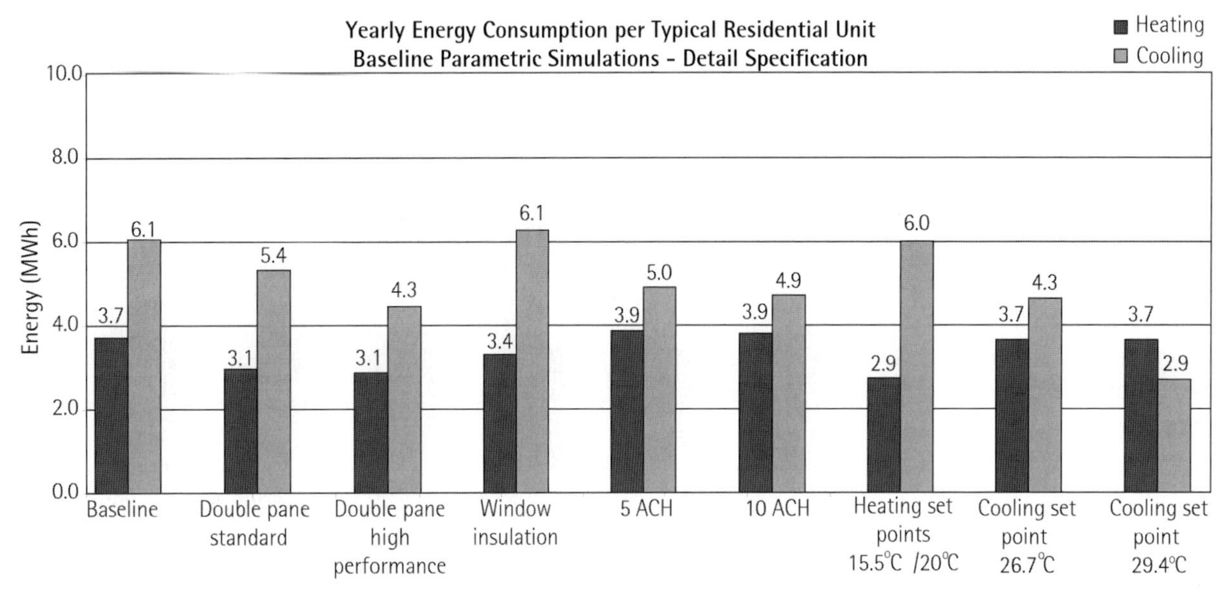

Figure 3 Energy use for detail specification upgrades versus the baseline case

Combined Feature Simulations

The next set of simulations (simulations 1-3) centered on combining the upgraded features for optimally inclusive energy savings. The most feasible results were then identified from information gathered from the baseline parametric tests. A comparison of the baseline case versus the simulation of multiple upgrades is displayed in Figure 6. Table 2 describes the simulations and combined upgrades.

It was found that by using a combination of the best upgrades, a savings of as much as 67 percent can be gained from these design features alone, as shown in the first three simulations in Figure 6. The next simulation (simulation 4) took a more feasible approach, using upgrades that are more cost effective in the construction process. This method was also effective, and created a 52 percent savings compared to the baseline case. The following three simulations (simulations 5-7) increased the window area to 50 percent WWA in order to provide architectural flexibility. The goal of these tests was to show how heating and cooling loads can be affected by solar heat gain and window conduction. The tests highlighted the importance of window design in this climate. Simulations 1-3 had night insulation and an overhang, with a standard double-pane window, while simulation 4 did not include the overhang, and simulations 5-7 included neither the overhang nor the night insulation. The combined upgrades also integrated ventilation at 5 ACH or 10 ACH into a simulation of the effects of the heating and cooling set points. The energy data from the simulations in this section were converted into cost data, as shown in Figure 7.

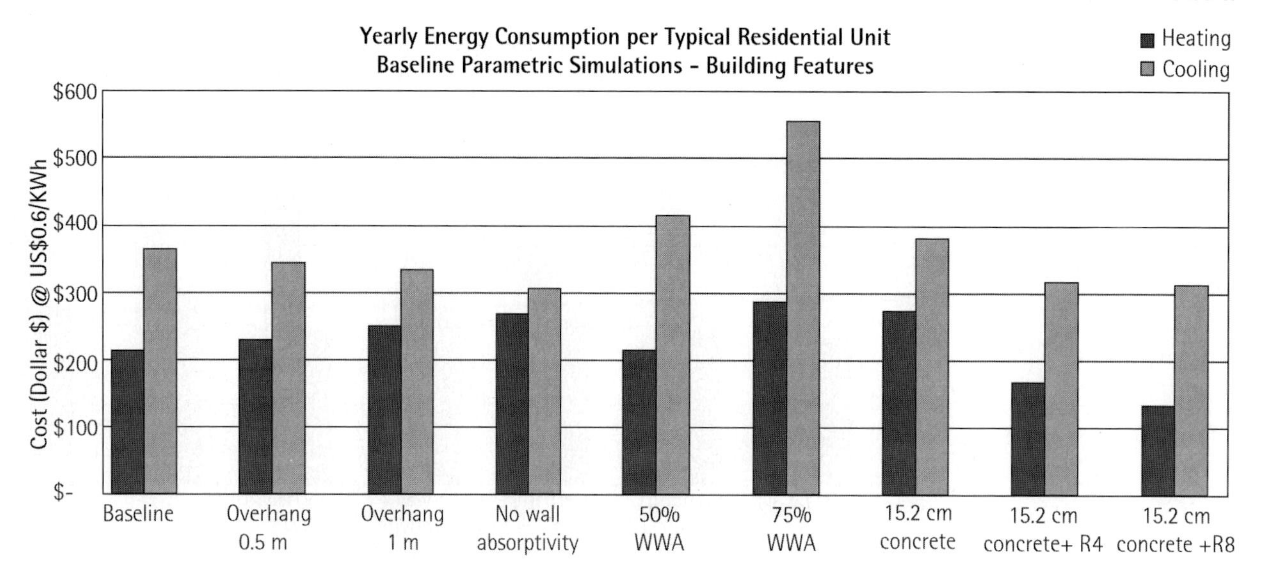

Figure 4 Cost savings for building feature upgrades versus the baseline case

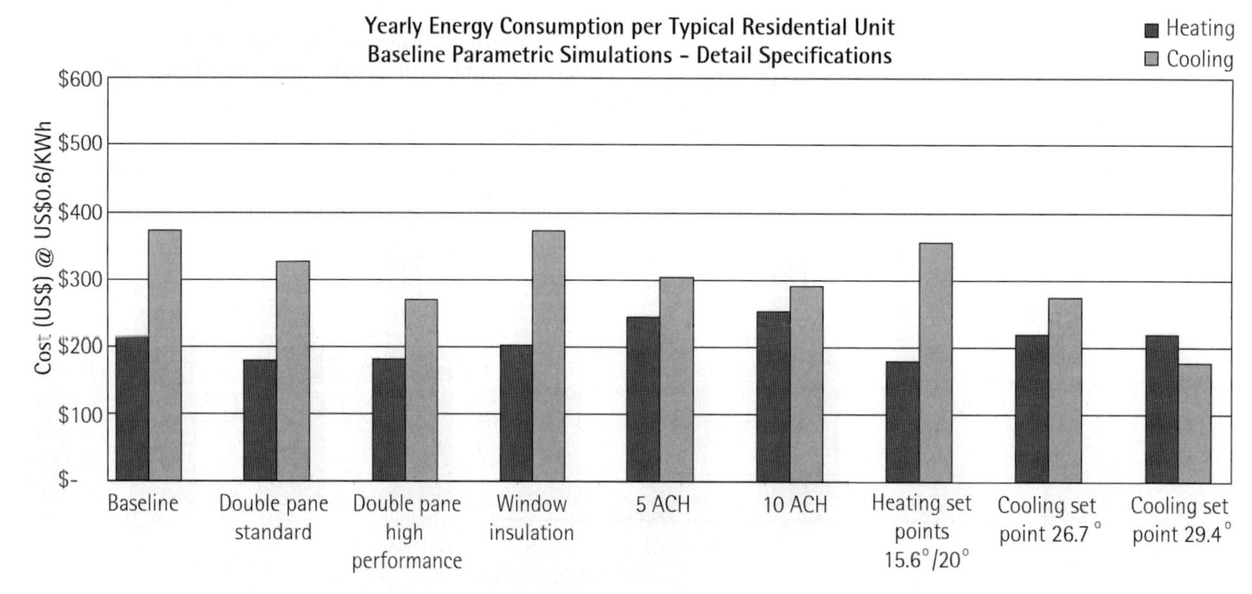

Figure 5 Cost savings for detail specification upgrades versus the baseline case

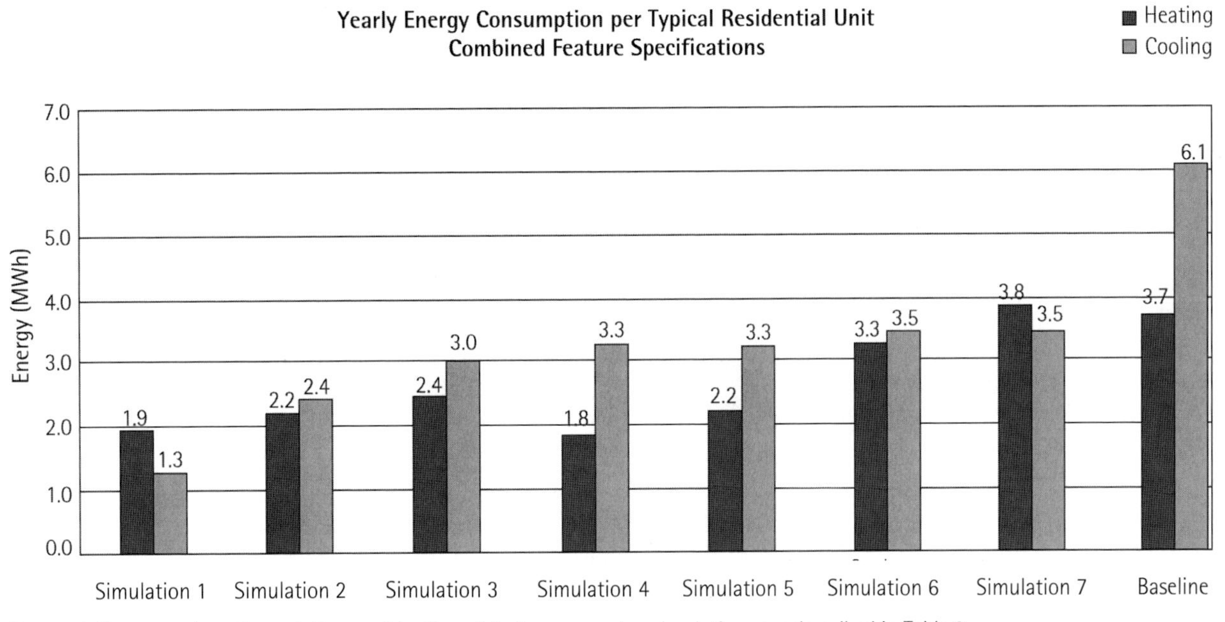

Figure 6 Energy savings through the combination of feature upgrades; simulations are described in Table 2

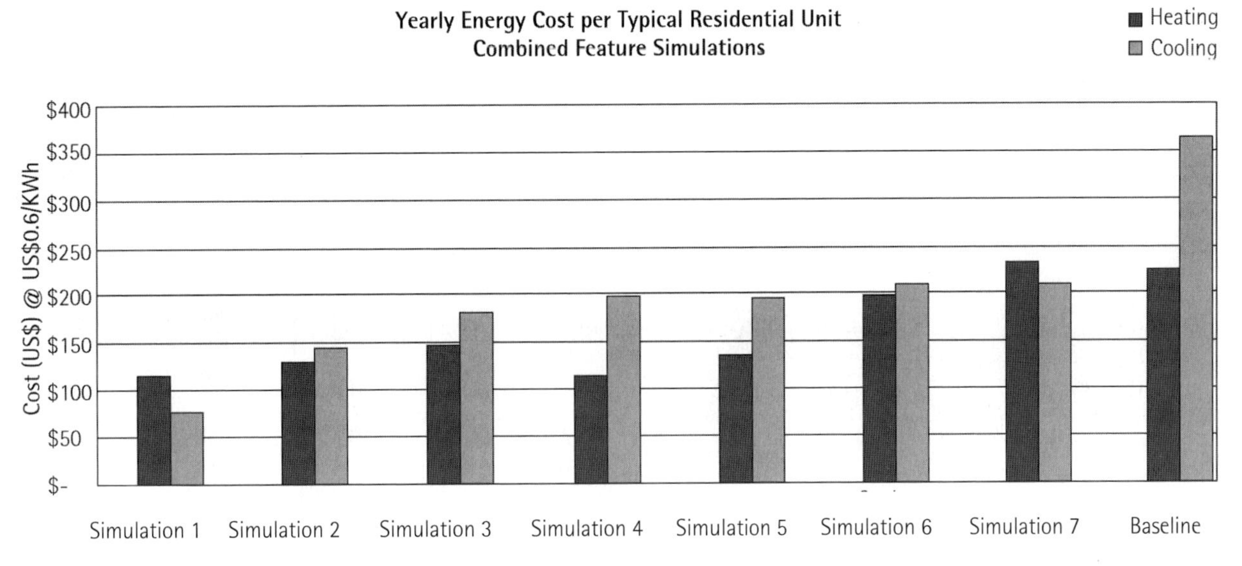

Figure 7 Cost savings through the combination of feature upgrades; simulations are described in Table 2

Simulation	Upgrade Combinations
1	Cooling set point 29.4°C Ventilation at 10 ACH during moderate seasons High performance double-pane window 15.2 cm concrete + R8 (5 cm foam insulation) 33% WWA Window night insulation One-meter overhang on the south façade Heating setback from 20°C (day) to 15.6°C (night) No wall absorptivity
2	Cooling set point 26.7°C Ventilation at 5 ACH during moderate seasons Standard double-pane window with clear glass 15.2 cm concrete + R4 (2.5 cm foam insulation) 33% WWA Window night insulation Half-meter overhang on the south façade Heating setback from 20°C (day) to 15.6°C (night) No wall absorptivity
3	Cooling set point 26.7°C Ventilation at 5 ACH during moderate seasons Standard double-pane window with clear glass 15.2 cm concrete + R4 (2.5 cm foam insulation) 50% WWA Window night insulation One-meter overhang on the south façade
4	Cooling set point 26.7°C Ventilation at 5 ACH during moderate seasons Standard double-pane window with clear glass 15.2 cm concrete + R4 (2.5 cm foam insulation) 50% WWA Window night insulation
5	Cooling set point 26.7°C Ventilation at 5 ACH during moderate seasons Standard double-pane window with clear glass 15.2 cm + R4 (2.5 cm foam insulation) 50% WWA
6	Cooling set point 26.7°C Ventilation at 5 ACH during moderate seasons Heating setback from 20°C (day) to 15.6°C (night)
7	Cooling set point 26.7°C Ventilation at 5 ACH during moderate seasons
	Baseline

Table 2 Table showing summary of combined upgrades

Design Tools

Following the simulations, communication tools were developed for designers to help them to integrate the technical findings into the design process. A shading worksheet was generated to allow architects to see immediate effects of building features and detailed specifications, such as of changing the orientation, overhang lengths, and shading coefficient during the duration of one year (Figures 8a and 8b). This allowed architects to make better decisions regarding windows and shading devices early on in the design process. The shading calculations worksheet for Shanghai, in addition to Beijing and Shenzhen worksheets, are included in the compact disc available with book.

Recommendations

This research shows that significant energy savings can be obtained through strategic design of both the building features and the detail components. The use of insulation on the exterior walls provided the greatest savings on the heating load. The use of upgraded window components provided the greatest reduction of cooling load. Important components to consider included items such as shading devices for the summer sun or a lower solar heat gain coefficient on window glazing, as in a low emissivity (low-e) window.

Also of great significance was taking advantage of natural ventilation, which helped the occupant to lower the heating and cooling loads. The equipment should promote the use of higher summer set points and lower winter set points. Taking advantage of these set points alone will both increase the comfort for the occupants and result in potential energy savings of 30 to 50 percent. Lastly, the key to such savings lies in the architects' knowledge base of such strategies for timely and strategic integration into the design process.

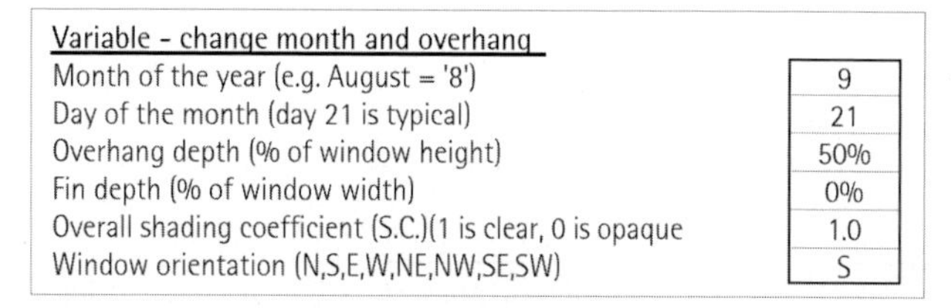

Variable - change month and overhang	
Month of the year (e.g. August = '8')	9
Day of the month (day 21 is typical)	21
Overhang depth (% of window height)	50%
Fin depth (% of window width)	0%
Overall shading coefficient (S.C.)(1 is clear, 0 is opaque	1.0
Window orientation (N,S,E,W,NE,NW,SE,SW)	S

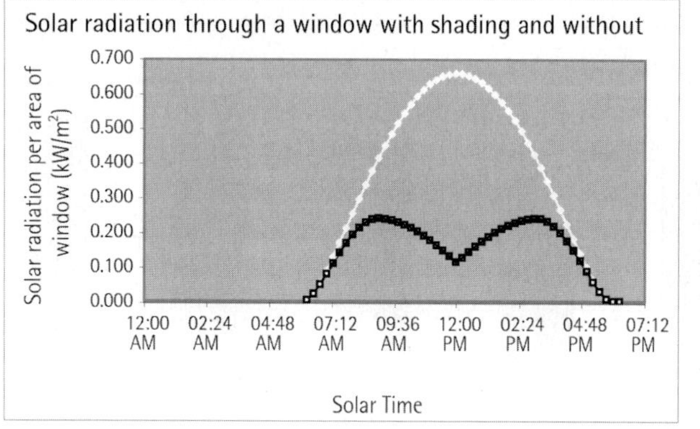

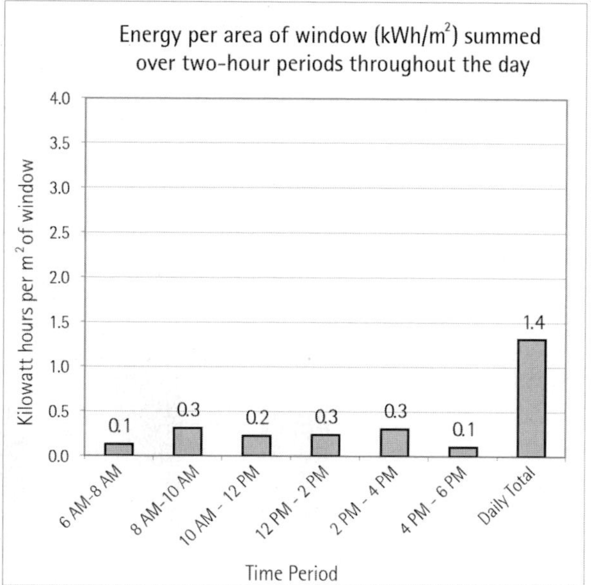

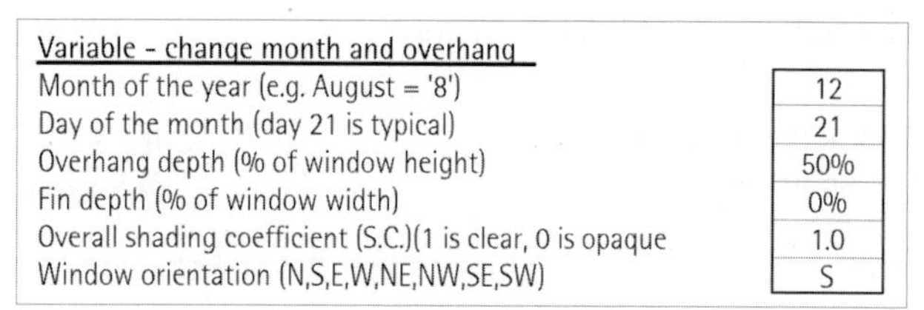

Variable - change month and overhang	
Month of the year (e.g. August = '8')	12
Day of the month (day 21 is typical)	21
Overhang depth (% of window height)	50%
Fin depth (% of window width)	0%
Overall shading coefficient (S.C.)(1 is clear, 0 is opaque	1.0
Window orientation (N,S,E,W,NE,NW,SE,SW)	S

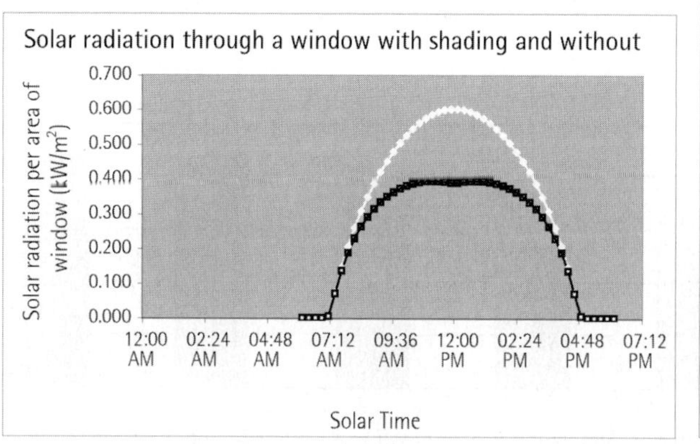

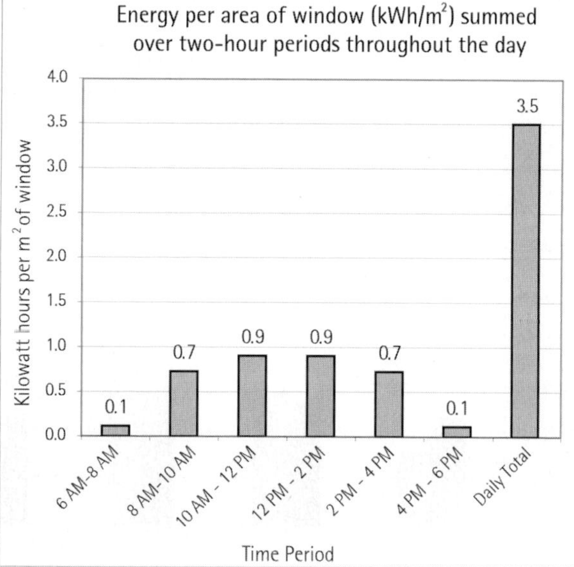

Figure 8a Shanghai shading calculations worksheet showing solar radiation and energy per area of window for a south-facing window with an overhang 50% of the window height in September and December

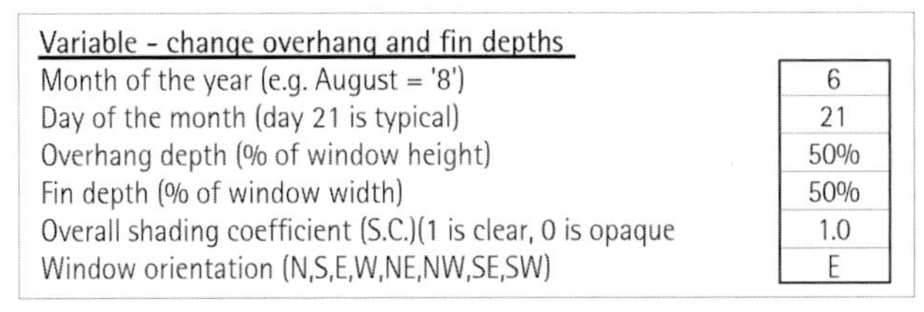

Variable - change overhang and fin depths	
Month of the year (e.g. August = '8')	6
Day of the month (day 21 is typical)	21
Overhang depth (% of window height)	50%
Fin depth (% of window width)	50%
Overall shading coefficient (S.C.)(1 is clear, 0 is opaque	1.0
Window orientation (N,S,E,W,NE,NW,SE,SW)	E

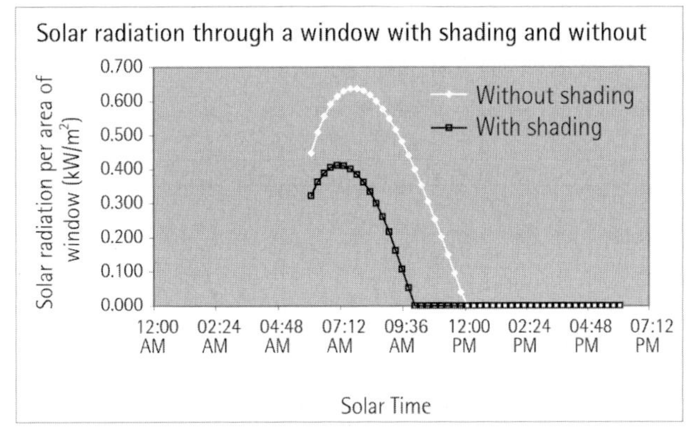

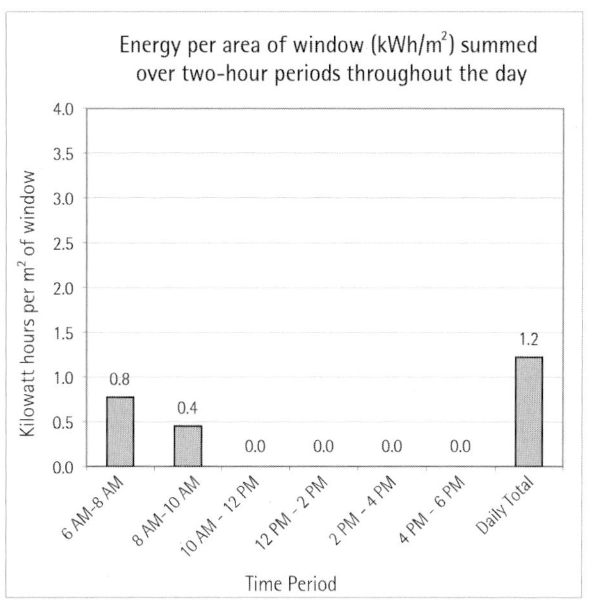

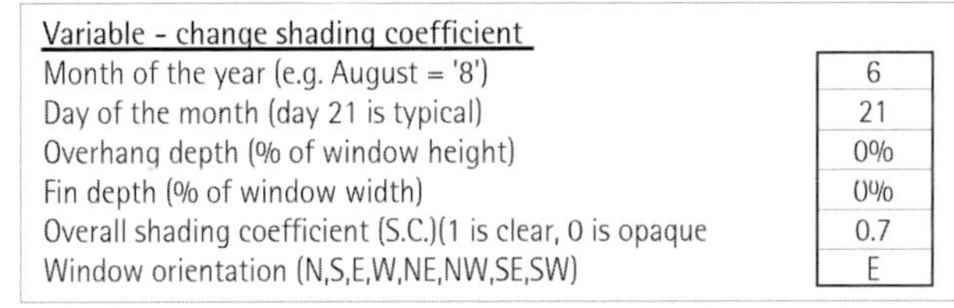

Variable - change shading coefficient	
Month of the year (e.g. August = '8')	6
Day of the month (day 21 is typical)	21
Overhang depth (% of window height)	0%
Fin depth (% of window width)	0%
Overall shading coefficient (S.C.)(1 is clear, 0 is opaque	0.7
Window orientation (N,S,E,W,NE,NW,SE,SW)	E

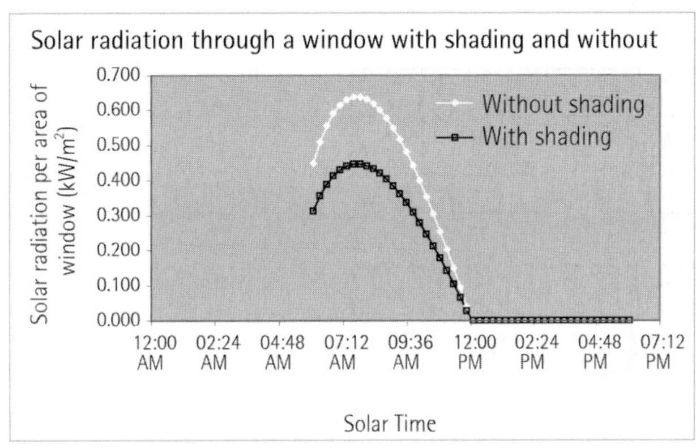

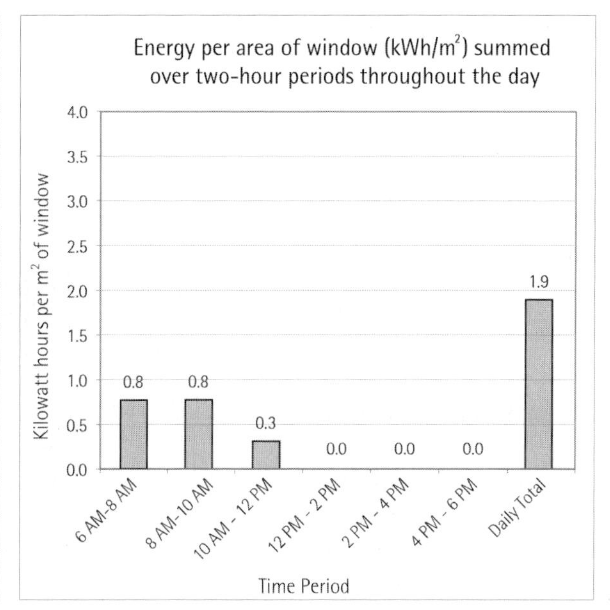

Figure 8b Shanghai shading calculations worksheet showing the effects of overhangs, vertical fins, and window shading coefficients on solar radiation and energy per area of window for a east-facing window in June

ARCHITECTURAL DESIGN

The design team worked on three schemes for the project based on the initial energy studies, climate research, and client considerations. The first scheme prioritized natural ventilation and an alternative to building massing; the second concentrated on building orientation with respect to sun angles and natural ventilation. Natural ventilation and shading were the focus of the third scheme, which also provided a unique alternative to building massing. The design goal for all buildings was to increase energy savings over the typical mid-rise building while maintaining or improving comfort levels. Equally important was to approach the urban design with a visual alternative to the monotonous block-type housing common in Shanghai.

Scheme I

This design was a response to the surrounding urban fabric and attempted to avoid building towers of equal massing commonly found in Chinese construction. The building was located on the periphery of the neighborhood facing a main city boulevard on the north side, which helped to shape the urban space and to provide identity of the neighborhood in the complicated urban fabric. The volume of the building gradually stepped back in an east-west direction along the height of the buildings (Figures 9-11). This helped to serve as the transition between large-scale urban development in the neighborhood and the existing small-scale neighborhood to the south.

Scheme I was composed of two buildings with a combination of single-story and duplex units that encouraged natural ventilation by eliminating the circulation corridor (Figure 12). The scheme created central vertical circulation cores for each cluster of units, thus exposing both north and south walls to the exterior,

Figure 9 Architectural elevations (from south) for scheme I

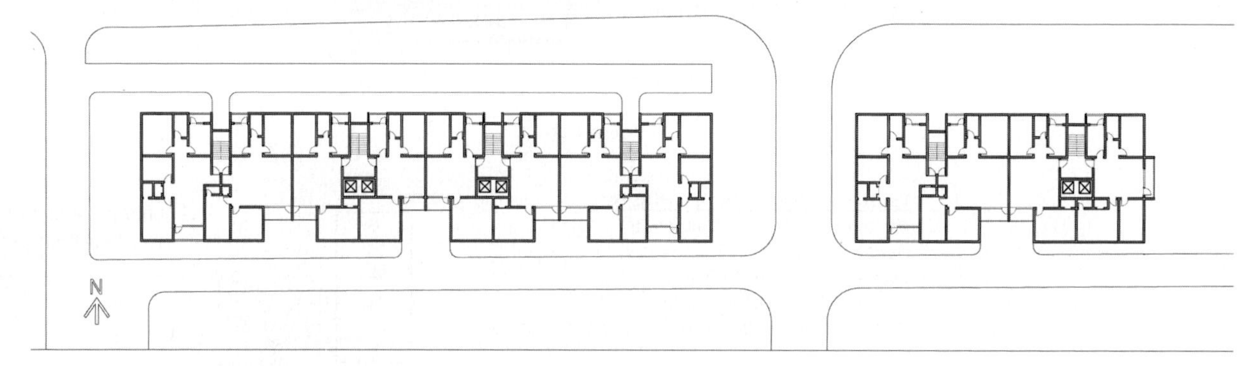

Figure 10a Typical floor plan for scheme I - levels 1-6

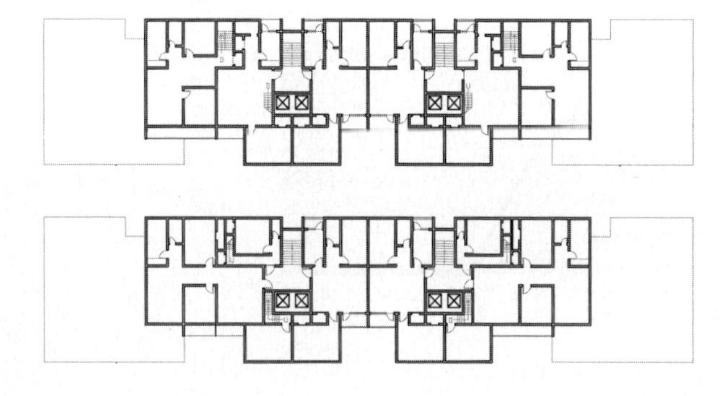

Figure 10b Typical floor plans for upper and lower floors for scheme I - levels 7-10

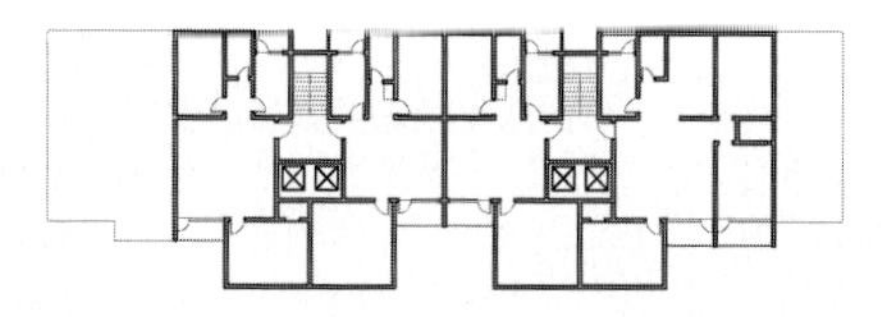

Figure 10c Typical floor plan for scheme I - levels 11-12

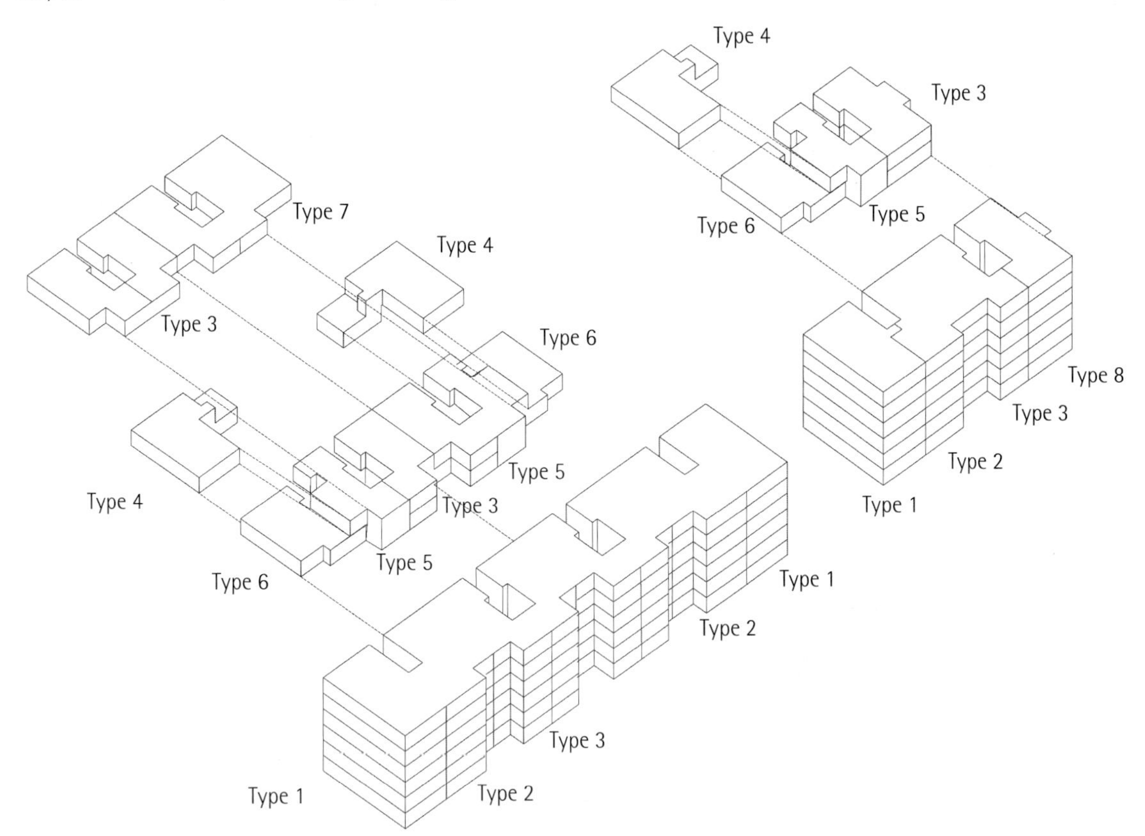

Unit Type	Area Per Unit (m^2)	Number of Units
Type 1	142	12
Type 2	146	12
Type 3	125	49
Type 4	152	7
Type 5	137	7
Type 6	143	7
Type 7	145	5
Type 8	125	15

Total Units	Two Bedrooms	Three Bedrooms
114	70	44

Total Floor Area (m^2)	Building Footprint Area (m^2)	Site Area (m^2)
16,771	1,605	4,686

Total Public Circulation Area (m^2)	Total Private Floor Area (m^2)	Ratio of Circulation to Private
1,559	15,211	1:10

Figure 11 Scheme I unit types - (left) unit layout and (right) and description of unit types

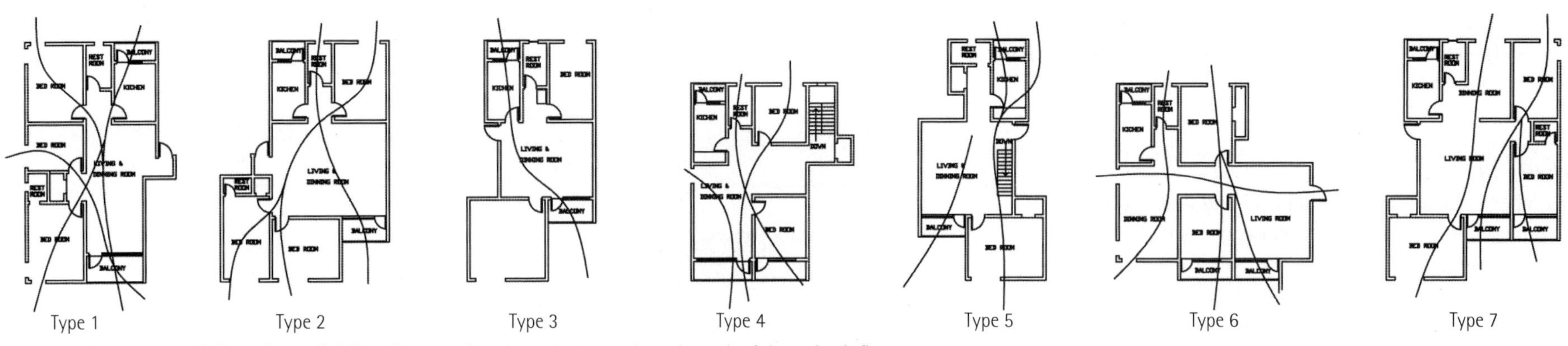

Figure 12 Preliminary unit floor plan and airflow diagrams for scheme I; arrows show the path of the main air flow

which allowed for better natural daylight and interior cross ventilation through the units.

Scheme II

A primary goal was to develop a strategy that could both keep hot temperatures out during the summer and allow solar gains to be optimized during the winter. In order to do this, the buildings were oriented 10° west of north to let the winter morning sun in (Figure 13). The design team then performed sun-angle light studies on this strategy (Figure 14). Scheme II was based on the findings of these studies and incorporated a south-facing sunspace and open balconies to aid in shading the building during summer months.

Light studies (Figures 15-19) were performed using ray-tracing software to simulate the effect of proper shading on a typical unit. Colors in false color maps correspond to illuminance levels; a level of ~ 500 lux was adequate for most tasks. Excessively high levels are associated with glare and unwanted heat gains. It was found that during the hotter summer months, little direct sunlight penetrated into the unit. During the cooler winter months, sunlight penetrated deep into the unit, warming it and making more efficient use of the more seasonally scarce daylighting. Such simple well-informed orientation not only can have a great influence on energy savings for heating and cooling, but also provides a comfortable, shaded, and well-lit environment for occupants.

Scheme III

This scheme differed from schemes I and II by dividing the program into three smaller buildings instead of two larger buildings. In doing so, it was able to interweave green space with public and semi-public activity areas. The developer selected this scheme, and the final design

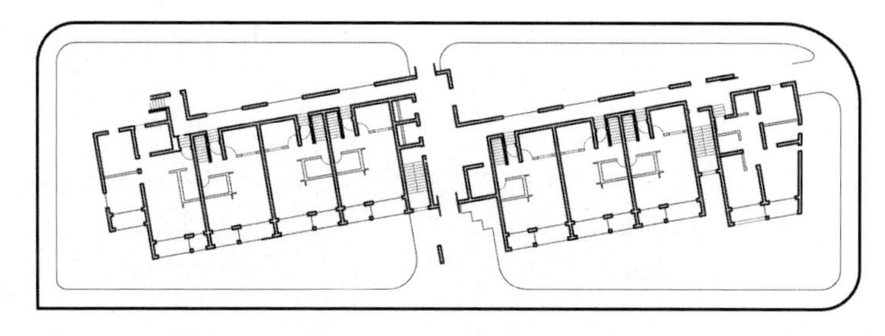

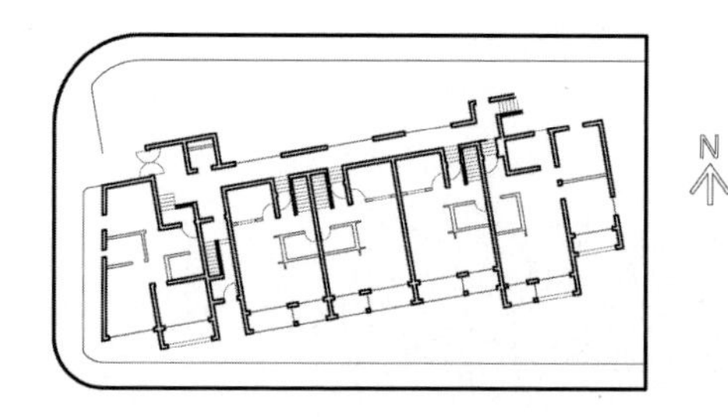

Figure 13 In the site plan of scheme II, buildings are oriented 10° west of north

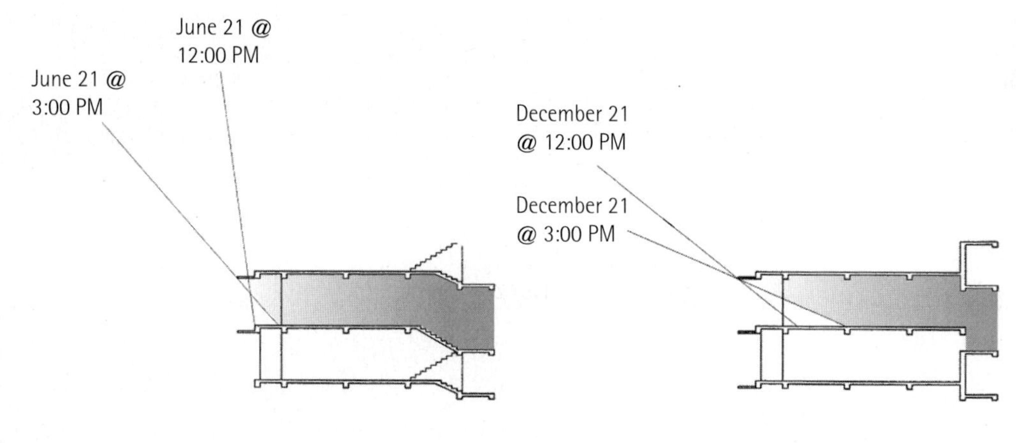

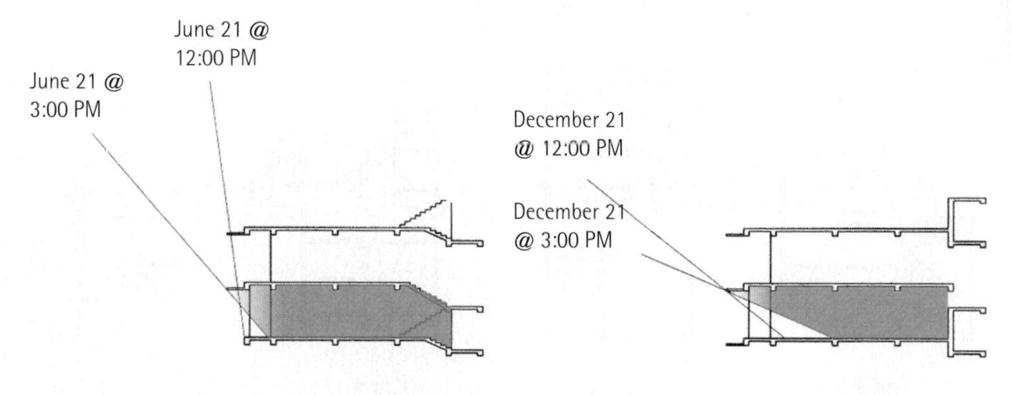

Figure 14 Sun angle studies for scheme II with 1 m overhang, showing upper floor and lower floors

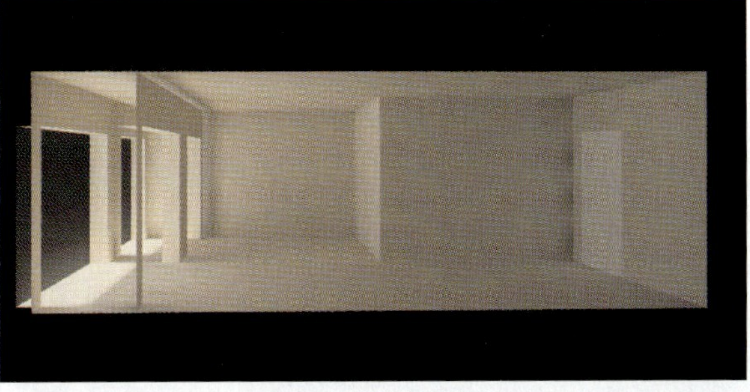

Figure 15 Photo-realistic renderings at 12 PM on 21 June without shading showing: (left) view of living room from entry, (middle) section view of south-facing living room, and (right) false-color maps for illuminance that offer a quantitative method for reading renderings. Illuminance levels range from dark blue = 500 lux, light blue = 4,000 lux, green = 8,000 lux, to red = 16,000 lux.

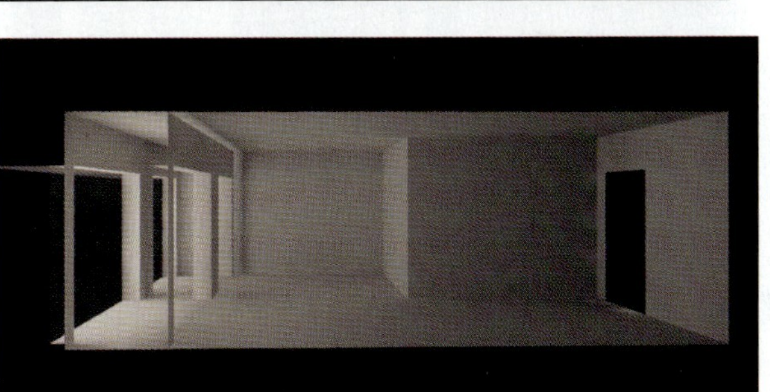

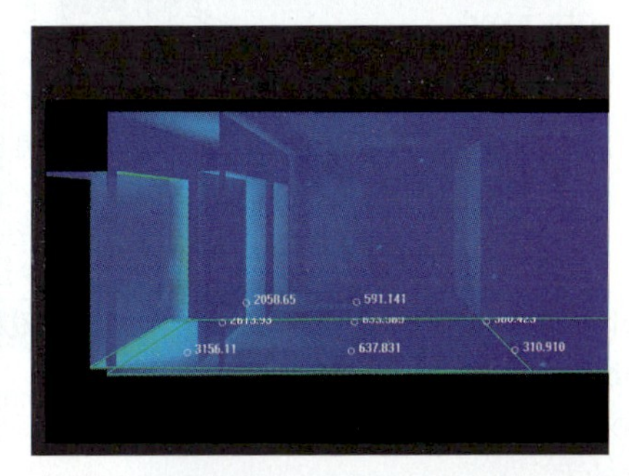

Figure 16 Photo-realistic renderings of 12 PM on 21 June with shading (1 m fixed overhang) showing: (left) view of living room from entry, (middle) section view of south-facing living room, and (right) false-color maps for illuminance that offer a quantitative method for reading renderings. Illuminance levels range from dark blue = 500 lux, light blue = 4,000 lux, green = 8,000 lux, to red = 16,000 lux.

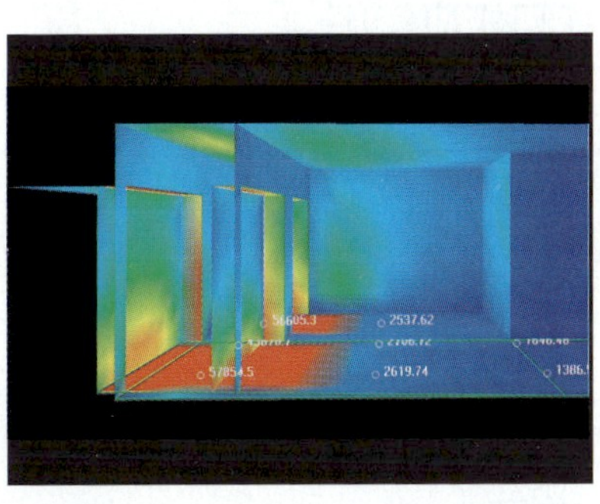

Figure 17 Photo-realistic renderings of south-facing living room with shading (1 m fixed overhang): (above left) at 9 AM on 21 December and (above right) false-color maps for illuminance that offer a quantitative method for reading renderings. Illuminance levels range from dark blue = 500 lux, light blue = 4,000 lux, green = 8,000 lux, to red = 16,000 lux.

Figure 18 Photo-realistic renderings of south-facing living room without shading at 12 PM on 21 December: (middle left) view from entry and (near left) section

Figure 19 Photo-realistic renderings of south-facing living room with shading (1 m fixed overhang) at 12 PM on 21 December: (middle left) view from entry and (near left) section

(shown in Figures 20-30) created a variety of experiences at the indoor and outdoor areas around the residences. These smaller building masses also promoted airflow through the site and permitted pedestrian and garden spaces between the housing.

The design priority for the generally hot and humid climate of Shanghai was to achieve a reduction of the summer cooling load and associated energy for cooling. This was accomplished in individual units primarily by reducing initial solar gains through the use of both fixed and operable overhang shading systems. An open plan and section design for each unit encouraged airflow that both cooled and removed excess moisture created by cooking and bathing. The scheme also used a skip-stop elevator, with corridors every other floor to economize on the cost of elevators and to create neighborhood units within the towers. Duplex units were oriented north/ south, and crossed over the corridor to promote natural ventilation (Figure 24-27).

Within the typical unit, natural ventilation was provided for the kitchen and dining rooms by utilizing vertical space over the corridor, which allowed airflow into the units. An open stairwell was provided for airflow to move between both floors of the unit (Figure 26). Within the single-floor units, located at the end of each corridor, all windows were on the north or south façade, with the exception some east- or west-facing windows, carefully located beneath balcony overhangs to provide shading. In all units, living rooms and master bedrooms faced south, which was an important cultural consideration when designing in China. The south-facing balconies had operable shading devices in order to change the effective overhang length and resultant shading benefits (Figure 28).

Figure 20 Street view (from north) of scheme III, final design

Figure 21 Site plan of scheme III, final design; this plan shows existing lower-rise buildings to the south

Figure 22 North elevation of scheme III, final design

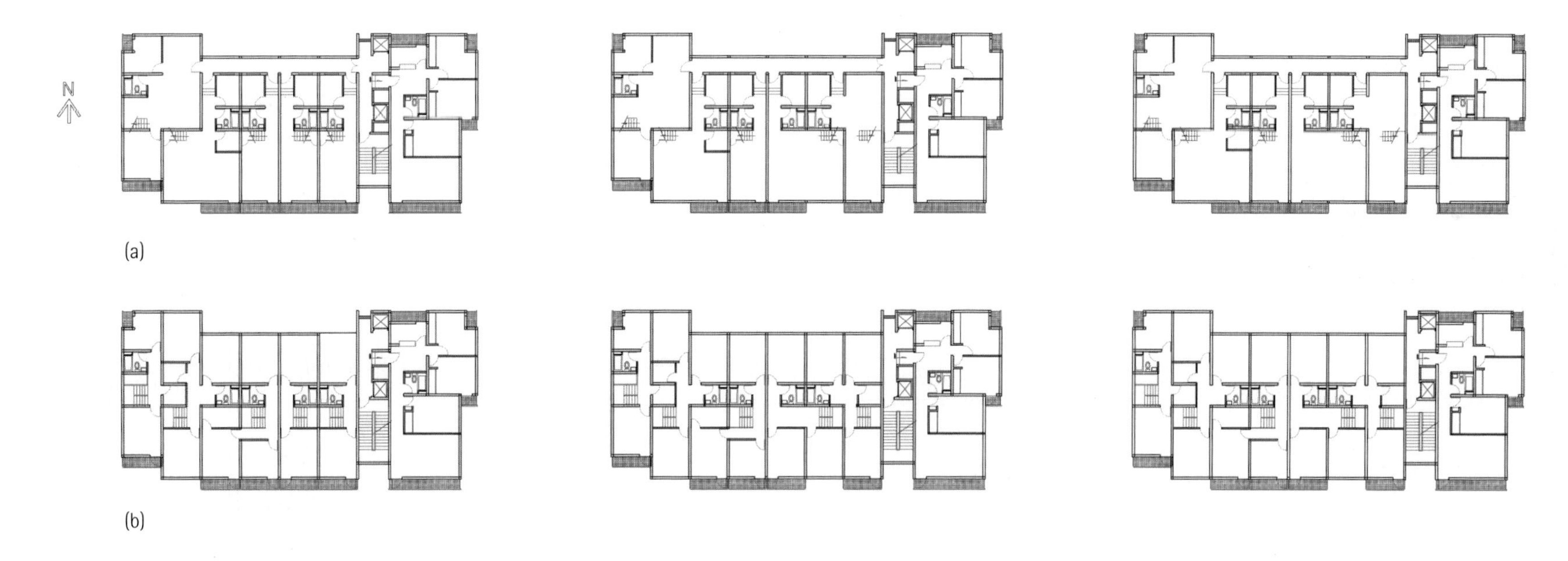

Figure 23 Architectural floor plans of: (a) lower, and (b) upper levels of duplex units in scheme III, final design

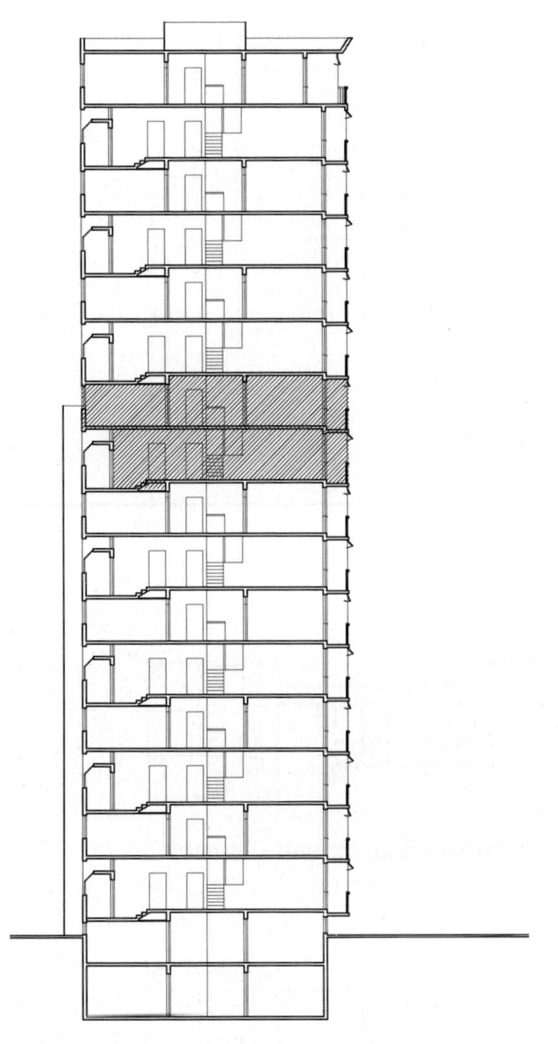

Figure 24 Building cross section illustrating duplexes (shown hatched) and corridors

Figure 25 Floor plans of units in a typical building showing both: (a) lower, and (b) upper floors of duplex

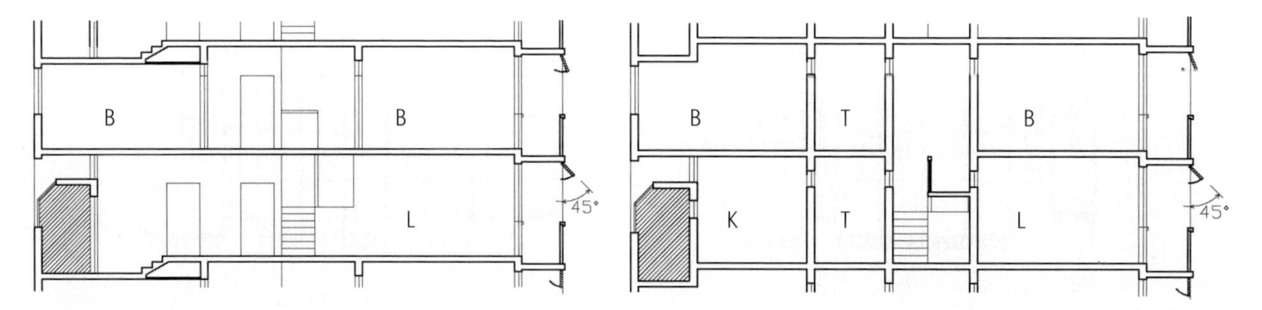

Figure 26 Enlarged sections of typical duplex unit - airflow is possible to the kitchen and dining room through high-level ventilation above the corridor (shown hatched)

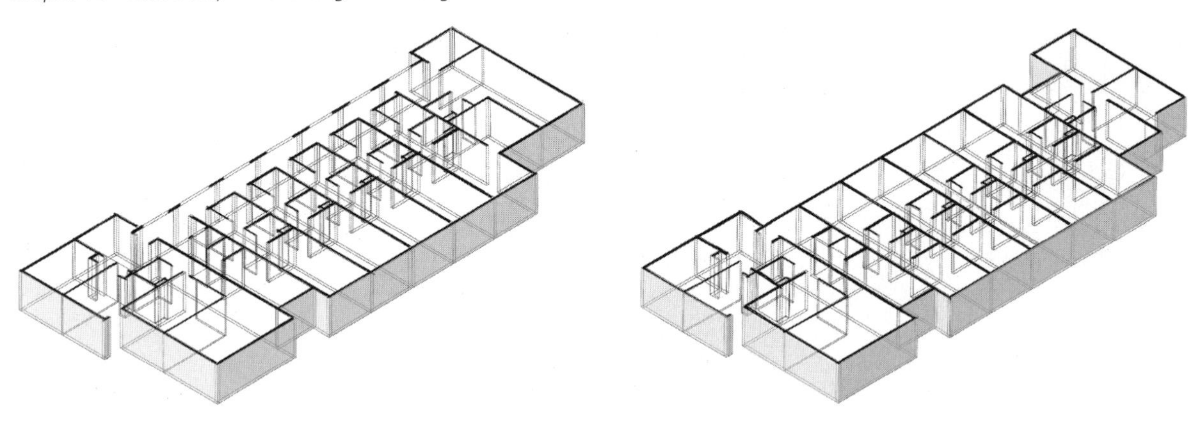

Figure 27 Axonometric of typical floors showing (left) lower and (right) upper levels; the corridor is located only at the lower level

Figure 29 Site model of scheme III, final design, showing building massing

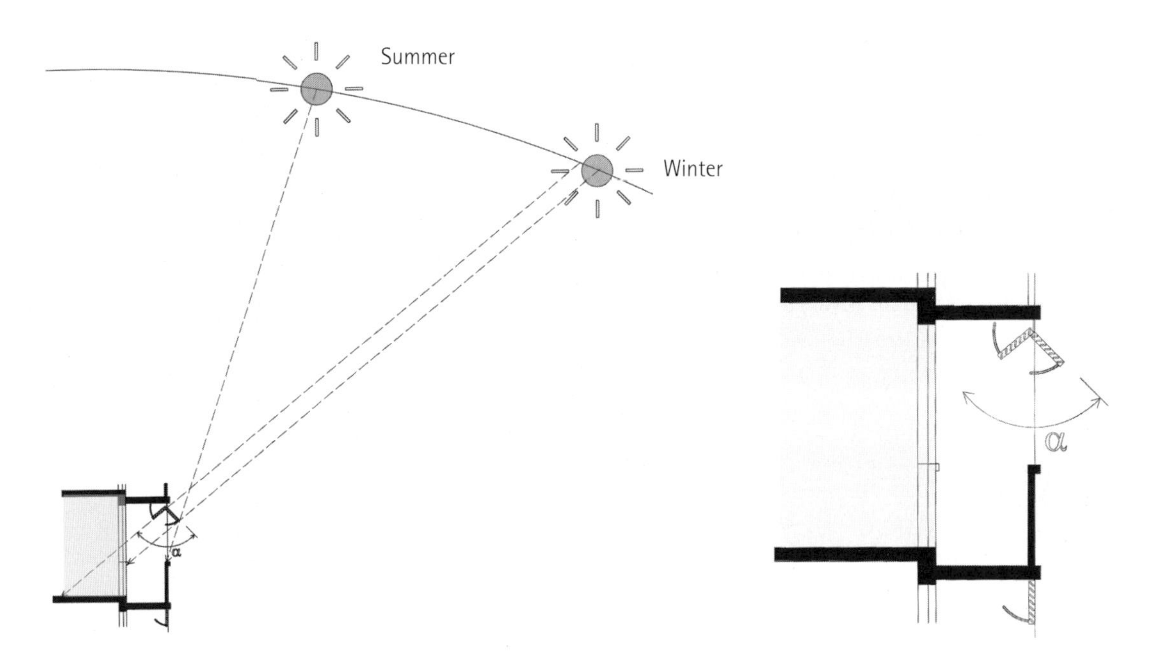

Figure 28 Shading diagram through (left) south sun-space and (right) diagram of moveable shading device for sun-spaces

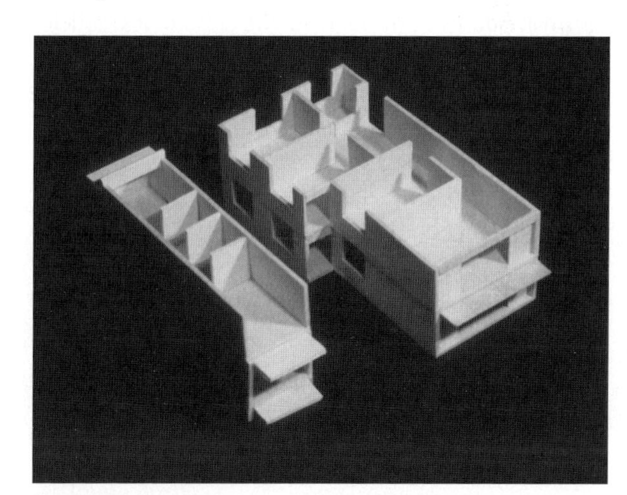

Figure 30 Model of scheme III, final design, showing duplex unit layout

NATURAL VENTILATION STUDIES

In addition to energy and daylighting studies, the team performed a series of ventilation studies at both the site and unit levels. The results of these studies affected window placement at the scale of both the building and individual units.

Site-Level Studies

The studies shown in this section involve the simulation of indoor and outdoor airflow by CFD. The developer preferred the final scheme (scheme III), but wished to convert the proposed design into two buildings rather than three. MIT simulated the effects of this change on natural ventilation within units, the results of which are shown in Figures 31 through 36. The CFD studies modeled surrounding buildings and were based on both the model illustrated in Figure 31 and the following assumptions of climate data for Shanghai:

- wind speed = 3 m/s at 10 meters above the ground;
- wind direction = southeast;
- air temperature = 24°C; and
- air relative humidity = 70 percent.

The studies found that the three-building scheme promoted better natural ventilation within the units studied (Figures 32-35). The three buildings promoted a more uniform condition across the width of the façade, whereas the wider two-building scheme resulted in areas with low air circulation impeded by the wide building. This was because the flow was completely blocked in the two-building scheme as shown in Figures 35 and 36. As a result, the airflow in some units of the two-building scheme could be very high, but in the rest the airflow was low. For natural ventilation design, it is preferable to use a scheme with more uniform flow, as

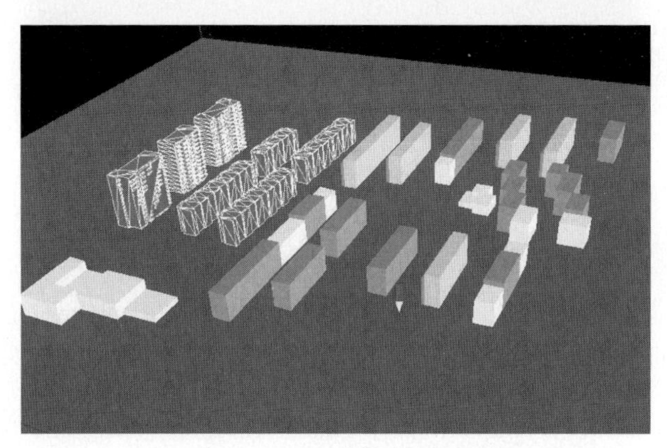

Figure 31 Axonometric showing area of site tested in CFD studies for (upper) 2-building scheme and (lower) 3-building scheme

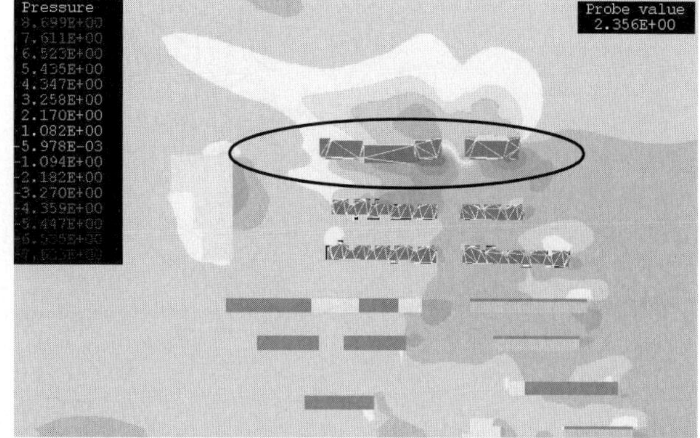

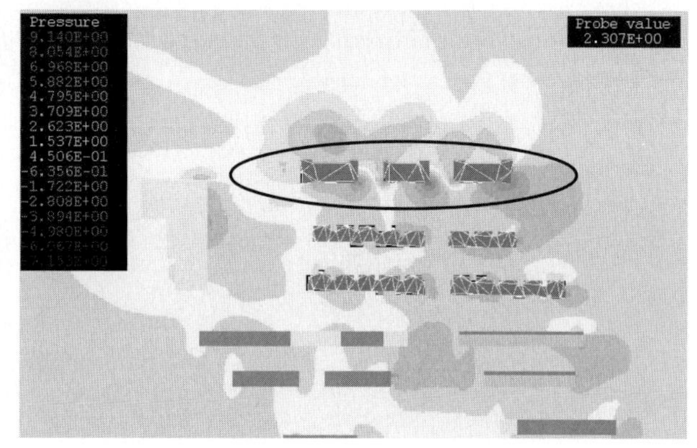

Figure 32 Images of pressure distribution around the buildings at 3 m above ground shown for: (upper) 2-building scheme, and (lower) 3-building scheme. Different gray shading levels show zones of pressure variation around building. The buildings studied are shown circled. (north is to the top in Figures 32, 33, 35, and 36)

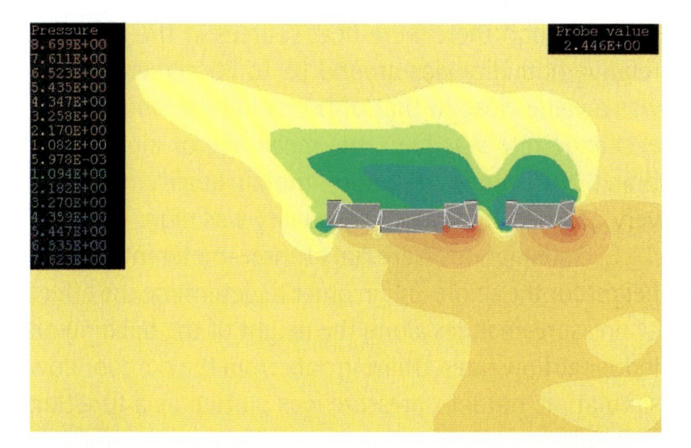

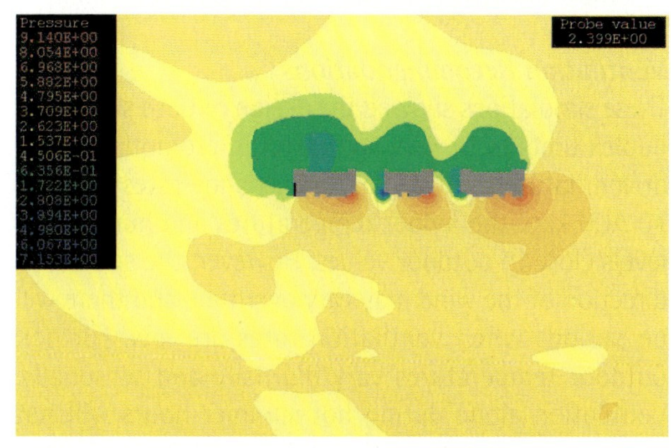

Figure 33 Images of pressure distribution around buildings at 25 m above ground shown for (upper) 2-building scheme and (lower) 3-building scheme (red – high pressure, yellow – moderate pressure, and green – low pressure)

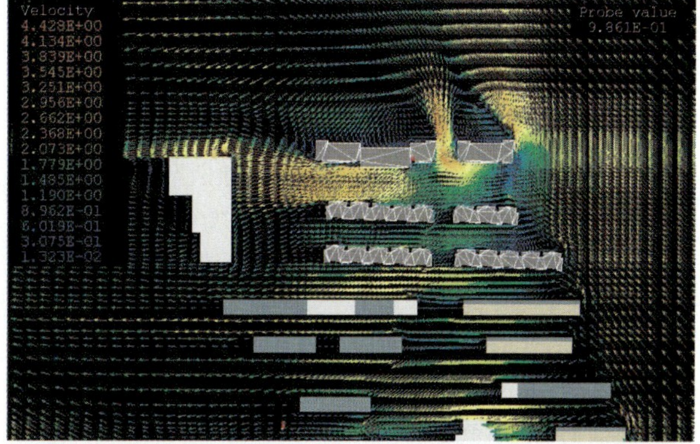

Figure 35 Images of wind speed at 3 m above ground shown for (upper) 2-building scheme and (lower) 3-building scheme

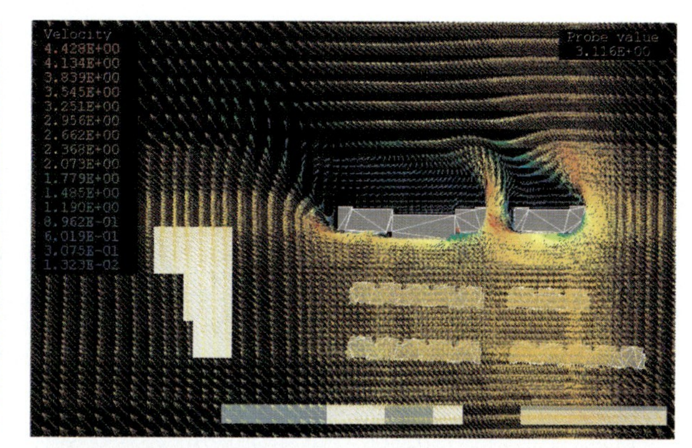

Figure 36 Images of wind speed at 25 m above ground shown for (upper) 2-building scheme and (lower) 3-building scheme

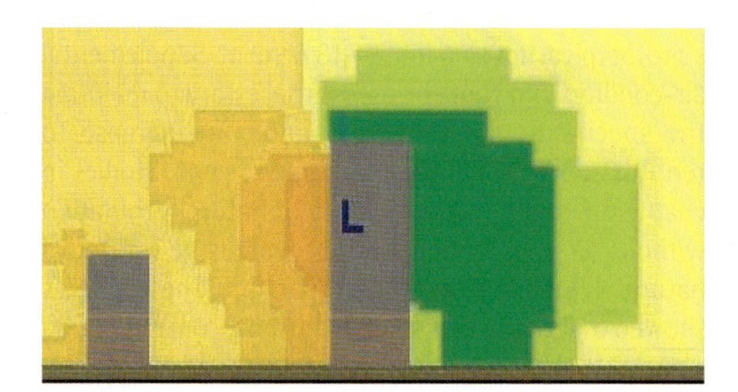

Figure 34 Sectional pressure distribution around the buildings for the 3-building scheme (red – high pressure, yellow – moderate pressure, and green – low pressure) (section looking west; north is to the left)

demonstrated by the three-building scheme. This study is valuable in demonstrating the influence of the actual architectural forms on the wind currents. With the natural climate data as a generator, we see that the number of buildings, the massing, and the orientation all play a role in creating pressure differentials that are influential in wind naturally entering and cooling individual units.

Ventilation Studies of Typical Units

In order to evaluate design decisions for window, door, and wall locations, indoor airflow within units was also simulated. These simulations were performed for all duplex units and single units. Because most results were similar, only the single-floor unit is described in this case study. The single units were located at the end of the circulation corridor, and thus were still able to have both north and south elevations exposed. Using the pressure distribution from site-level ventilation studies, along with the three-building CFD model shown in Figure 38, the design team was able to calculate the distributions of airflow, air temperature, relative humidity, predicted percentage dissatisfied (PPD), and the mean age of air. These tests were done with an assumption of an outside air temperature of 24°C and a relative humidity of 70 percent. Further input parameters for the single and duplex unit simulations are shown in Table 3.

The single-floor unit studied was unit G (Figures 37 and 38). Indoor airflow simulations and computed results by CFD are shown in Tables 3, 4, and Figures 38 through 46. These indicate that the maximum air velocity in the unit was less than one meter per second, which is a comfortable value for cross ventilation. The air exchange rate varies from 16 ACH on the first floor to a maximum of 40 ACH at approximately two-thirds of the maximum height of the building. With the air exchange rate of 16 ACH, the indoor air temperature increased less than 1°K, although there were heat sources in the unit. The relative humidity was around 65 to 70 percent, which was a value close to that of the outdoors. Since the air exchange rate was high, the mean age of air was less than 120 seconds. Therefore, the air quality would be very good when outdoor air quality was high.

CFD simulations were run at three-meter intervals in height for the single unit in order to determine the effect of pressure changes along the height of the building on indoor airflow rates. Utilizing data from the outdoor flow simulation, outdoor pressure was plotted as a function of height at each of the openings (Figure 47).

Ventilation Recommendations

These simulations showed the design of both single and duplex units to be very successful in promoting wind-driven natural ventilation. Air exchange rates of 16 to 40 ACH kept the indoor temperatures and humidity at levels close to outdoor values. However, the speed and direction of the wind may vary over time and there will be periods when ventilation rates are low. Further, outdoor temperatures vary diurnally and seasonally. Ventilation alone during hot summer hours will not consequently produce thermally comfortable indoor conditions.

The energy studies presented earlier in this case study showed that 5 ACH natural ventilation during April, May, September, and October could reduce cooling energy in a prototypical housing unit by 18 percent. Supplemental air-conditioning could be used as necessary to maintain the specified thermostat set point. An increase to ten ACH yielded minimal additional savings. Studies in chapters 5 and 6 focused on use of natural ventilation without air-conditioning and estimated the impact of natural ventilation on thermal comfort. These studies showed that occupants achieve the most comfort during summer months by ventilating at night and minimizing

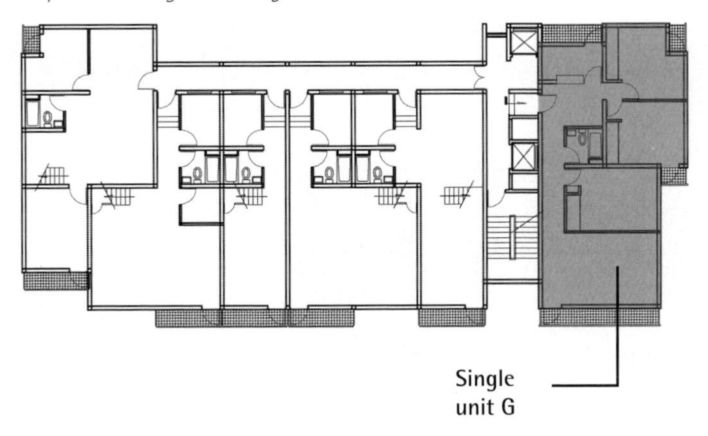

Figure 37 Floor plan diagram, showing location of unit G

Input Parameters	Inlet Value	Source Value	
Temperature/ heat flux	24° C	Occupant**	100 W
		Television	300 W
		Refrigerator	400 W
Pressure	*		
Humidity ratio	13 g water /kg air	Occupant**	55 g water /hour
Relative humidity	70%		
CO_2 concentration	400 ppm	Occupant**	5 mL/s
Age of air	0	1 (everywhere)	

* Both inlet and outlet pressures were input and vary between the duplex and single units

** Values refer to each of the 4 occupants

Table 3 Input parameters for CFD simulations

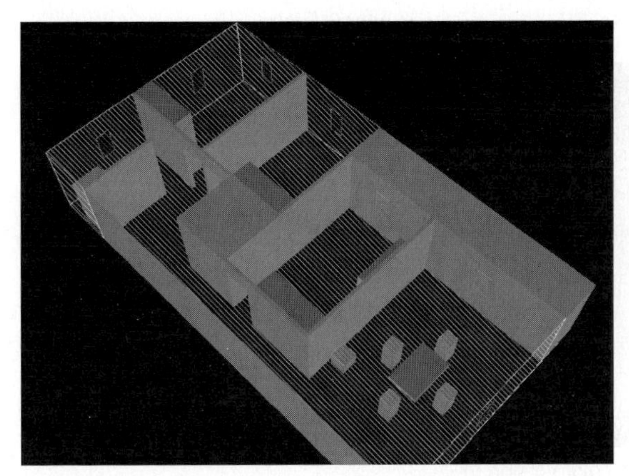

Figure 38 Single residential unit (G) modeled in CFD

Solved Parameter	Range of Values
Temperature	24-25°C
Velocity	0.01-1.2 m/s
Pressure	3-5 Pa
Percentage of persons dissatisfied	5-30%
Persons dissatisfied due to draft	0-100%
Humidity ratio	13-14.5 g water/kg air
Relative humidity	70-75%
CO_2 concentration	400-500 ppm
Ventilation effectiveness	20-100%
Age of air	0-102 s

Table 4 Solved parameters for CFD simulation of single unit (G)

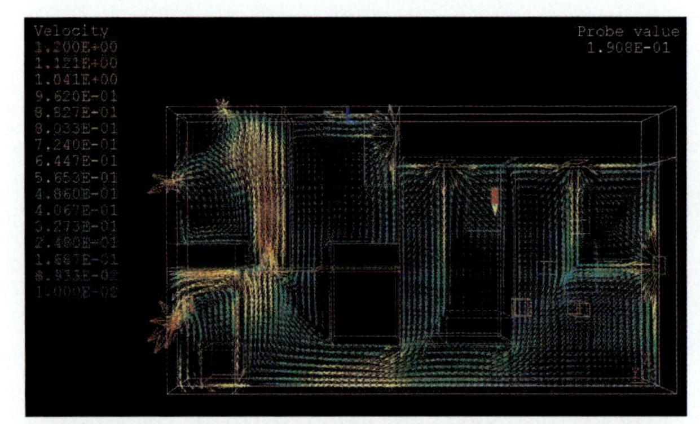

Figure 39 Air velocity distribution (m/s) at z = 1.5 m (see color version in chapter 6, Figure 14d)

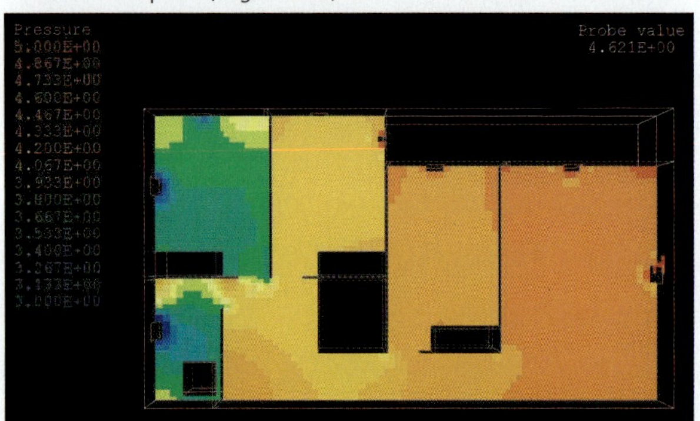

Figure 40 Pressure distribution - colors from dark to light grey indicate pressure levels from maximum to minimum

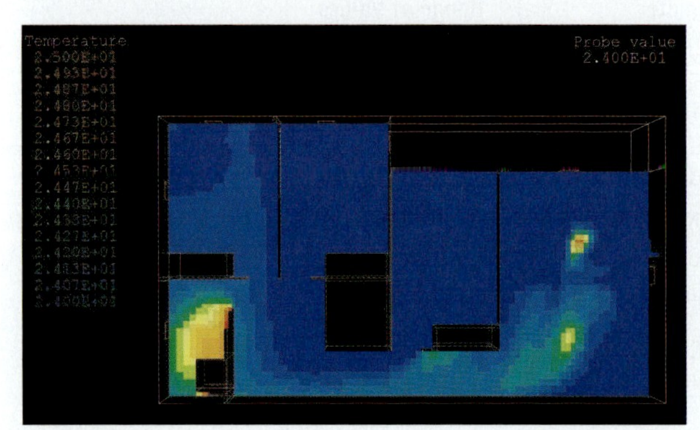

Figure 41 Air temperature (°C) at z = 1.5 m (see color version in chapter 6, Figure 14b)

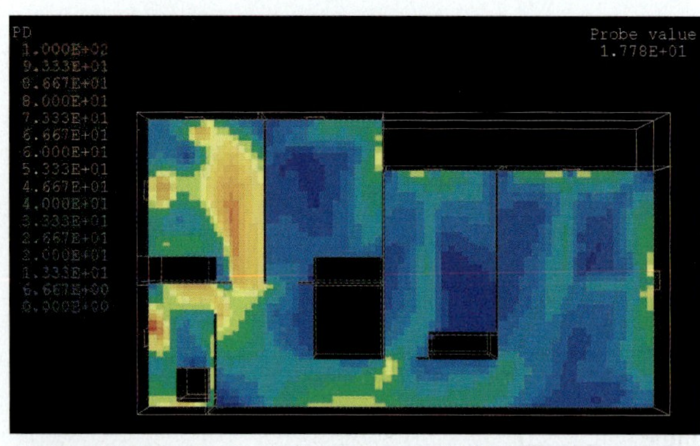

Figure 42 Persons dissatisfied due to draft (PD) contours at z = 1.5 m

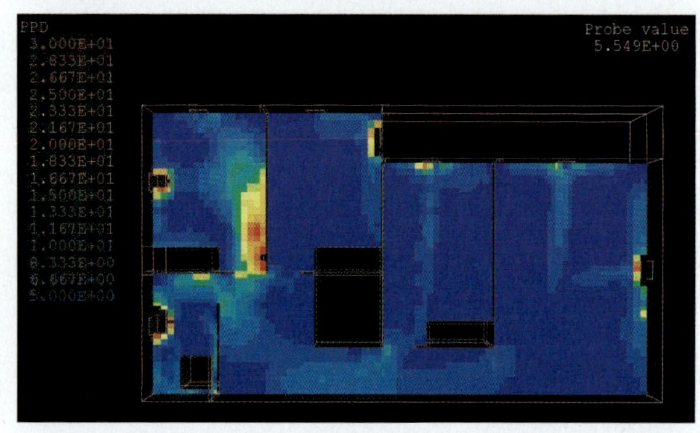

Figure 43 Percentage of persons dissatisfied (PPD) contours at z = 1.5 m (see color version in chapter 6, Figure 14c)

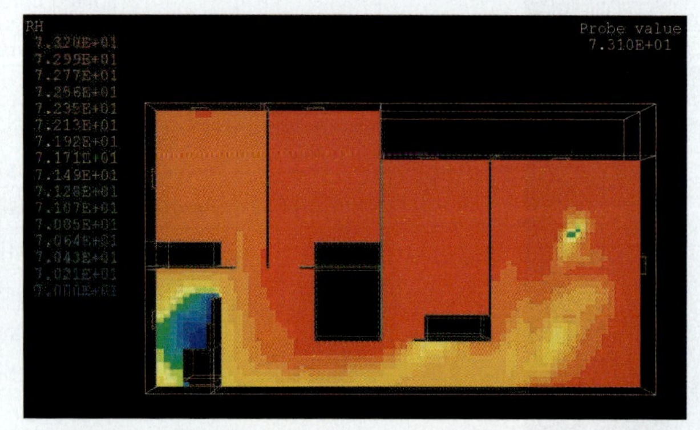

Figure 44 Relative humidity (ϕ) at z = 1.5 m (see color version in chapter 6, Figure 14e)

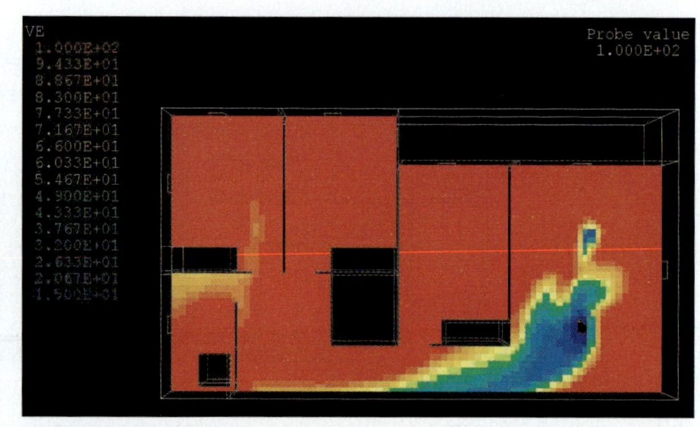

Figure 45 Ventilation effectiveness contours at z = 1.5 m

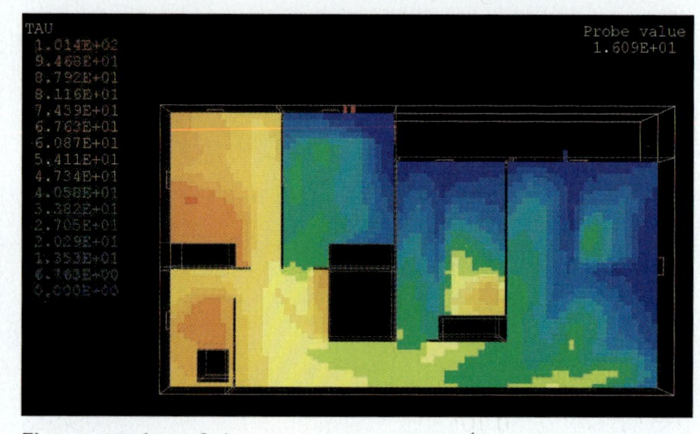

Figure 46 Age of air contours at z = 1.5 m (see color version in chapter 6, Figure 14f)

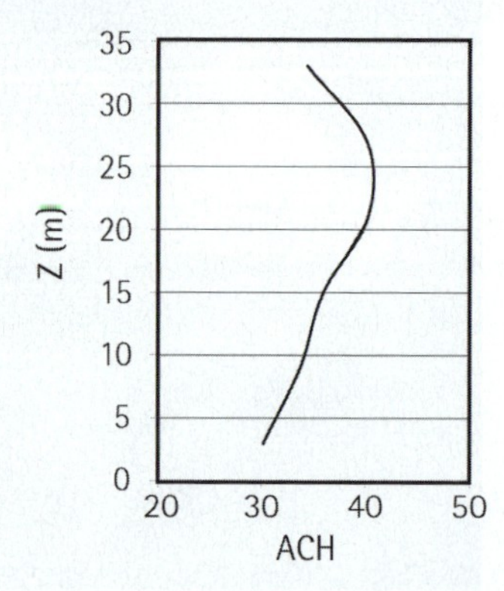

Figure 47 ACH versus height for the single unit (G)

airflows during the day. In practice, energy-conscious occupants who have air-conditioning should use natural ventilation when possible and use air-conditioning sparingly to maintain indoor conditions at the upper bound of what they consider to be comfortable.

SUMMARY

This project sought to propose a mid-rise building design that was both energy-efficient and unique at an urban scale. Initial design options all proposed to reduce the appearance of building massing, utilize natural ventilation, and implement shading techniques to reduce the energy consumption of the building, as well as provide recommendations for building feature and detailed specification upgrades. Through the initial analyses of the energy-reduction features to ventilation studies analyzing the temperature and relative humidity of typical units, the design team was able to create a design that improved the occupant comfort levels and microclimate conditions at each building while reducing the overall need for mechanical systems. These studies further highlighted the ability to increase building standards through the integrated use of energy-efficient detailing.

The design process was completely holistic in that both design and technical findings were carried out simultaneously, creating a dynamically informed process that was true to the contextual urban conditions, the natural environment, and the needs of the occupants.

ACKNOWLEDGMENTS

Tongji University provided invaluable assistance with client coordination and was instrumental in the design development of this project.

The *Natural Ventilation Studies* section of this chapter is adapted from: Allocca, C. 2000. Sustainable Buildings for China: Indoor Airflow Analysis of Duplex and Single Apartment within Shanghai Building; paper written for the Sustainable Urban Housing in China Project.

The *Energy Studies* section of this chapter is adapted from: Norford, L., B. Dean, L. Glicksman, A. Scott and Q. Chen. 2000. Energy Simulation Results to Better Design High-Rise Residential Buildings in Shanghai China; paper written for the Sustainable Urban Housing in China Project.

REFERENCES

AEC. 2006. *VisualDOE 4.0.* Architectural Energy Corporation. <http://www.archenergy.com/products/visualdoe>

Carrilho da Graça, G., Q. Chen, L. Glicksman and L. Norford. 2002. Simulation of Wind-Driven Ventilative Cooling Systems for an Apartment Building in Beijing and Shanghai. *Energy and Buildings* 34(1): 1-11.

EnergyPlus Weather Data. 2006. Shanghai/Hongqiao. Available from <http://www.eere.energy.gov/buildings/energyplus/cfm/weather_data.cfm>.

CHAPTER THIRTEEN

CASE STUDY FOUR - SHENZHEN WONDERLAND PHASE IV

Juintow Lin

PROJECT DESCRIPTION

Shenzhen, located in southern China, is in a region with a warmer climate than Beijing or Shanghai. Vanke Architecture Technology Research Center in Shenzhen invited MIT to work on the site plan and architectural design for energy-efficient housing for phase IV of the Shenzhen Wonderland development (Figure 1). When this project was initiated in 1999, the first two phases of housing were already completed, while phase III and IV were in design development. The program for phase IV included 259 units on 20,000 square meters with a floor area ratio (FAR) of 1.45–1.65, for a total of 30,000 square meters of gross internal area. Unit areas varied from 45 to 135 square meters, and building heights ranged from five to eleven stories.

Several factors influenced the final scheme designed by MIT. These included energy recommendations based on research of energy saving scenarios as well as urban design concepts, such as an interest in providing a variety of building and unit types. Due to the large size of the development, which was approximately five acres, the team sought to create clusters of areas, so that occupants would have a sense of neighborhoods within the context of a large gated development. The existing Shenzhen Wonderland development has ample landscaped areas, which provide places for the community to enjoy time outdoors (Figures 2-4). First-floor units have garden

L. Glicksman and J. Lin (eds), Sustainable Urban Housing in China, 182-211

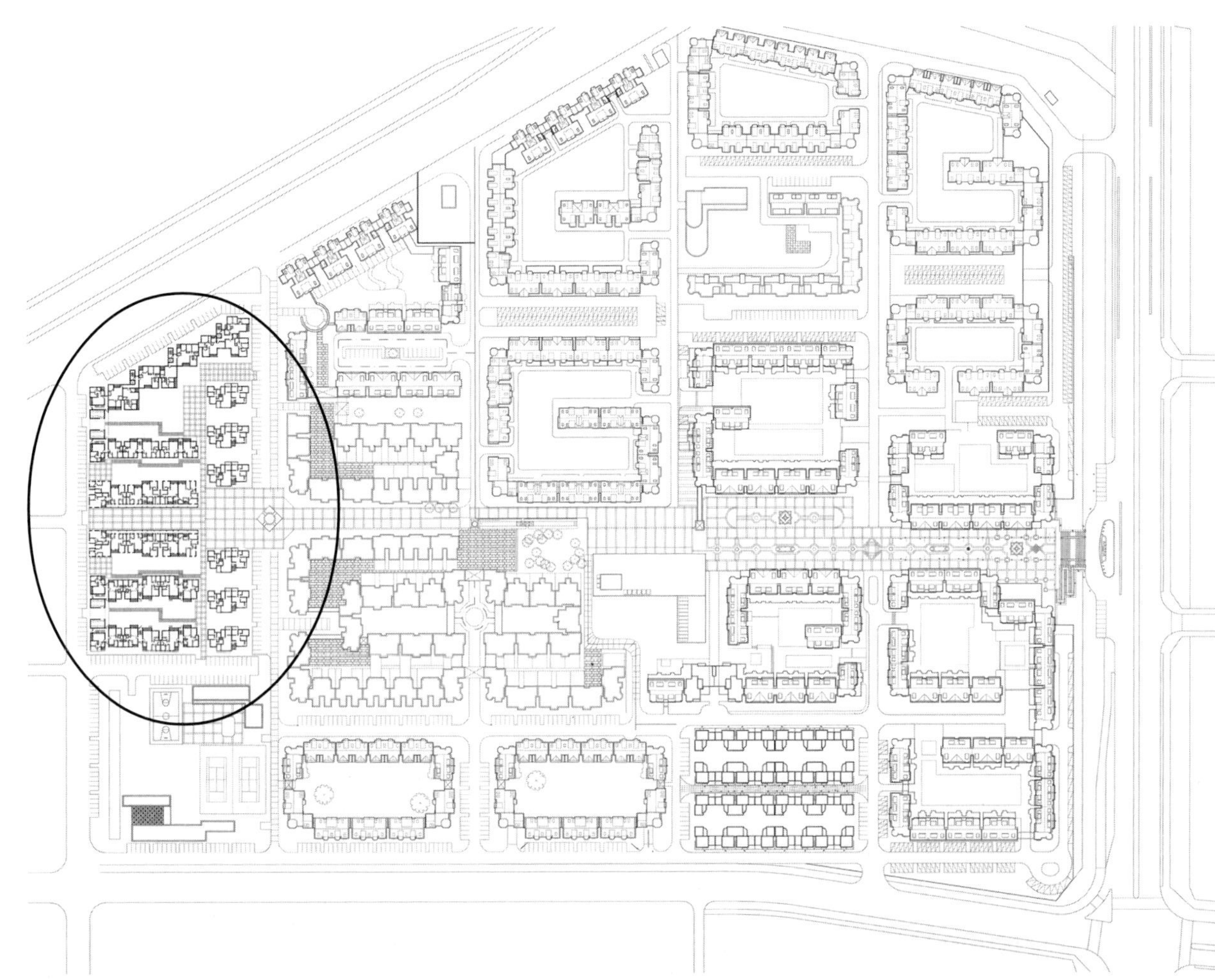

Figure 1 Site plan of Shenzhen Wonderland development; preliminary site layout for phase IV is shown circled

Figure 2 Image typical of the Shenzhen Wonderland development

Figure 3 First-floor unit garden patios

Figure 4 Covered outdoor walkways provide shaded areas

Figure 5 Preliminary site plan of phase IV

patios with direct connections to the outdoors. The landscape in general was designed with a great attention to detail across an inclusive range of scales.

As with previous case studies, the goal was to minimize the energy use of mechanical equipment and improve indoor air quality. A secondary goal of the MIT team was to create a site design (Figure 5) that encouraged more community interaction than the designs for phases I and II.

TECHNICAL RECOMMENDATIONS

The design of Shenzhen Wonderland phase IV began with in-depth technical studies analyzing energy-saving scenarios. These preliminary technical simulations, developed using the DOE-2 building energy simulation program (AEC 2006), shading studies, CFD for prediction of airflow, and sound engineering information, were based on detailed analyses of site conditions and building design. Results from the analyses were summarized to include recommendations regarding building orientation, layout, façade design, ventilation, noise, thermostat, and air conditioners. Annual energy usage based on the various simulations was compared to the base case (Table 1), and the findings are summarized in Table 2, as well as in the text that follows. These findings showed that solar heat through windows dominated summer heat gains.

The technical recommendations for this project were further reinforced by encouraging residents to follow guidelines on temperature controls, shading devices, and natural and mechanical ventilation. These guidelines were summarized into a residential user manual, described later in this chapter.

Passive Solar Building Design

Solar energy travels through the windows of a building, increasing the energy used by air conditioners and degrading occupant comfort. An ideal passive solar building minimizes internal energy use by optimizing its design to suit the annual path of the sun and the local climate. A primary goal in energy-efficient design for buildings in warm climates such as Shenzhen is that they must minimize solar heat gain during much of the year.

Based on solar energy calculations for each window direction (north, south, east, and west), recommendations were made regarding optimum window shape and orientation, as well as shading options. The passive design incorporated knowledge of the sun's path, including sun angles during different days and azimuth angles (sun-path from a bird's-eye view) throughout each day of the year. Figure 6 shows elevations at noon on three dates for Shenzhen (latitude 22.18°N), and Figure 7 shows azimuth angles for Shenzhen throughout the day on those dates.

People living in "northern" cities immediately note that the sun looks very different in Shenzhen when compared to cities such as Boston or Beijing. At 42°N latitude (Boston), the maximum noontime sun elevation is 72° in summer, and the minimum noontime elevation is 25° in winter. The sun is north of a building (absolute azimuth angle greater than 90°) only during early morning and late afternoon in June and July. However, in Shenzhen, at 22° N latitude; the maximum sun angle is 91° in summer; the minimum noontime elevation is 45° in winter; and the sun is north of a building throughout the entire day from 4 June to 10 July (and during much of the day the rest of summer).

In addition to understanding sun angles and paths, one must also consider the impact of solar energy contact. Figure 8 illustrates the amount of solar energy that hits each surface (north, south, east, and west). The

Building Component	Specification
Floor area	6 m x 12 m
Orientation	North/south facing façades
Wall type	Concrete block, no insulation
Window type	Clear single-pane, aluminum frame
Window area	40% window-to-wall area
Overhang type	No overhang
Electric heat pump	10 SEER / 7.8 HSPF
Heating set point	20°C
Cooling set point	24°C
Ventilation	No ventilation

Table 1 Specifications for base case simulated building

Window solar heat	40 percent
Infiltration	15 percent
Occupants	15 percent
Window conduction	8 percent
Wall conduction	8 percent
Lights, cooking, appliances	14 percent

Table 2 Breakdown of summer heat gains in base case dwelling

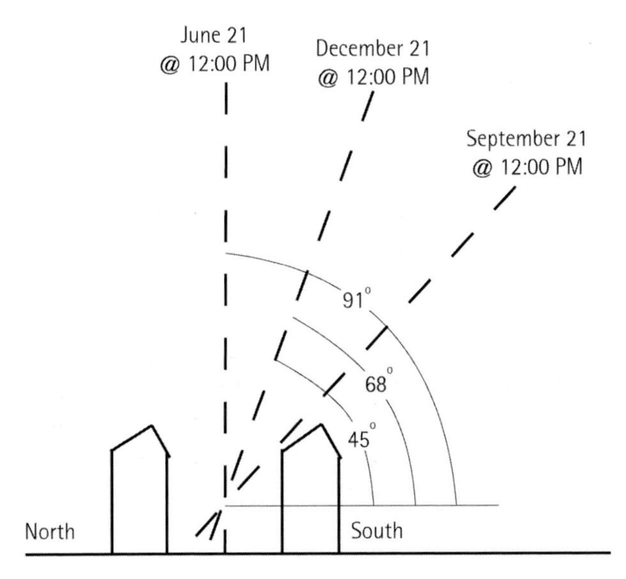

Figure 6 Sun angles at solar noon on 21 June, 21 September, and 21 December for Shenzhen

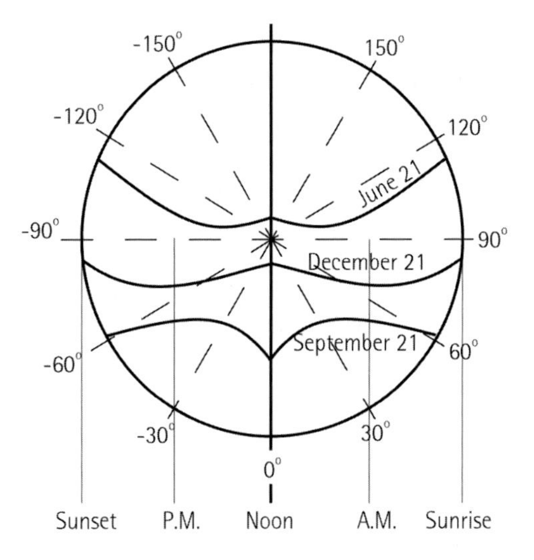

Figure 7 Azimuth sun angles throughout the day on June 21, September 21, and December 21 for Shenzhen

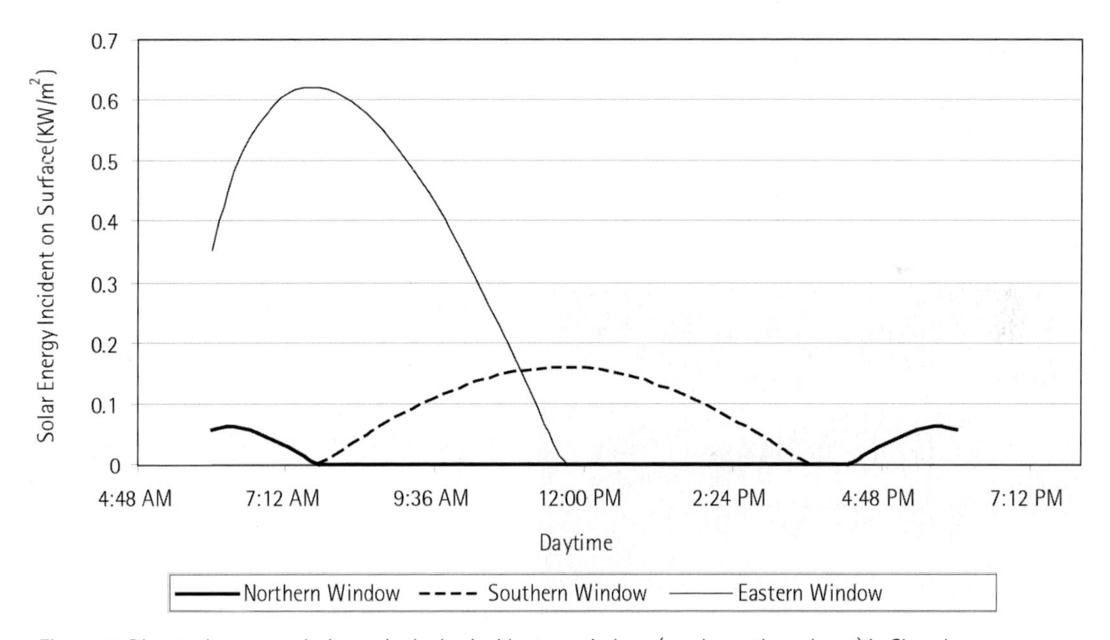

Figure 8 Direct solar energy during a single day incident on windows (north, south, and east) in Shenzhen on 21 August as a function of daytime (the values for a western window are the mirror image of the eastern window)

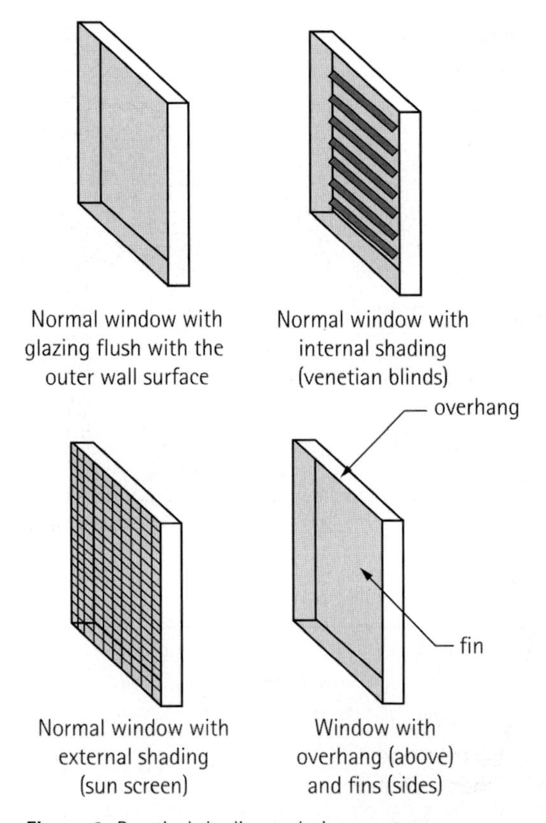

Figure 9 Practical shading techniques

Figure 10 This MIT building uses inset windows to create deep overhangs and fins

plot shows the solar power per area of window (kW/m^2) incident on each surface on 21 August. The western window curve (not shown) is a mirror image about solar noon of the eastern window curve (shown). Note also that near 21 June there will be no southern exposure and more northern exposure, but near 21 December, the opposite is true.

Shenzhen's climate is hot and humid during most of the year. Therefore, minimizing solar heat gain through windows, especially during the warmest months, will both reduce air-conditioning loads and improve the energy efficiency of a building. Eastern and western windows contribute significantly to solar gain during the summer months. Southern windows have very large solar gains even during the winter months. Opening windows during winter (when air is cooler than comfort levels) balances the excess wintertime solar gains and can easily solve this problem. Buildings should have minimum window (and surface) exposure on their eastern and western faces. Therefore, a slender building with its long sides facing north and south was desirable.

As orientation affects site planning, proper consideration of shading techniques will clearly influence architectural design. Two fundamental shading strategies exist for reducing solar energy that passes directly into a building. The first method places obstructions between the sun and the window (e.g., trees, external shades, or adjacent buildings). The second method uses the window itself to reflect excess sun (e.g., reflective coatings or interior blinds). The four practical techniques illustrated in Figures 9 and 10 to shade windows are: exterior shading screens (similar to insect screens), interior shades (such as venetian blinds), window overhangs (above the window), and window fins (on the sides of the window). Other practical shading techniques use vegetation, adjacent structures, or other natural obstructions.

The effectiveness of a window's shading properties is measured by the window's shading coefficient. The amount of sun energy (Q in Joules) that a window lets in is given by:

$$Q = 0.87 * SC * E,$$

where E is the solar energy hitting the window (in Joules) and SC is the window's shading coefficient. The shading coefficient compares how much sun a specified glazing allows through compared to a standard single layer of clear glass without shading. For example, a clear piece of glass has a shading coefficient of 1.0, while an opaque wall has a shading coefficient of zero. Shading coefficient values are derived experimentally. Overhangs and fins cannot be quantified using shading coefficients because the amount of shading they provide changes dramatically throughout the day and year.

To find the best method for shading windows in Shenzhen, we studied eight shading cases (using combinations of the above techniques). Table 3 describes each case while Figure 11 show the performance of each case throughout a year for each window direction (north, south, and east). The total solar energy gain per area of window (MWh/m^2) for the year (and for the nine-month period excluding winter shown in Figure 12) further shows how each design performs.

The graphs shown in Figures 13 through 15 present the results of each case as the solar energy per area of window area (kJ/m^2) over one day (the twenty-first day of each month) for a northern window, a southern window, and an eastern-western window. Northern windows saw minimal solar gains, and only during the warmer months (April through August). Southern windows saw large gains during winter months but

Case	Shading Coefficient	Overhang %	Fin %	Description
0 (base)	1.00	0	0	3 mm clear, single-glazed
1	0.58	0	0	3 mm clear, single-glazed with venetian blinds
2	0.28	0	0	3 mm clear, single-glazed with exterior shade or 6 mm reflective double-glazed
3	1.00	30	0	Base case with overhang extending 30% of window height
4	1.00	30	30	Base case with overhang extending 30% of window height and fin extending 30% of window height
5	1.00	100	0	Base case with overhang extending 100% of window height
6	1.00	100	100	Base case with overhang extending 100% of window height and fin extending 100% of window height
7	1.00	0	30	Base case with fin overhang extending 30% of window width
8	1.00	0	100	Base case with fin overhang extending 100% of window width

Table 3 Description of window shading cases

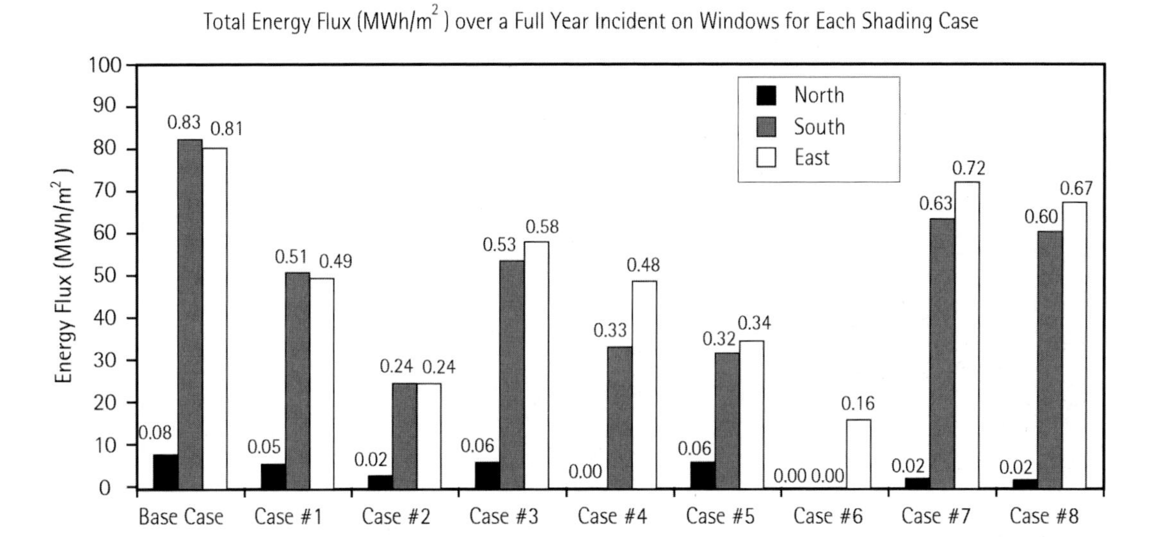

Figure 11 Annual energy gain per unit area of window

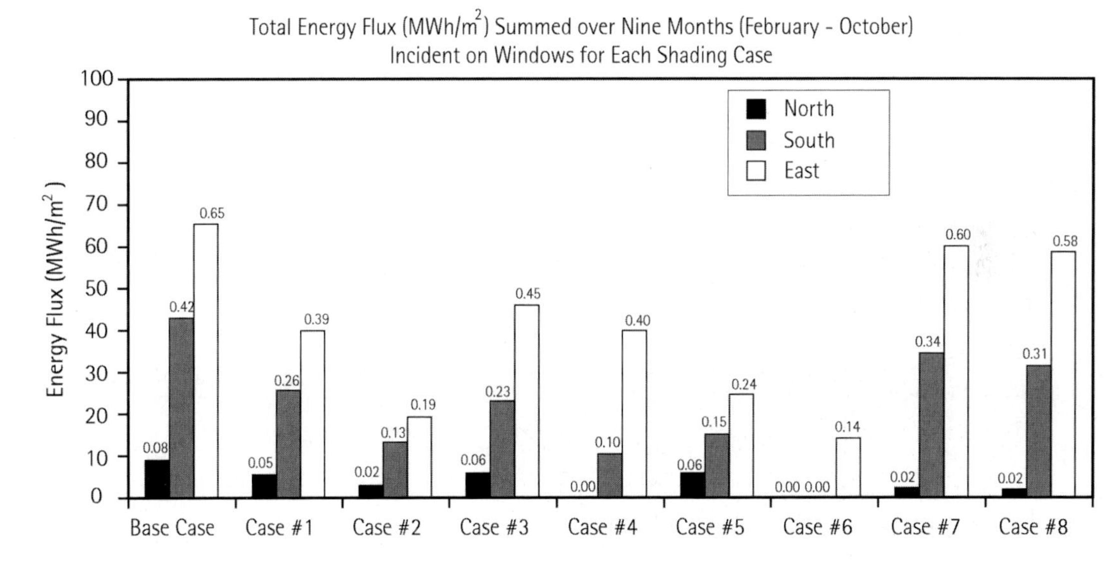

Figure 12 Nine-month energy gain per unit area of window

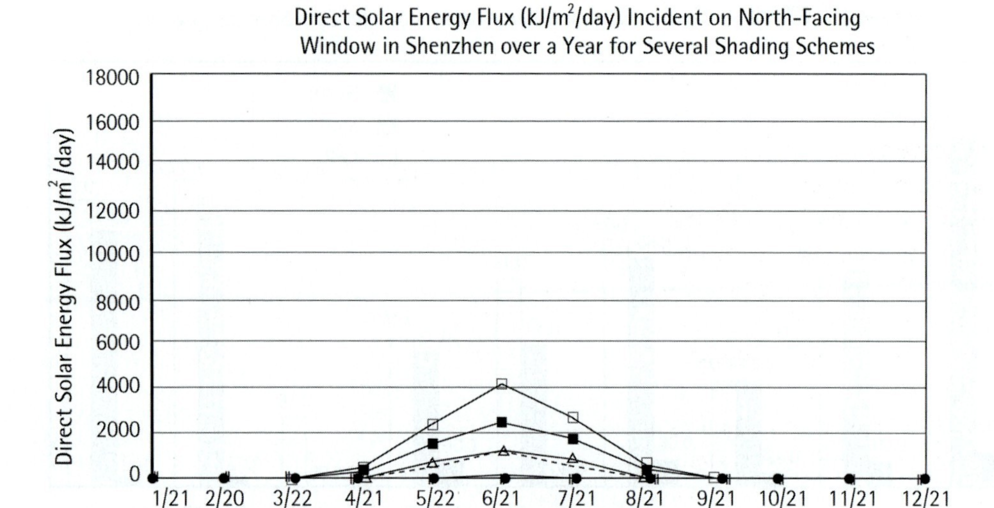

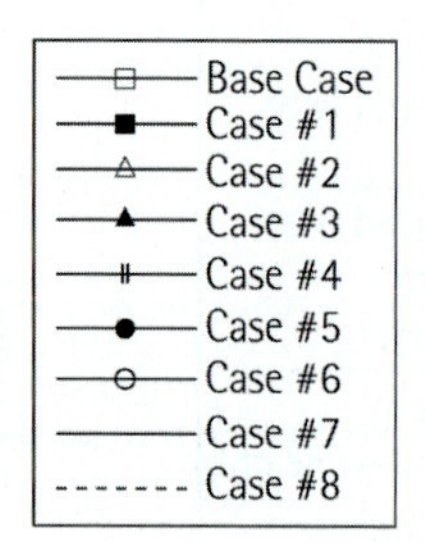

Figure 13 Daily average solar energy through north-facing windows for each month and for a variety of window shading cases (refer to Table 3 for description of cases)

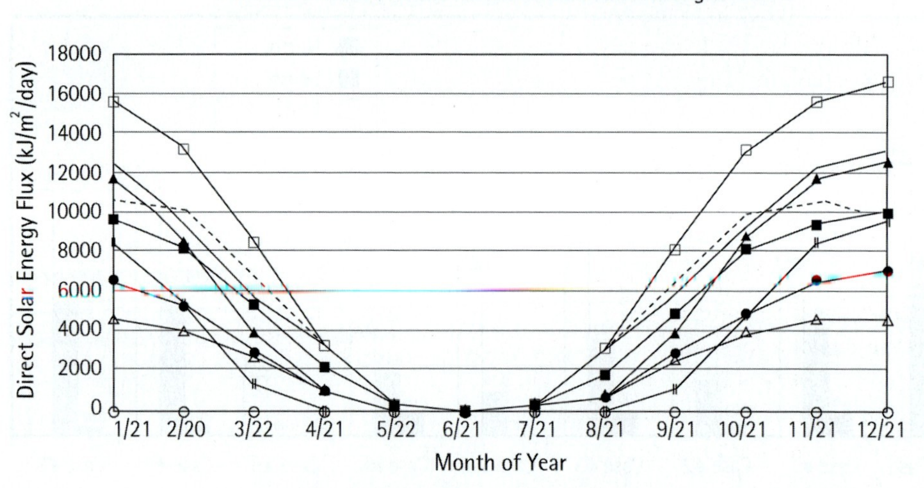

Figure 14 Daily average solar energy through south-facing windows, for each month and for a variety of window shading cases (see Figure 13 legend and Table 3 for description of cases)

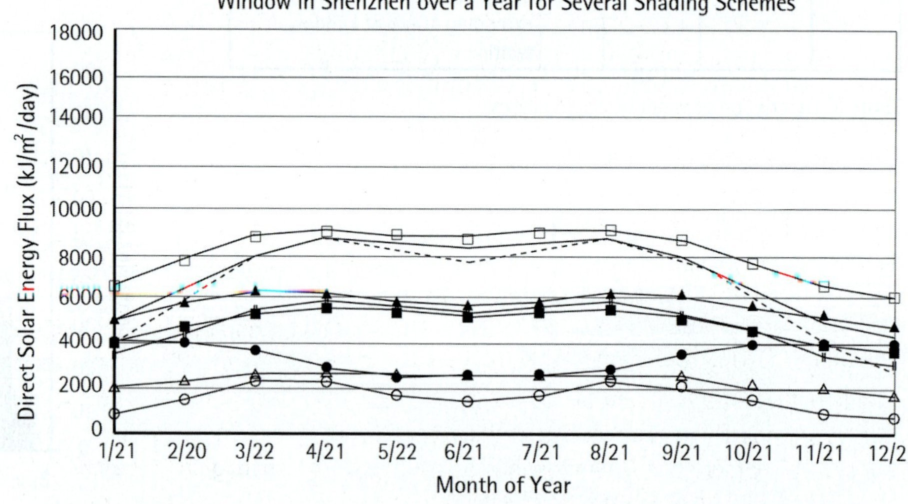

Figure 15 Daily average solar energy through east- or west-facing windows for each month and for a variety of window shading cases (see Figure 13 legend and Table 3 for description of cases)

relatively little during the summer. East-west windows saw relatively even gains throughout the year.

Recommendations for Windows and Walls

As shown in Table 2, solar gain through windows accounts for approximately 40 percent of the total cooling energy load for a base case unit in Shenzhen. Over a year, the energy from direct sunlight per area incident on all building surfaces (north, south, east, and west) totaled nearly 2.53 $MWh/^2$. Northern windows received 3 percent of the total direct annual solar energy per area, while southern windows received 33 percent, and western and eastern facing windows each received 32 percent (64 percent combined). As a result, it was desirable that building designs minimized western and eastern exposures in order to maximize energy-efficient designs. Recommendations for window design in Shenzhen follow in the next several paragraphs.

Northern windows should have small overhangs and vertical fins, or architectural projections, for the purpose of shading. Although shading the northern windows does not have a large impact on energy savings (less than one percent annually), it will greatly improve comfort during the summer months when the sun is to the north. Overhangs should be sized with a depth equal to ten percent of the window height and fins with depths equal to ten percent of the window width. This shading plan will block 100 percent of direct sunlight from August to April, 70 percent in May and September, and 50 percent in June.

Southern windows should have overhangs and fins. Shading the southern window has a large impact on energy savings. The recommended overhang has a depth equal to 20 percent of the window height and fins with depths equal to 10 percent of the window width. This shading plan will block 100 percent of direct sunlight from April through August, 60 percent in March and September, 40 percent in February and October, and 30 percent from November through January.

Western and eastern windows must be "self-shaded" to block late-afternoon and early-morning sunlight. Overhangs and fins are ineffective because the sun angles are so low in the morning and afternoon. As shown in Figure 15, the only overhang and fin configuration that reduces energy levels more than the solar heat gain coefficient described in Case #2 is overhangs and fins at 100 percent of the window width and height, or Case #6. This is impractical, so we suggest reducing the solar heat gain coefficient (SHGC) of the window to less than 0.4. This may be achieved by using double-pane windows with low-e or reflective coatings or mounting an exterior screen (similar to an insect screen) on a standard single-pane window. Similar to the Shanghai case study, a shading worksheet was prepared for designers to test various combinations of window shading techniques. Figure 16a shows the difference between a window facing south in June and September. The June window receives no solar radiation. Figure 16b shows the difference between overhangs and fins sized at 100 percent and changing the SHGC to 0.4 on solar radiation through a window in August.

Walls with western exposure should have R=1.5 Km^2/W or greater (e.g., block wall with six centimeters of glass fiber insulation inside the weather barrier). This will help keep the west-facing units cooler late in the day during the hot summertime months. This upgrade will increase the comfort of occupants late in the day during the warmest months, but does not save much energy over the entire year (less than one percent).

The color of the exterior walls was found to have negligible impact (less than one percent) on the annual energy use (comparing red brick with white paint). Therefore, the aesthetics of the color scheme was more important than the potential energy savings impact.

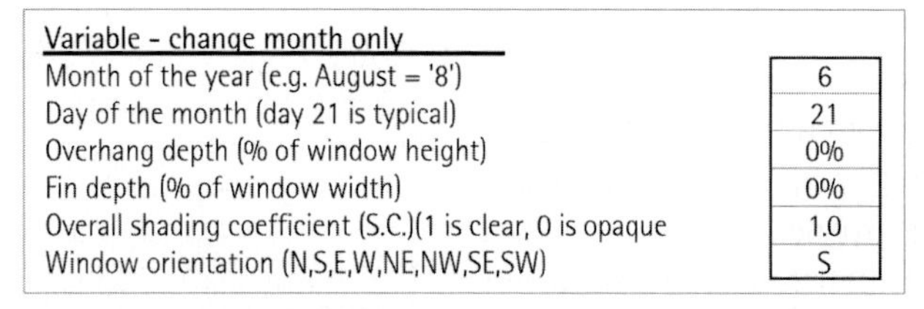

Variable - change month only	
Month of the year (e.g. August = '8')	6
Day of the month (day 21 is typical)	21
Overhang depth (% of window height)	0%
Fin depth (% of window width)	0%
Overall shading coefficient (S.C.)(1 is clear, 0 is opaque	1.0
Window orientation (N,S,E,W,NE,NW,SE,SW)	S

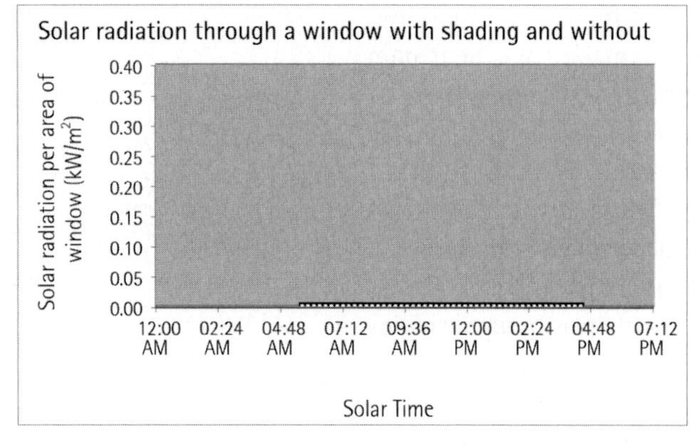

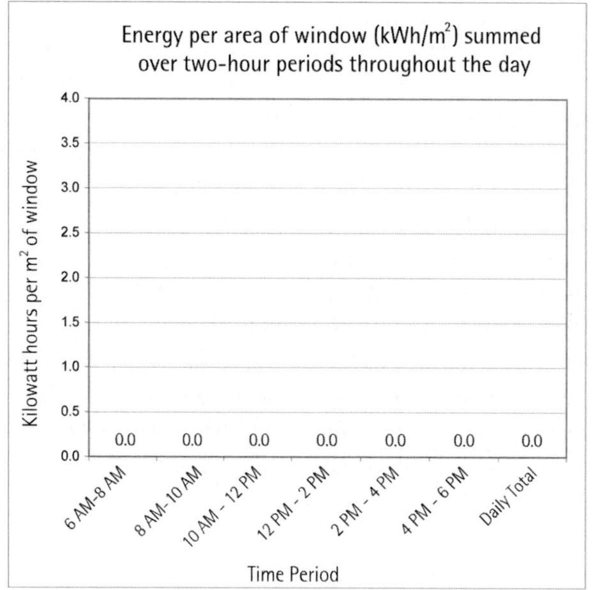

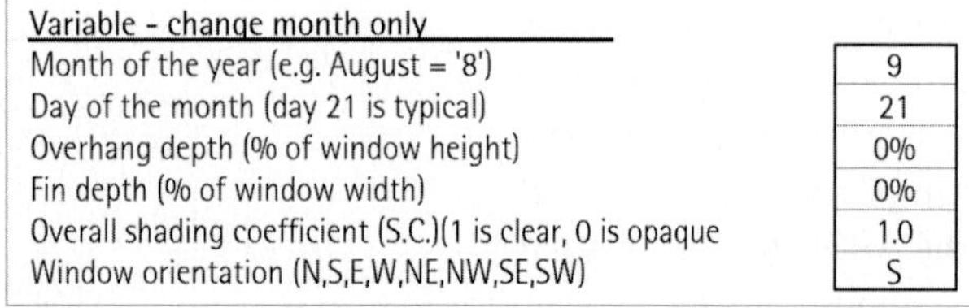

Variable - change month only	
Month of the year (e.g. August = '8')	9
Day of the month (day 21 is typical)	21
Overhang depth (% of window height)	0%
Fin depth (% of window width)	0%
Overall shading coefficient (S.C.)(1 is clear, 0 is opaque	1.0
Window orientation (N,S,E,W,NE,NW,SE,SW)	S

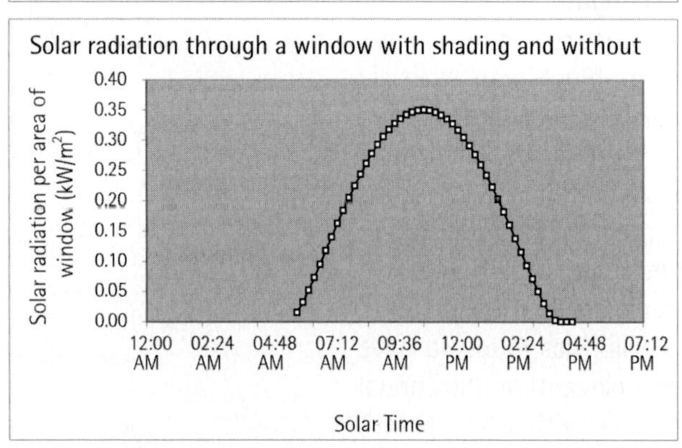

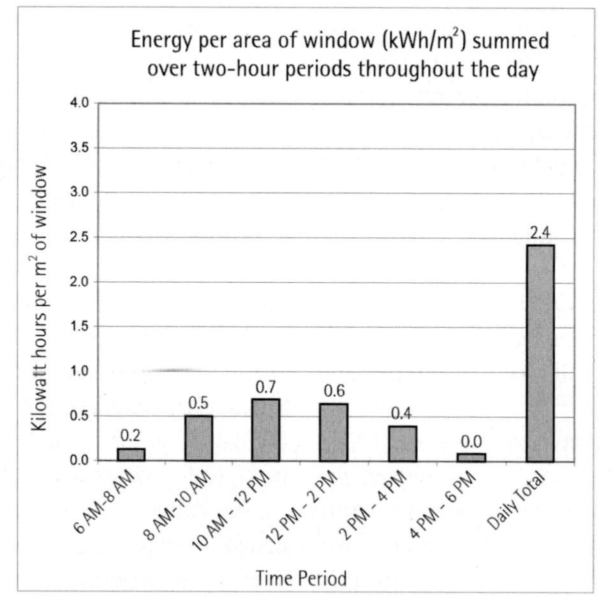

Figure 16a Shenzhen shading calculations worksheet showing solar radiation and energy per area of window for a south-facing window in June and September

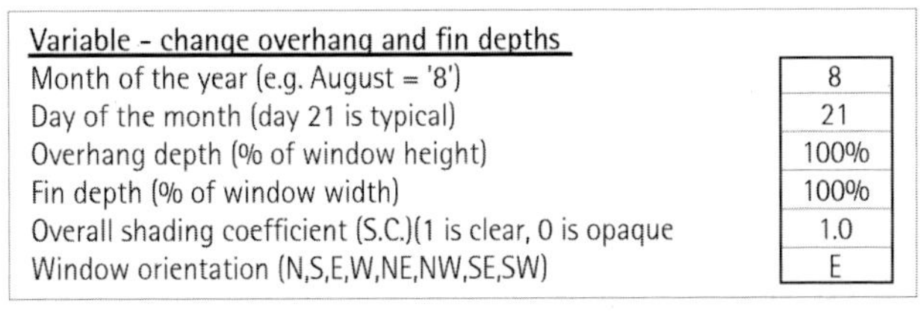

Variable - change overhang and fin depths	
Month of the year (e.g. August = '8')	8
Day of the month (day 21 is typical)	21
Overhang depth (% of window height)	100%
Fin depth (% of window width)	100%
Overall shading coefficient (S.C.)(1 is clear, 0 is opaque	1.0
Window orientation (N,S,E,W,NE,NW,SE,SW)	E

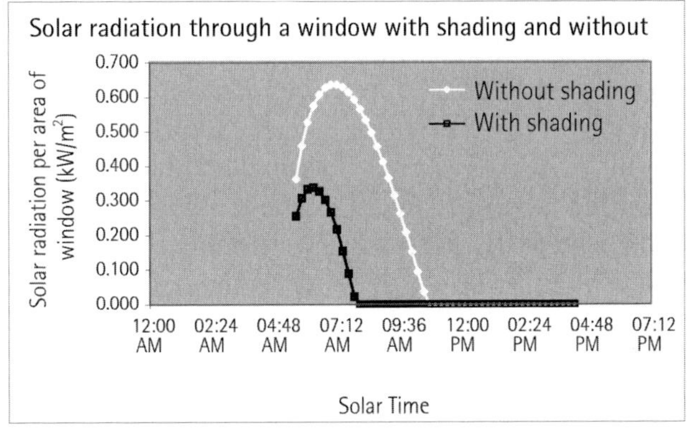

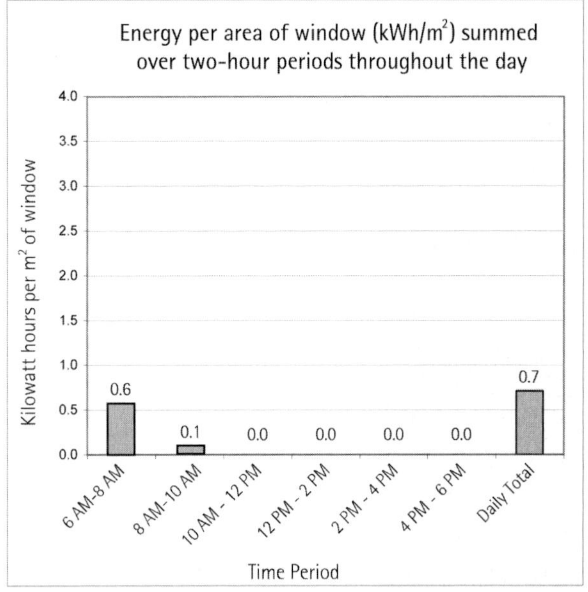

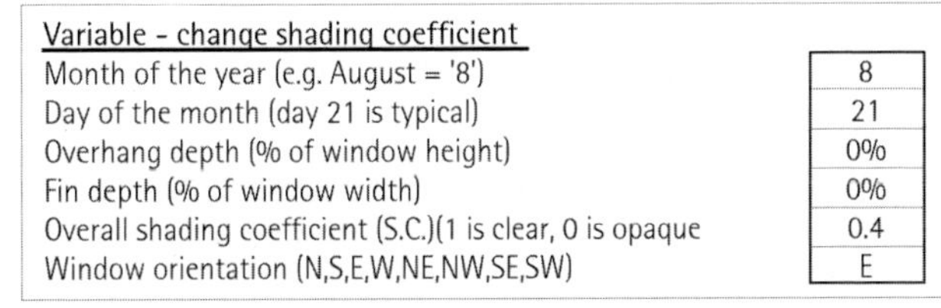

Variable - change shading coefficient	
Month of the year (e.g. August = '8')	8
Day of the month (day 21 is typical)	21
Overhang depth (% of window height)	0%
Fin depth (% of window width)	0%
Overall shading coefficient (S.C.)(1 is clear, 0 is opaque	0.4
Window orientation (N,S,E,W,NE,NW,SE,SW)	E

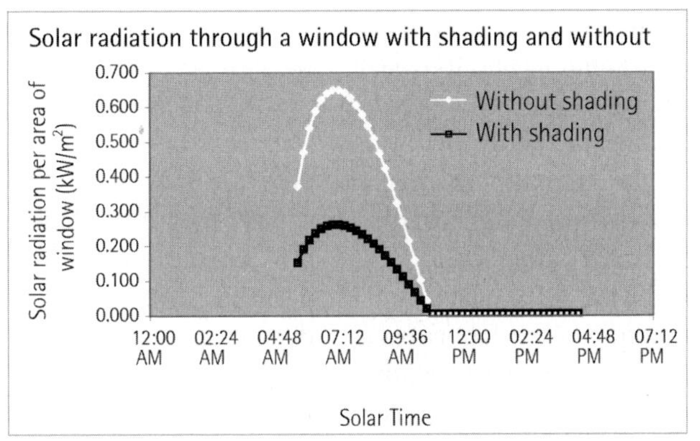

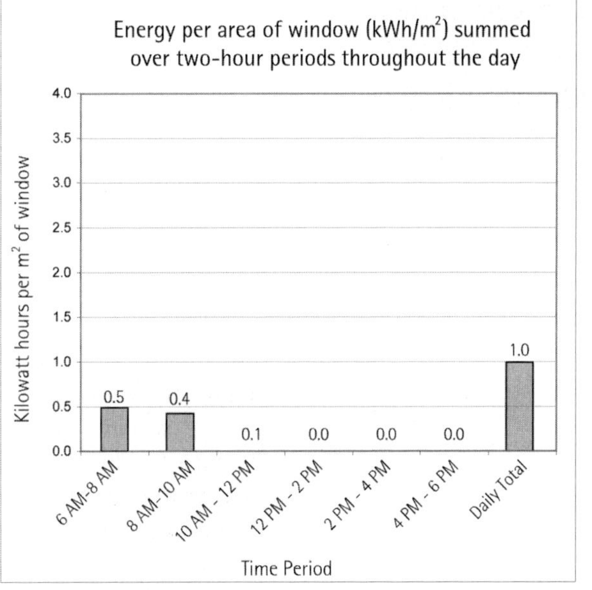

Figure 16b Shenzhen shading calculations worksheet showing results of overhangs, vertical fins, and window shading coefficients on solar radiation and energy per area of window for a east-facing window in August

Figure 17 Solar water heaters mounted on roofs in Xi'An

Infiltration

Infiltration accounted for approximately 15 percent of the total cooling energy load for a base case unit. Therefore, construction and detailing techniques should ensure that infiltration is less than 0.35 ACH in each unit when air-conditioning is used. All windows, doors, fans, and cracks should be sealed thoroughly to reduce leak. One method to test infiltration values is to execute a blower door test on a random sample of units.

Roof

The roof should be insulated with at least R=3.67 Km^2/W. For example, 14 centimeters of typical fiberglass or comparable insulation inside the weather barrier would accomplish an R-value of 3.67. This improvement only provided less than 100 kWh (less than 3 percent) of energy savings over the year, but should be helpful in keeping the top-level units cooler year-round when air-conditioning is not used.

Solar hot water collectors (Figure 17) could provide domestic hot water to many units (if not all), while simultaneously removing excess heat from the rooftop. Solar hot water collectors, if used, should be integrated architecturally with aesthetic considerations as well. The roof pitch in most designs was set at 22° toward the south because this was the optimum incline to collect the sun throughout the entire year for Shenzhen. Solar hot water heaters use direct sun radiation to collect heat and transfer it to water for use in the home. Several designs exist, ranging from glass plate collectors to evacuated tube solar collectors. The former were more practical for this application. They could be custom built and integrated into a roof structure or pre-fabricated in a factory and installed on the roof after construction.

The U.S. Department of Energy states that every square meter of solar collector area on a roof will heat roughly 60 liters of water per day throughout the year in a southern climate (average throughout the year) (U.S. DOE 2006). Solar hot water heaters installed on every roof would greatly contribute to a more sustainable building. Although they may only supply a portion of the total hot water demand, the solar water heaters also function to remove heat from the roof area, leaving top-floor units more comfortable.

Ventilation

Energy simulations showed that proper ventilation during spring and fall could save up to 30 percent in annual energy usage if used at the correct times. Wind-driven ventilation alone will give between 0 and ~30 ACH, depending on the direction of the wind during spring and fall months (shown in the wind rose in Figure 18) ad the unit's location (Figure19 and Table 4).

The wind rose indicated that prevailing winds are primarily from the southeast and northeast. Therefore it is important not to block the eastern side of the site; instead it should be open to allow wind movement through the site. Wind should flow freely through courtyards, balconies, and units in order to keep people more comfortable during the hot summer months through the use of natural ventilation.

The site design for phase IV left large openings to the east, for winds to enter. The schematic site plans for phase III, located directly to the east indicated an east-west orientation for buildings, and allowed for winds to penetrate well into phase IV. Revised plans, however, showed that several buildings in phase III were redesigned and blocked the wind from entering phase IV. This is discussed further in the section on computational fluid dynamics, *CFD Simulations*.

Also, in light of the benefits of the prevailing easterly wind flow, phase IV should not block the wind from the future development to the west. The design of phase IV created several openings in the west side. If no development was planned to the west of phase IV, site-level ventilation could have been improved by closing off the western openings.

Natural ventilation is a valuable tool for an energy-efficient residential unit. Unfortunately, wind is often unreliable, and is also often inhibited by adjacent buildings. Large exhaust fans should therefore be installed in each unit to limit the need for air-conditioning when wind alone does not produce enough natural ventilation. These fans would provide an average hourly air change rate of 36 (for a 100 m^2 x 3 meter high unit, that means 3 m^3/s through the fan). The cooling load was calculated as an average for the six months of March through May and September through November.

A typical exhaust fan must be about 0.5-0.7 minimum diameter to provide the required airflow rate for a pressure loss of 150 Pa (20 Pa static pressure). We suggested placing the exhaust fan in the kitchen of most units for aesthetic reasons. In units located along the railroad tracks, the placement of fans on the side exposed to the railroad was avoided, because they could leak noise into the units. For top-floor units, we suggested placing the fan in the wall near the top of the cathedral ceiling to expel the hot air. In most installations, and for most units, the total static pressure loss was near 20 Pa.

Natural and fan-assisted ventilation provided fresh air. It was crucial to use ventilation at the proper times (when the outside air is cooler than the cooling set point temperature) in order to save energy. This is discussed further in the residential user manual. To ensure good indoor air quality during the summertime (when air conditioners are used and windows are closed), occupants should ventilate for a few minutes each morning using the exhaust fan.

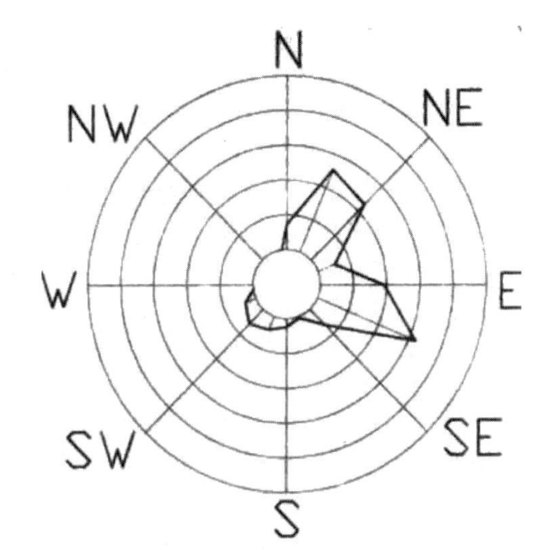

Figure 18 Shenzhen wind rose

Apt. #	Delta P (Pa)	Q (m^3/s)	ACH
1	0.0	0.00	<1
2	0.7	0.38	13
3	3.0	0.79	28
4	0.3	0.25	9
5	0.3	0.25	11
6	0.0	0.00	<1
7	2.0	0.65	23
8	0.7	0.38	10
9	1.0	0.46	16
10	0.7	0.38	14
11	0.3	0.25	11
12	0.0	0.0	0.0
13	0.0	0.0	0.0
14	0.0	0.0	0.0

Table 4 Phase IV units - ACH levels when east wind is blowing at 3 m/s

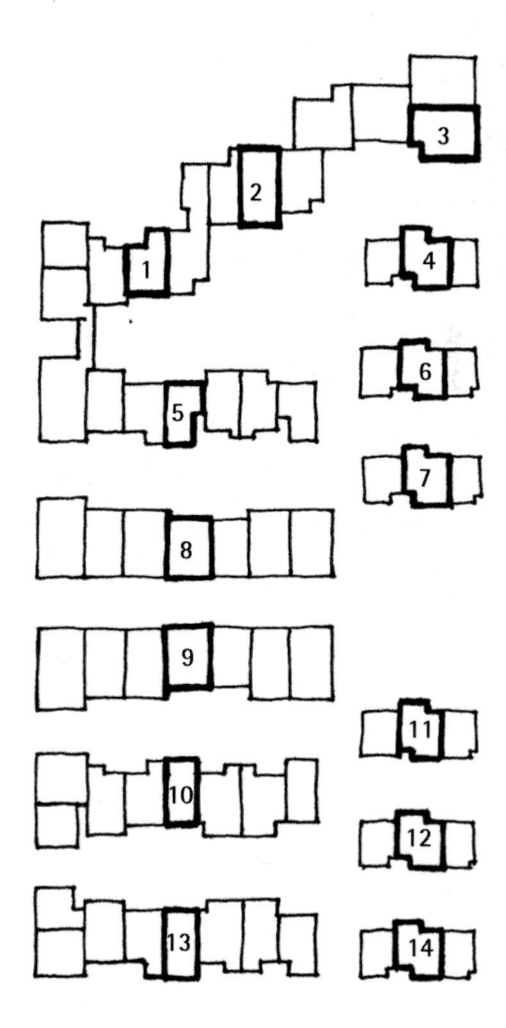

Figure 19 Key of selected phase IV units simulated in ventilation studies

Thermostat

It was extremely important for the occupants to be aware of the temperature and humidity both indoors and out so that they could operate the windows, fan, and air conditioner efficiently. A residential user manual was created in order to guide occupants on how to use natural ventilation, mechanical ventilation, and air-conditioning based on the temperature difference between indoors and outdoors. It was recommended that all the units have digital, programmable, clock-thermostats that monitor and display the indoor and outdoor temperature.

Air Conditioner

Occupants should be encouraged to buy efficient air conditioners. The units should have a minimum COP of three (COP = heat extracted (W) / electricity input (W)). Air conditioners that have too much capacity (W) operate inefficiently. Energy calculations indicated that air conditioners should be sized at ~100 W/m^2. The required capacity was based on floor area of the unit, insulating qualities of the windows and walls, and the degree of shading of the windows from direct sun during the hottest part of the day.

Noise

Of primary concern on the north side were the railroad tracks and the large amount of consequent noise generated. The north edge of phase IV site bordered the railway to the north.

People expect an outside noise infiltration of less than 35 dBA in bedrooms, less than 40 dBA in living and dining rooms, and less than 45 dBA in bathrooms and kitchens (Figure 20). Trains produce ~90 dBA of noise, mostly at low frequencies. A traffic barrier can provide up to 15 dBA noise reduction; the distance away from the tracks (~20 meters) will give 6 dBA reduction; and the proper window/wall/door combination can provide 20-40 dBA reduction when doors and windows are closed (Figure 21). A 20 dBA reduction is possible if doors and windows have gasket seals, and up to 40 dBA reduction is possible if all noise recommendations listed below are used.

The northern row of buildings could not avoid the train sounds, but could block the sounds from reaching the rest of the site. In order to accomplish this, no openings were placed between the courtyard and the train tracks through the northern row of buildings. Even a small opening in the northern row of buildings, such as a narrow walkway, would greatly increase the noise levels in the courtyard. The northern row of units worked to shield the rest of the site from noise and also protected its own occupants from noise. In the following two paragraphs are the recommendations for all exterior surfaces with an unobstructed view of the railroad tracks.

Windows adjacent to the railroad tracks should have tightly sealing gaskets. These windows should have thick glass (each pane should be at least six millimeters thick) (Table 5). The most important consideration is to eliminate air gaps between windows and the wall. For example, a 1 millimeter x 1 meter air gap would reduce the transmission loss (TL) through a 1 m^2 window from 30 dBA to 26 dBA. If this gap were increased to 1 centimeter x 1 meter, the equivalent TL would further be reduced to 19 dBA.

Exterior doors adjacent to the railroad tracks should be heavy (such as metal with foam core, or 100 percent solid wood), and should have gasket seals around the door and weather stripping at the bottom of the door (Table 6). Again, eliminating air gaps is crucial. In addition to upgrades for the windows along the north wall, the mass of exterior walls adjacent to the railroad tracks should be increased to help reduce noise inside. One

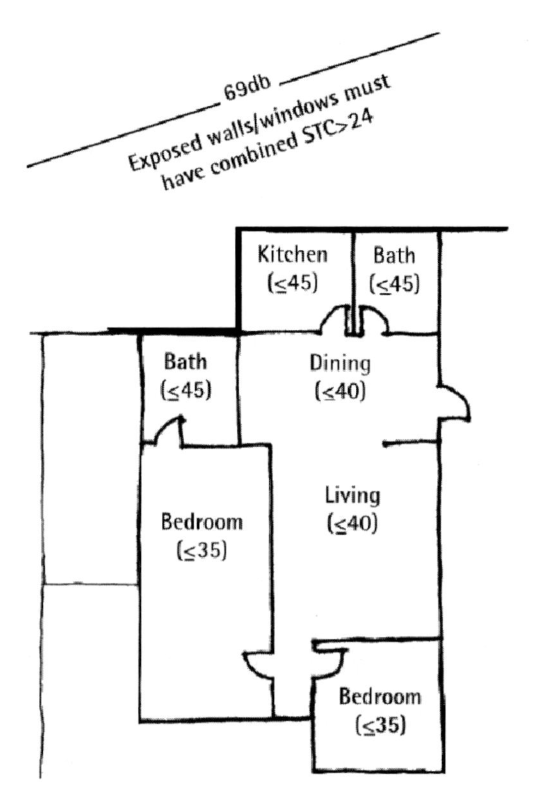

Figure 20 Typical acceptable indoor noise levels (STC stands for sound transmission class)

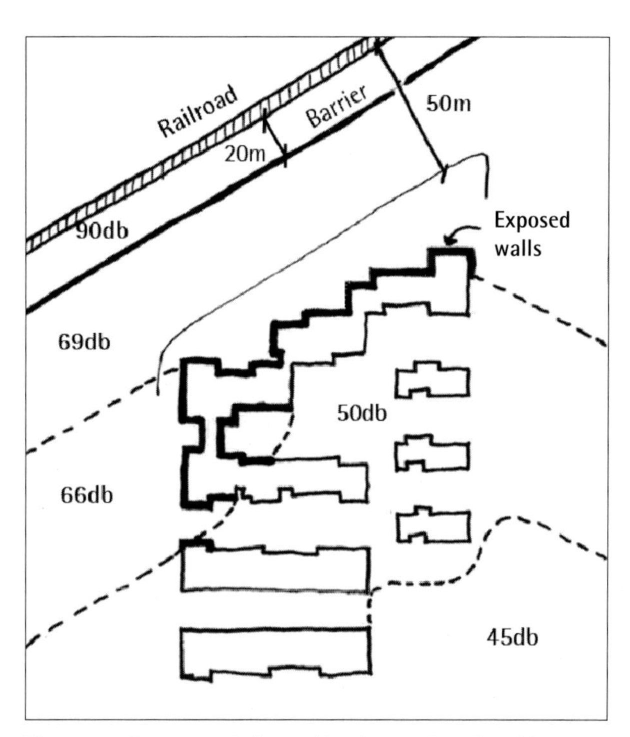

Figure 21 Recommended sound levels near the railroad in Shenzhen Wonderland phase IV. Walls shown with dark lines should follow recommendations for windows, doors, and wall construction.

Window Type	Transmission Loss (dB @ 500 Hz)
Single pane (3.2 mm glass with gasket seal)	28
Single pane (6.4 mm glass with gasket seal)	31
Single pane (13 mm glass with gasket seal)	36

Table 5 Window type versus transmission loss

Door Type	Transmission Loss (dB @ 500 Hz)
Hollow-core metal (no gaskets)	16
Solid-core wood (no gaskets)	26
Solid-core wood (with foam gaskets)	30

Table 6 Door type versus transmission loss

Units Primarily Facing North and South

Building Design Features			
Windows	North	Overhang: 10% of window height Fin: 10% of window width Solar heat gain coefficient less than 0.6 Clear double-pane window (vinyl frame if possible)	**
	South	Overhang: 20% of window height Fin: 10% of window width Solar heat gain coefficient less than 0.6 Clear double-pane window (vinyl frame if possible)	***
	Noise control	Tightly sealed windows (double-pane, non-operable if possible) on walls adjacent to railroad	**
Walls	Insulation	Block wall plus 6 cm (R=1.5 or greater) of insulation on west-facing exterior walls	*
	Noise control	Increase mass of walls adjacent to railroad by filling block wall cavities with dense concrete	**
Infiltration	Blower door test (random sample units)	Less than 0.35 ACH (fresh air by natural or mechanical ventilation); all joints and cracks should be sealed well	***
Roof	Insulation	Have at least 14 cm (R-3.67) of typical fiberglass insulation or comparable insulation rating inside weather barrier to keep the top floor units from being too hot in the summer	***
Door	Noise control	Exterior door adjacent to railroad should be heavy, such as metal with foam core or solid wood and should have gaskets and weather stripping to minimize infiltration of air or noise	**
Building Equipment Features			
Cooling	Air-conditioning	3 COP minimum, higher COP is recommended	**
Ventilation	Off peak ventilation	Closable through-wall exhaust fan (ensure that they are properly sealed to have minimal infiltration when ventilation is not in use). For top floor units, the exhaust fan should be placed near the roof.	***
Thermostat	Programmable digital clock thermostat	Install a programmable digital clock thermostat that controls the air conditioner; install a thermostat that reads indoor and outdoor temperatures	***
Hot water	Solar collector	Significant energy can be collected from the sun to heat the water used for bathing and washing.	*

*** Essential for sustainability ** Recommended feature * Additional feature

Table 7 Technical summary of recommendations for residential units facing north/south

Units Primarily Facing East and West

Building Design Features			
Windows	East and west with overhang	Overhang: 100% of window height Fin: 0% of window width Solar heat gain coefficient less than 0.8 Clear single-pane window (vinyl frame if possible)	***
	East and west no overhang	Overhang: 0% of window height Fin: 0% of window width Solar heat gain coefficient less than 0.4 See window specification sheet for window types	
	Noise control	Tightly sealed windows (double-pane, non-operable if possible) on walls adjacent to railroad	**
Walls	Insulation	Block wall plus 6 cm (R=1.5 or greater) of insulation on west facing exterior walls.	*
	Noise control	Increase mass of walls adjacent to railroad by filling block wall cavities with dense concrete.	**
Infiltration	Blower door test (random sample units)	Less than 0.35 ACH (fresh air by natural or mechanical ventilation); all joints and cracks should be sealed well	***
Roof	Insulation	Have at least 14 cm (R-3.67) of typical fiberglass insulation or comparable insulation rating inside weather barrier to keep the top floor units from being too hot in the summer	***
Door	Noise control	Exterior door adjacent to railroad should be heavy, such as metal with foam core or solid wood and should have gaskets and weather stripping to minimize infiltration of air or noise	**
Building Equipment Features			
Cooling	Air-conditioning	3 COP minimum, higher COP is recommended	**
Ventilation	Off peak ventilation	Closable through-wall exhaust fan (ensure that they are properly sealed to have minimal infiltration when ventilation is not in use). For top floor units, the exhaust fan should be placed near the roof.	***
Thermostat	Programmable digital clock thermostat	Install a programmable digital clock thermostat that controls the air conditioner; install a thermostat that reads indoor and outdoor temperatures	***
Hot water	Solar collector	Significant energy can be collected from the sun to heat the water used for bathing and washing	*

*** Essential for sustainability ** Recommended feature * Additional feature

Table 8 Technical summary of recommendations for residential units facing east/west

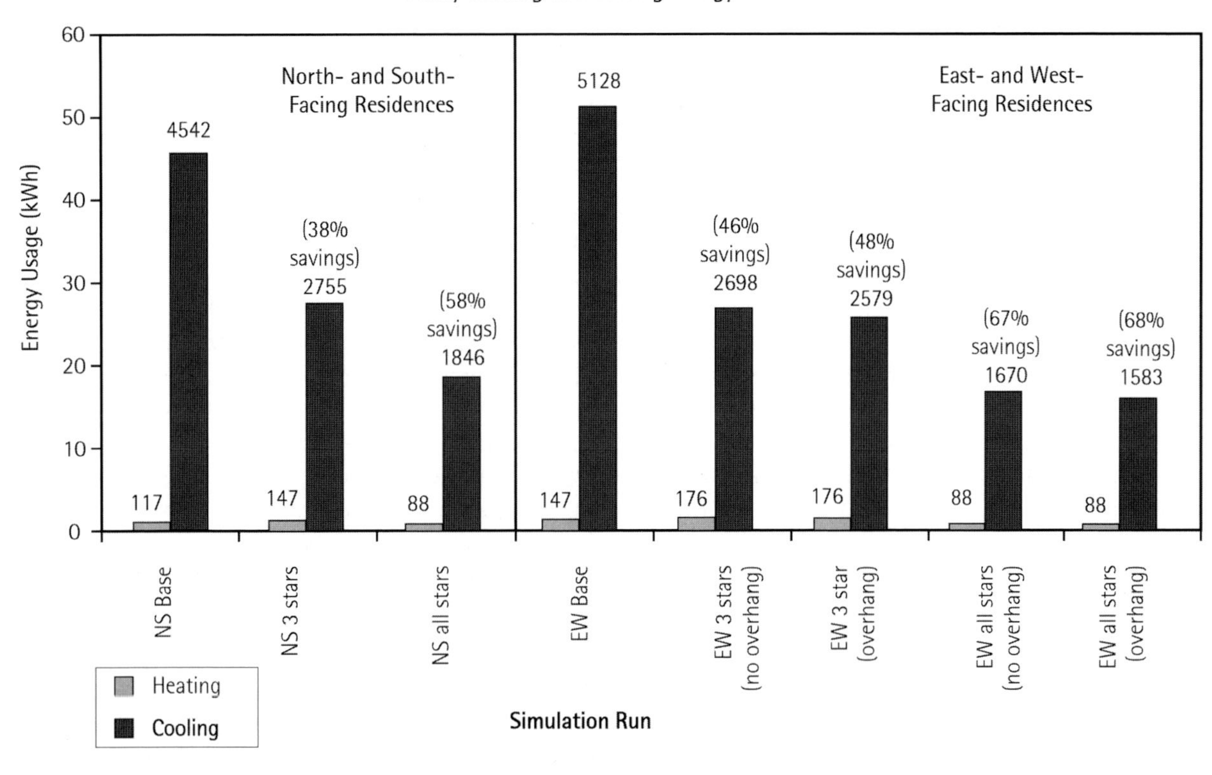

Figure 22 Shenzhen yearly heating and cooling energy use for single-story units. This graph summarizes recommendations listed in Tables 7 and 8.

	KWh (annually)	% Savings
Base Case	4500-5200	-
3 Star recommendations	~2700	40-50%
1, 2, & 3 Star	1500-1800	60-70%

Table 9 Possible savings resulting from incorporating recommendations

possible solution could be filling in concrete blocks with denser concrete.

Summary of Technical Recommendations

The technical recommendations are summarized in Tables 7 and 8. Each table ranks each recommendation with one, two, or three stars, depending on its importance to the sustainability of the design. Energy simulations were run for various combinations of upgrades, and these are summarized for north-south and east-west oriented buildings in Figure 22. If all recommendations are implemented, energy savings of up to ~2000 kWh (~60 percent) annually are predicted compared to the base case, which was based on existing units constructed in the development (Table 9). Some items that have only one or two stars are still important, especially for occupant comfort during the hottest months, though they may not be essential to the reduction of annual energy use.

The design strategy was to reduce cooling loads by using the specification upgrades discussed earlier, and integrate them into the design, from site planning to architectural design. In addition, the site plan provided a scheme that prioritized cooling strategies, including shading, building orientation, and natural ventilation.

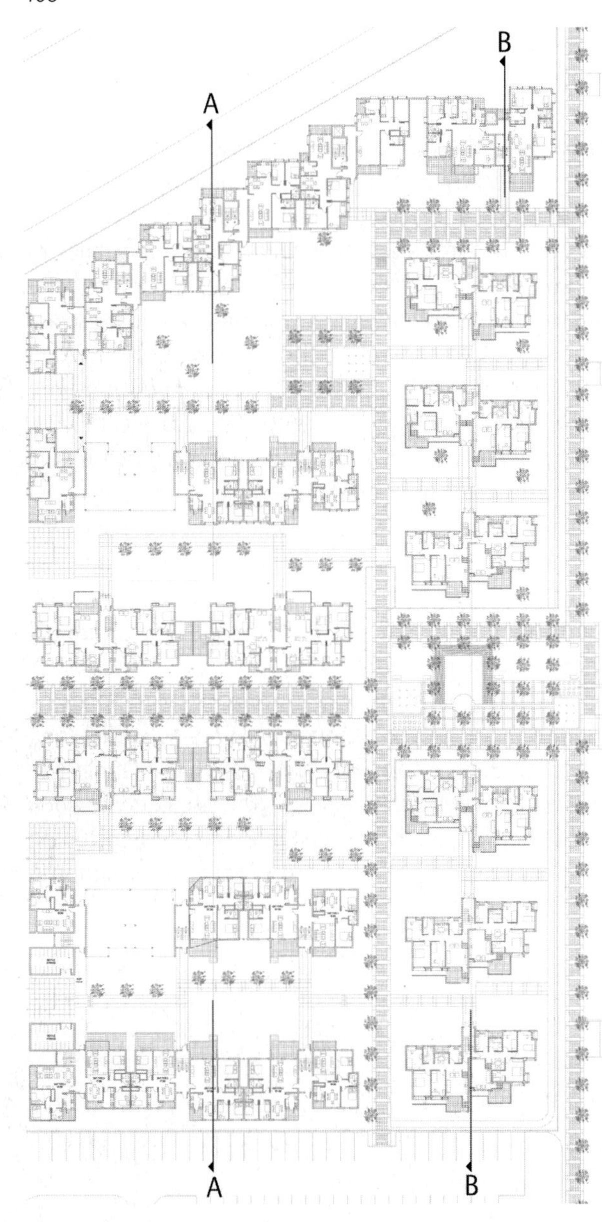

Figure 23 Final site plan of phase IV

ARCHITECTURAL DESIGN

Site Considerations

The site design illustrated in this section provides a number of strategies for reducing energy consumption. Like many of the buildings in the other Shenzhen Wonderland phases, the buildings were oriented to provide high levels of north-south exposure in order to maximize shading. This orientation increased the amount of solar gain in the winter, and the high sun angle in the summer required only slight overhangs to counteract the hot summer sun. Along the east and west axis, the edges were perforated, to allow for natural airflow ventilation from the east and to decrease exposure from the intense afternoon sun from the west. Some units within buildings with north-south orientation were also staggered, affording instances of localized shading and cross-ventilation. The planning of the site took into consideration density as well as the regulatory winter sunlight code, which requires that each unit must receive at least one hour of direct sunshine every day of the year, a rule that had significant repercussions on site planning.

The second major effort of the design sought to address issues of sustainability at the urban scale. Efforts were made to create lower buildings at a higher density that translated to a more humane scale of development. One notable exception was the northern edge that required taller buildings to provide a barrier against noise pollution from the trains. Units were staggered in order to expose them to the exterior environment and take advantage of the natural daylight. A central goal was to increase opportunities for interaction with natural exterior conditions. This relationship was mediated through the implementation of small, intimate courtyards. Such spaces provided for more meaningful individual experiences and personal interactions than the singular large courtyard strategy prevalent in the existing plans. In addition, a large plaza was designed at the eastern edge of phase IV. This plaza served as the focal point of the east-west axis set up by previous phases of development. The final site plan and sections are shown in Figures 23 and 24, while Figure 25 conveys the general aesthetic of the buildings.

Architectural Form

The goal of the formal architectural expression was to create a contextually unique environment that fosters a diverse community. This was achieved by juxtaposing units of different sizes and qualities within many of the buildings. With respect to sustainable design, specific attention was paid to directional differences within the building composition. The east side, though sometimes hot, exploits the primary natural airflow. The south and north sides were more interchangeable with respect to natural ventilation issues. Also, special consideration was given to the relationship between courtyard and non-courtyard sides of the buildings where a coupling of unit pairs fostered a sense of both a front and backyard.

Clusters of different building types defined the development of phase IV, all of which were connected visually through courtyards and openings in buildings, as well as a series of pedestrian paths. There were several openings in the courtyard buildings that allowed winds to penetrate the site, and also allowed a pedestrian path network to connect all of the clusters of the site. Parking was provided at the periphery of the site, so that the experience of occupants could be primarily pedestrian.

Figure 24a Shenzhen Wonderland phase IV final design, section A-A

Figure 24b Shenzhen Wonderland phase IV final design, section B-B

Figure 25 Sketches of Shenzhen Wonderland phase IV

Courtyard Buildings

At the western edge of the site were courtyard building types, A, B, C, F, G, and H, that consisted of C-shaped buildings with the eastern edge left open to allow winds to penetrate into the courtyards, and subsequently through the units. These buildings utilized shared stairs between a pair of units in order to provide both north and south exposures, thus increasing the effectiveness of natural ventilation. Lower-level garden patios and upper-level balconies were provided for all units. The balconies provided opportunities for shading, as well as private outdoor areas for the inhabitants. The angle of the site allowed for an increasing courtyard size to the east. Buildings F, G, and H stepped up in height toward the northeast corner while still satisfying the winter sunlight code described earlier, and allowing daylight to penetrate the units. Unit areas varied from 68 to 123 square meters (Figures 26-29).

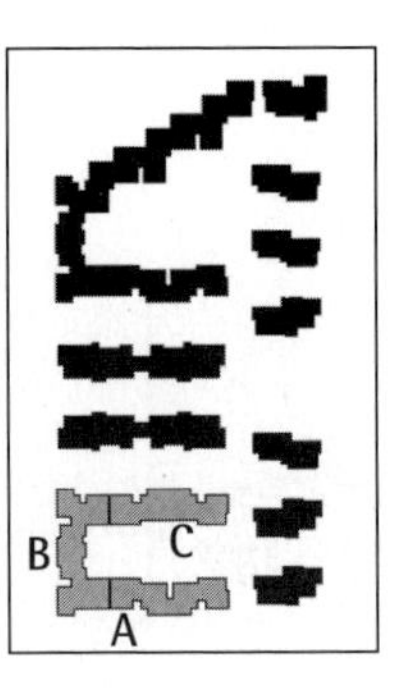

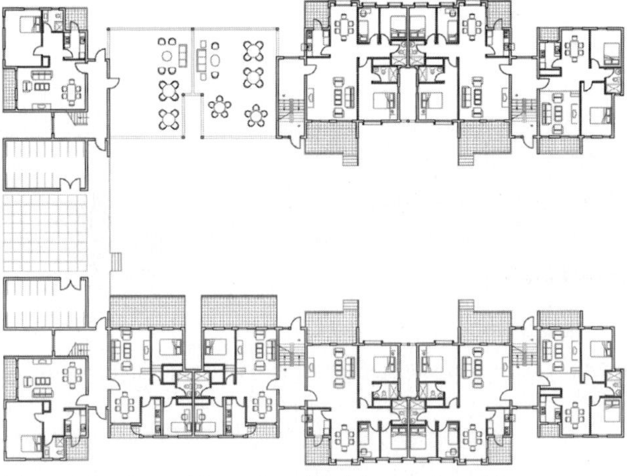

Figure 26a Ground floor plan of buildings A, B, and C

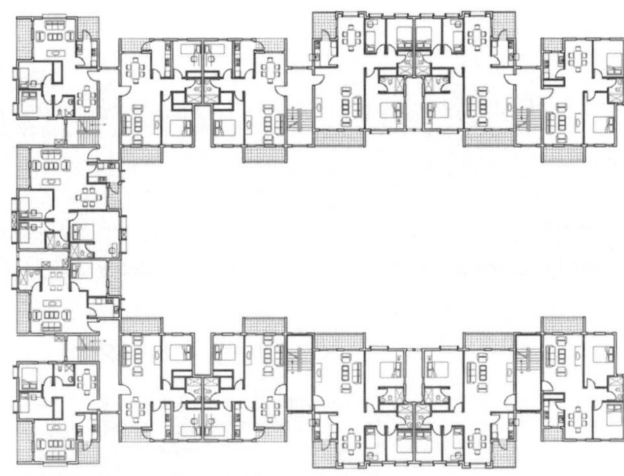

Figure 26b Typical floor plan (levels 2-4) of buildings A, B, and C

Figure 27 South elevation of buildings A, B, and C

Figure 28a Ground floor plans of buildings F, G, and H

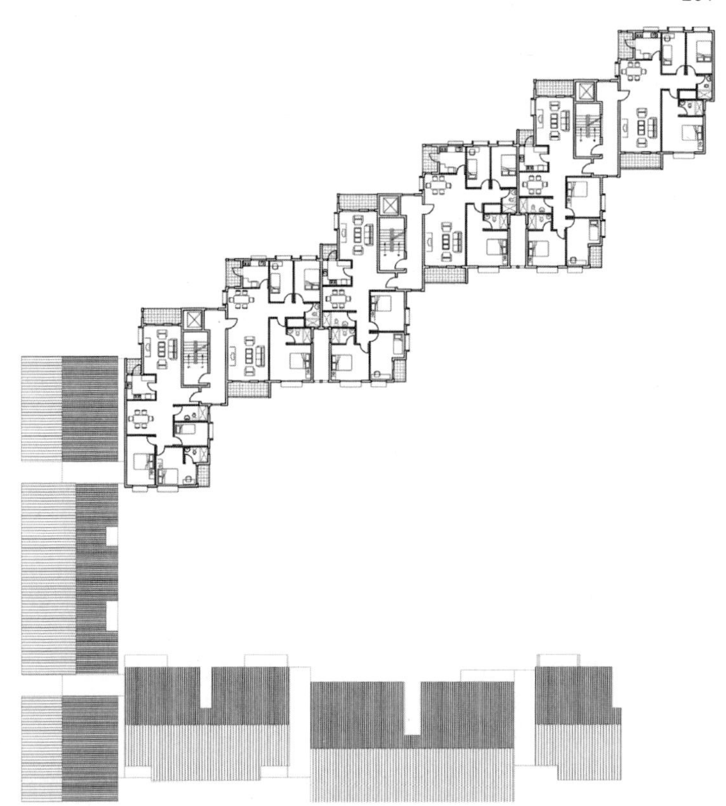

Figure 28b Typical upper level floor plans of buildings F, G, and H (lower level plans are similar to buildings A, B, and C)

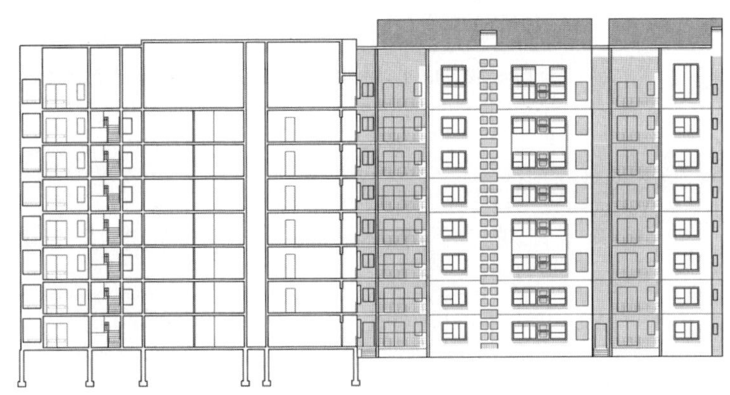

Figure 29 Elevation of building H (buildings F and G have similar elevations to A, B, and C)

Rowhouse Units

The east-west axis continued west from the plaza and provided access to buildings D and E. Because the height of buildings D and E was lower than the rest of the site (four stories), the distance between buildings could be reduced and still satisfy the winter sunlight code. The result was the creation of an urban pedestrian path. Top-level units had large roof terraces overlooking the pedestrian path below. Lower level units had garden patios or balconies facing the other side of the building. Unit sizes varied between 116 and 142 square meters (Figures 30 and 31).

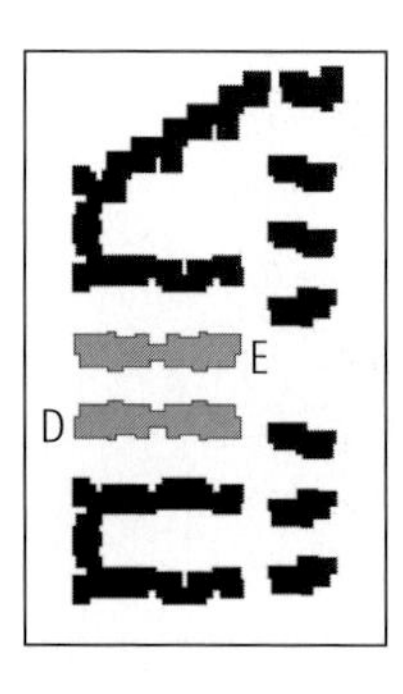

High-Rise Building

Located at the northeast of the site was a high-rise building that consisted of two units per floor, the sizes of which ranged from 113 to 135 square meters. This building type provided an alternative to the walk-up buildings in the rest of the site. Vertical access was provided by means of a central circulation space with an elevator and stairs. All rooms had ample access to light and ventilation. Parking was provided at the lowest level of the building (Figures 24b, 32 and 33).

Multi-Family Villas

Buildings J, K, M, L, N, and O were located at the eastern edge of the site. These were small buildings with a height of four stories each, oriented east-west, consisting of two units each. They played an important role in defining the eastern edge of the phase IV development, while allowing winds from the east to penetrate through the rest of the site. Areas ranged from 96 to 112 square meters (Figures 34 and 35).

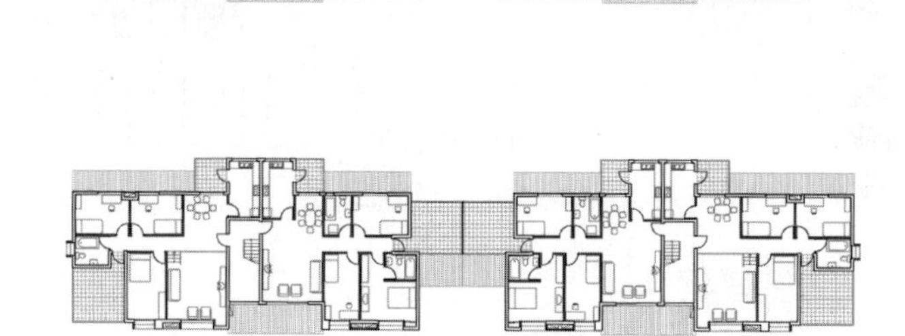

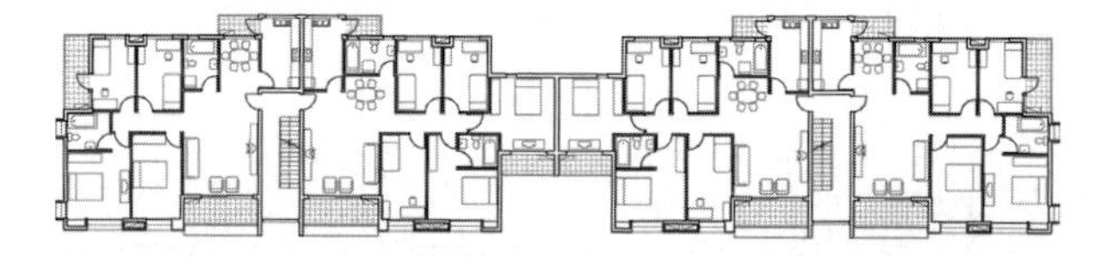

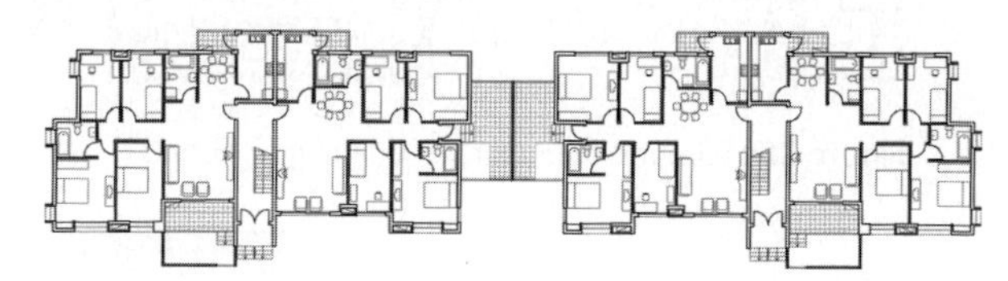

Figure 30 Plans of building D (building E similar), from top to bottom: roof plan, level 4 floor plan, levels 2-3 typical floor plan, and ground floor plan

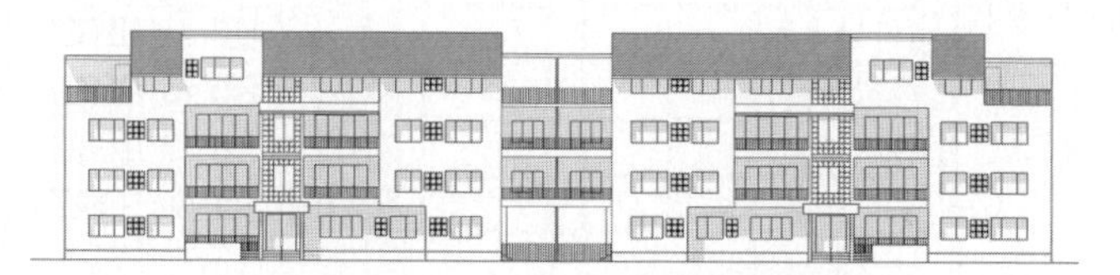

Figure 31 Elevation of building D (building E similar)

Figure 33 Elevation of building I

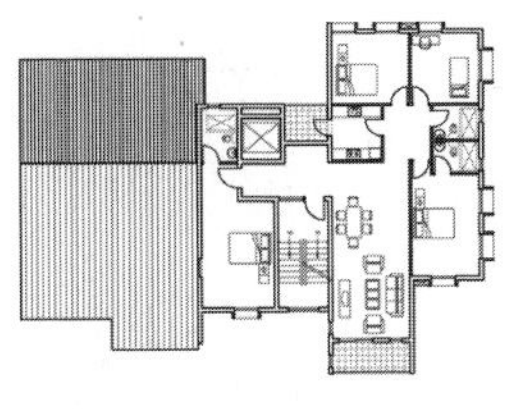

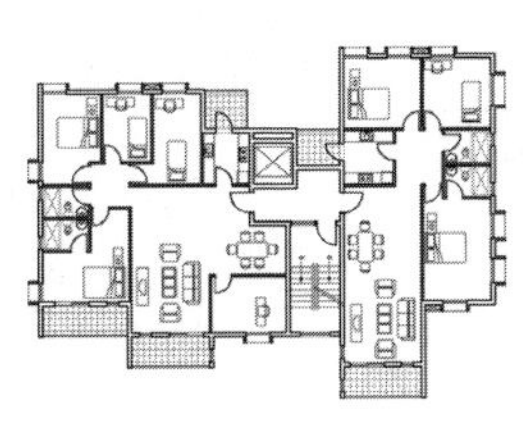

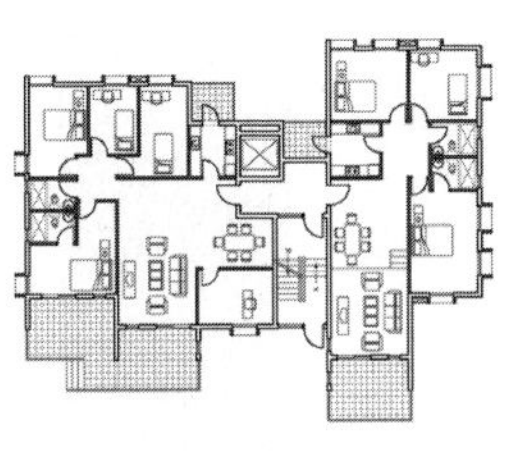

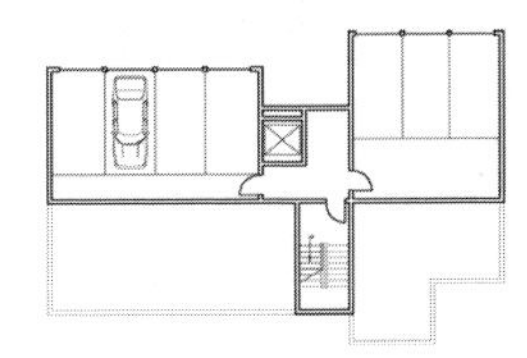

Figure 32 Plans of building I from top to bottom: top level (penthouse) floor plan, typical floor plan, ground floor plan, and basement level floor plan

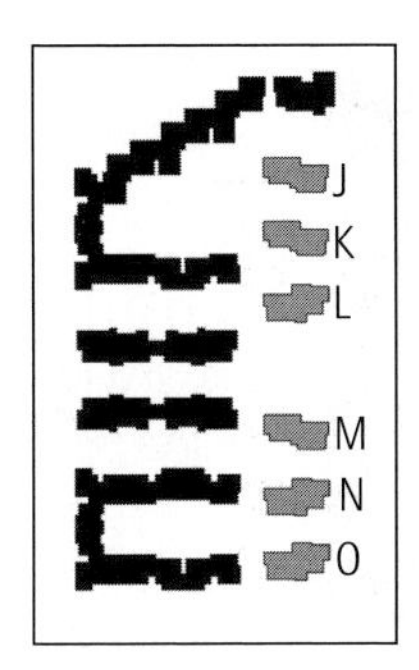

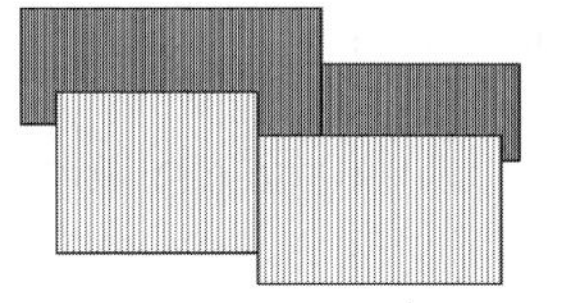

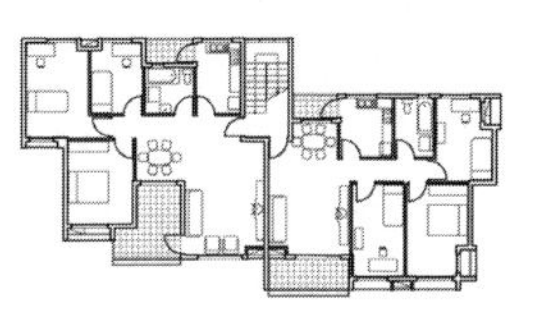

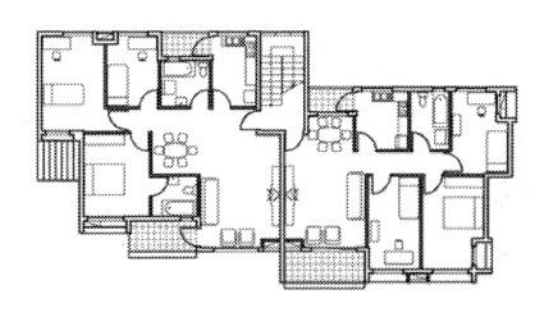

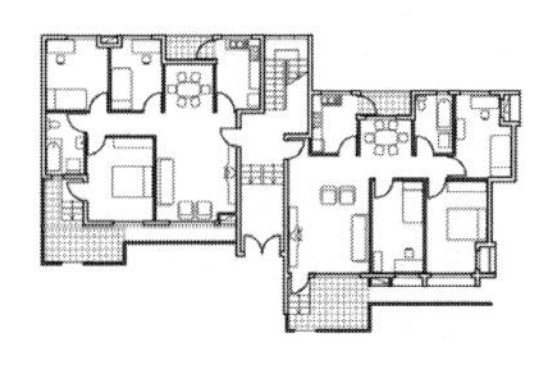

Figure 34 Plans of buildings J, K, and M (L, M, and O similar) from top to bottom: roof plan, level 4 floor plan, levels 2-3 typical floor plan, and ground floor plan

Figure 35 Buildings J, K, and M elevation (buildings L, M, and O similar)

CFD SIMULATIONS

The design of the site plan for Shenzhen Wonderland phase IV was informed regularly with exterior CFD and sun studies that measured the performance of the different design alternatives in terms of wind flow and sun movement. The primary goals were to optimize the use of wind and daylighting while minimizing heat gain. The high-density low-rise solution, which was adopted from the beginning, was appropriate for the achievement of both of these goals.

This final site plan design performed well in terms of wind control by allowing the prevalent winds to pass through the site from east to west, providing continuous natural airflow (Figure 36). In the CFD simulations of the preliminary design schemes, it was most notable that the wind distribution of the site had a more even flow throughout (Figures 37 and 38). The modifications implemented in the design of phase III included the addition of the north-south oriented buildings on the west that blocked the wind to the phase IV courtyards. It was recommended to Vanke to consider opening passages on the ground floors of the buildings in phase III at places that corresponded to the openings between buildings J, K, L, M, N, and O of phase IV, in order to improve air circulation conditions both to this phase of Shenzhen Wonderland as well as in the future areas to be developed in the west.

Pressure - wind from east at 3 m height

Pressure - wind from east at 0.5 m height

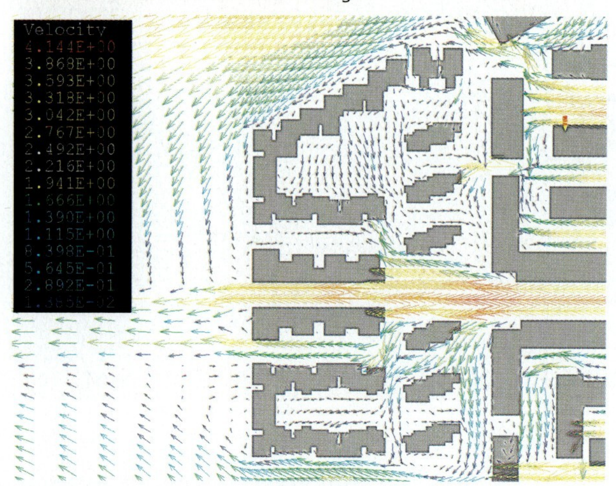

Velocity - wind from east at 3 m height

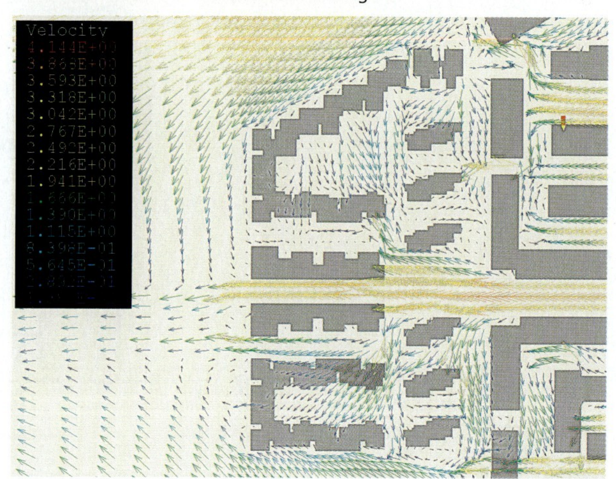

Velocity - wind from east at 0.5 m height

Figure 36 CFD results for the final site design of phase IV (north is to the top)

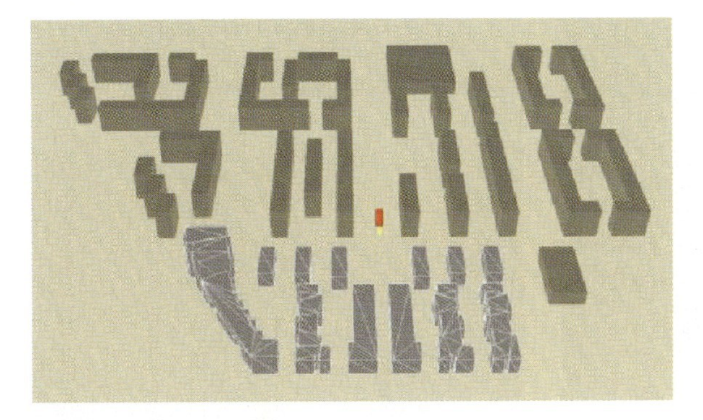

Figure 37 (upper left) 3-D view of site - model used for CFD studies (north is to the left)

Figure 38 (right) Preliminary CFD studies of the entire development and phases III and IV when phase III had all buildings oriented east/west (north is to the top)

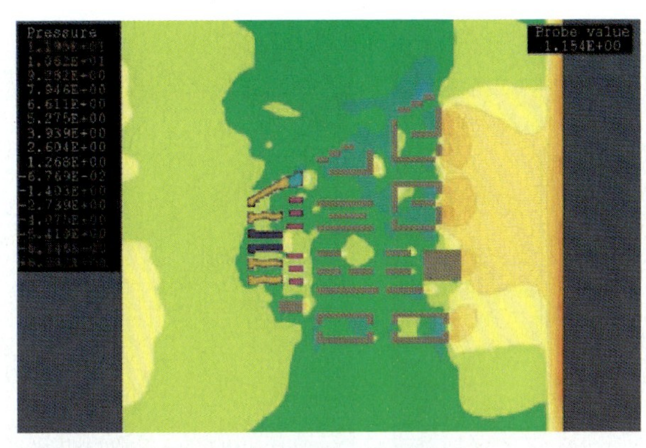

Pressure - wind from east, showing the full site

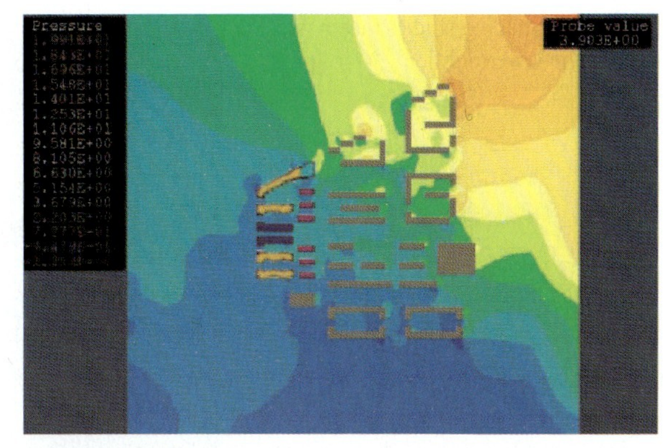

Pressure - wind from northeast, showing the full site

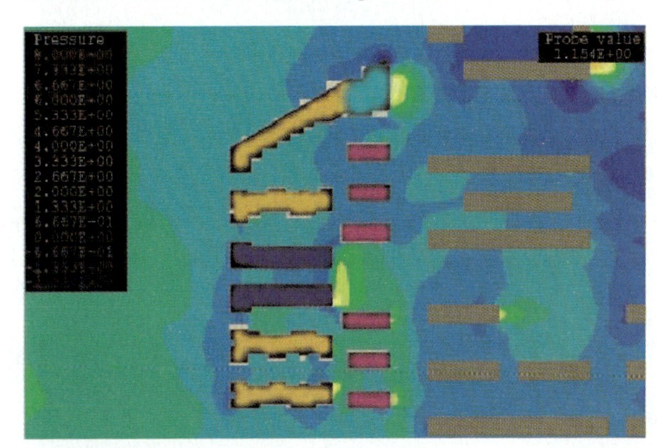

Pressure - wind from east

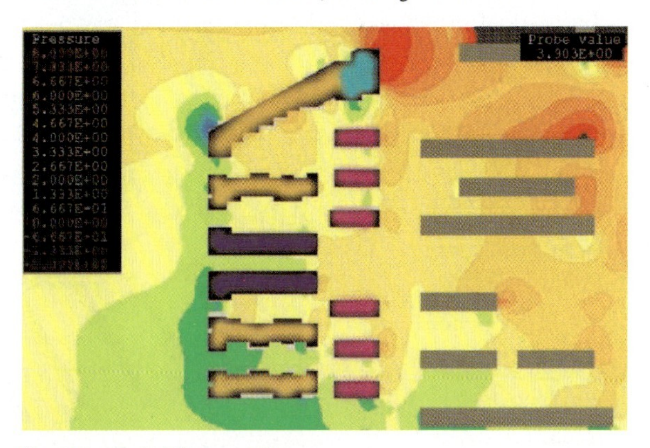

Pressure - wind from northeast

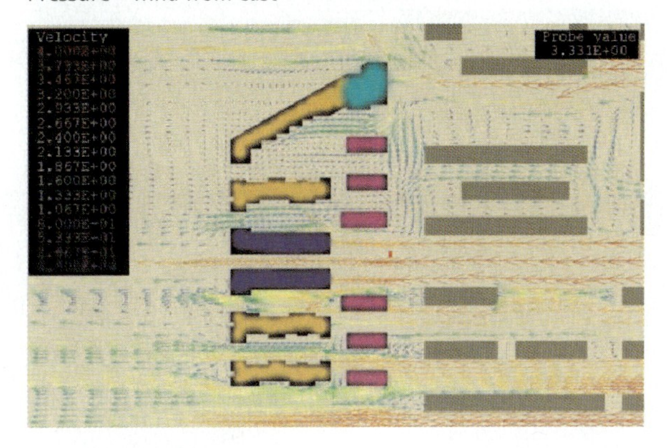

Velocity - wind from east

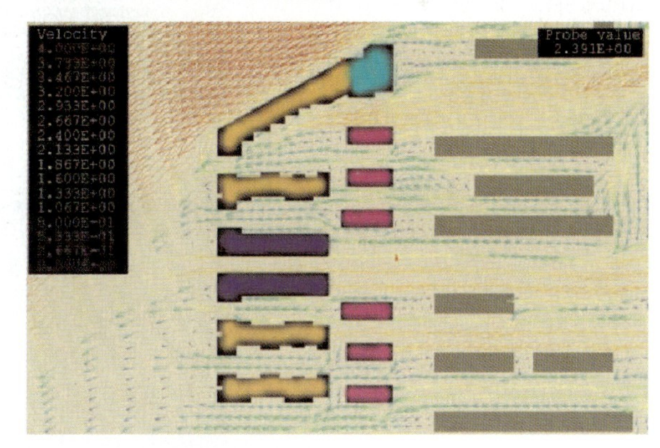

Velocity - wind from northeast

Figure 39 Shading studies (north is toward the lower right)

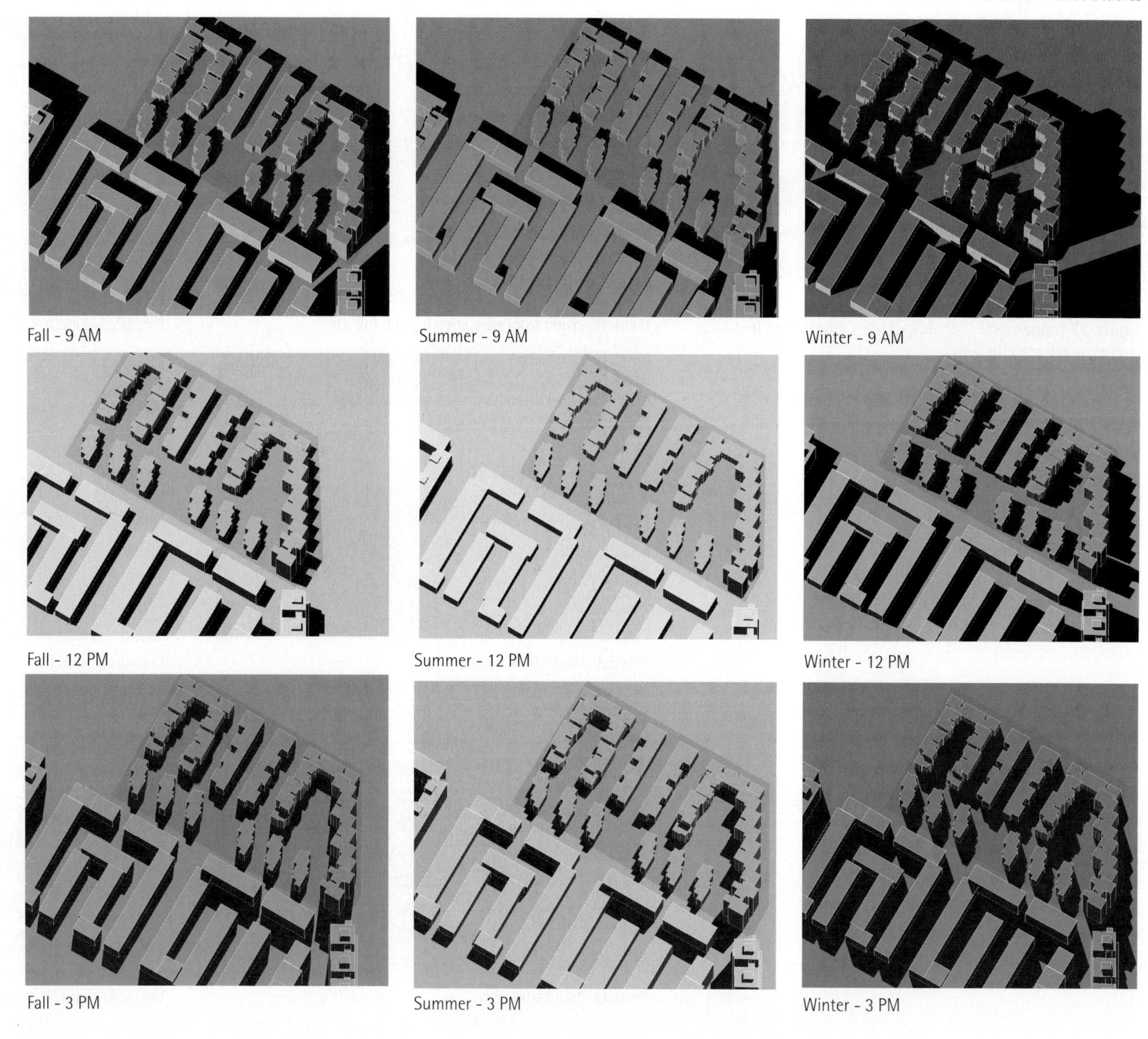

SOLAR SHADING STUDIES

The site design was optimized for shading, as the technical recommendations had shown the inadequacy of site wind conditions alone to provide enough ventilation to raise comfort standards to adequate levels. This response was achieved by minimizing the surface area of the western and eastern façades, resulting in buildings oriented along the east-west axis. The strategy served to optimize solar gain during the winter against the need for shading during the summer.

Shading studies showed that most of the sun falls on the roof during summer months. The north-south spacing between buildings considered the sun angle in order to allow direct sunlight to penetrate units during the winter (Figure 39).

Determining Sizes of Overhangs and Fins

As shown in the technical recommendations (Tables 7 and 8), one of the most important upgrades was to add overhangs to windows. The MIT design team explored alternative overhang solutions and made recommendations to Vanke.

For a window on the southern façade, the optimal overhang length was 20 percent of the height of the window, and the recommended length of vertical fins was 10 percent of the window width. Buildings built in previous phases of Shenzhen Wonderland already featured a design element similar to that, as illustrated in Figure 40. In those cases, the window could be converted to an energy-efficient design by placing the glass flush with the interior wall of the unit rather than with the exterior of the box. Some additional options to achieve the desired results are illustrated in Figure 41.

For a window on the eastern or western façades, the optimal overhang length was 100 percent of the height of the window. Some additional options to achieve this are illustrated in Figure 42. An alternate solution, as discussed in the passive solar building design section of this case study, was to use windows with a solar-heat-gain coefficient of less than 0.4.

Integrated Design of Window, Shading Device, and Air Conditioner

The design for phase IV provided design measures such as self-shading of buildings and shading devices on windows that would, if augmented, significantly reduce the need for air-conditioning. However, given the high levels of humidity in Shenzhen during summer months, some use of air conditioners would be necessary to maintain comfort conditions inside residences. Split air-conditioning units, which are used most frequently in housing, require the placement of the condenser unit outside the building to minimize noise disturbance to occupants.

It is important that designers specify the appropriate placement of these units and the route of connection pipes in order to avoid haphazard modifications by occupants that can become eyesores on the façade as well as potentially damage the outer protective envelope of the building if they are not properly sealed. As an alternative to making the condenser units visible from the outside, the design team developed a window that integrates both shading devices and air conditioners. Two options were developed. The first option used the space between protruding fins to house the condenser unit of the air conditioner (Figure 43). The second option created a smooth façade by thickening the wall and depressing both the window and the condenser unit into the face of the building. The leftover spaces were then integrated into the casework design so that space was not wasted. In addition, this area could further be used as a vertical shaft to accommodate downspouts and other in-wall utilities alongside the condenser units (Figure 44).

Figure 40 Existing overhangs and air-conditioner units at Shenzhen Wonderland

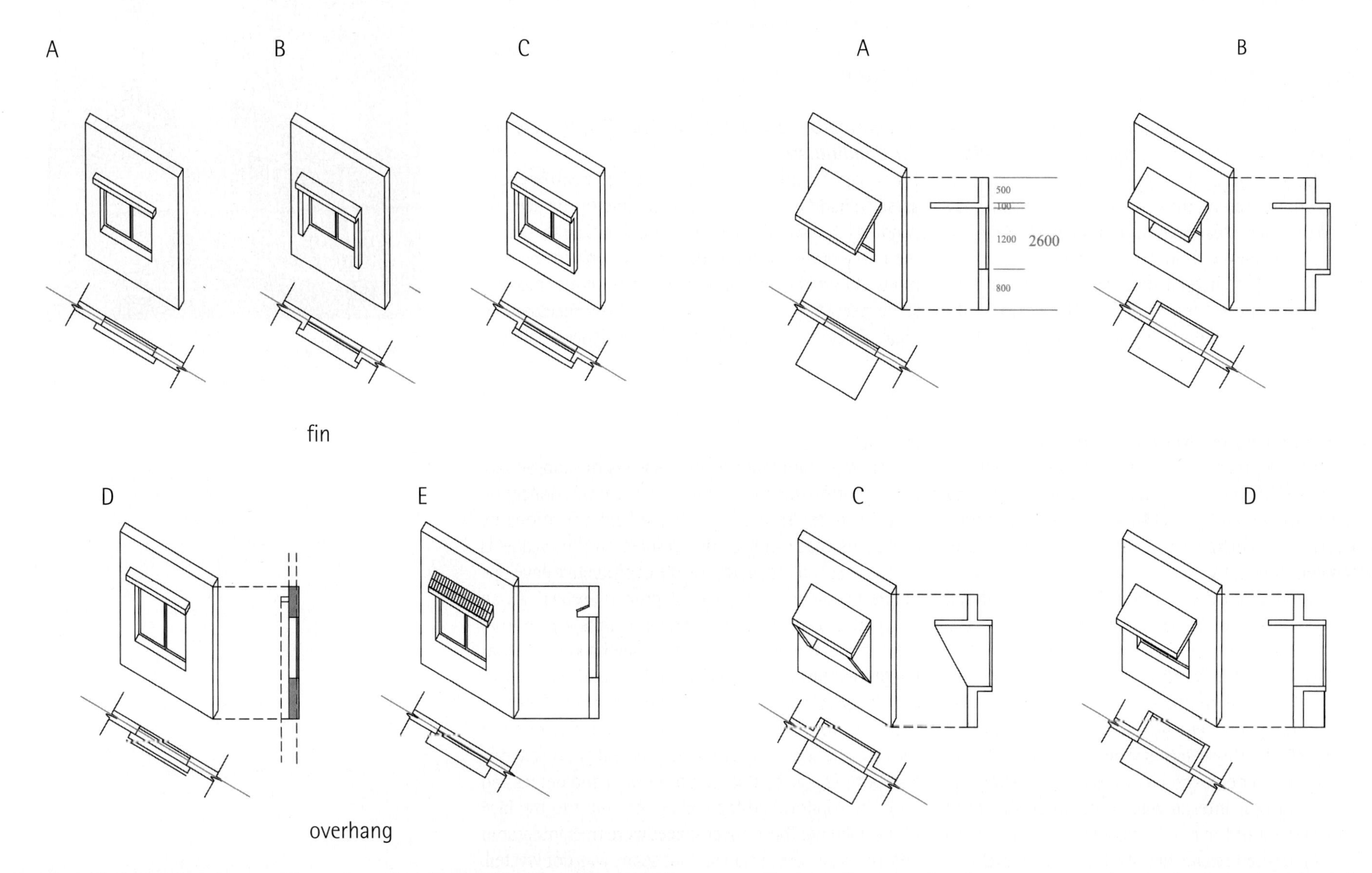

Figure 41 Illustration of overhangs and fins on the southern façade; 20% of the window height is recommended for overhangs, while 10% of the window height is recommended for vertical fins

Figure 42 Illustration of overhangs on western and eastern façades; 100% of the window height is recommended for overhangs

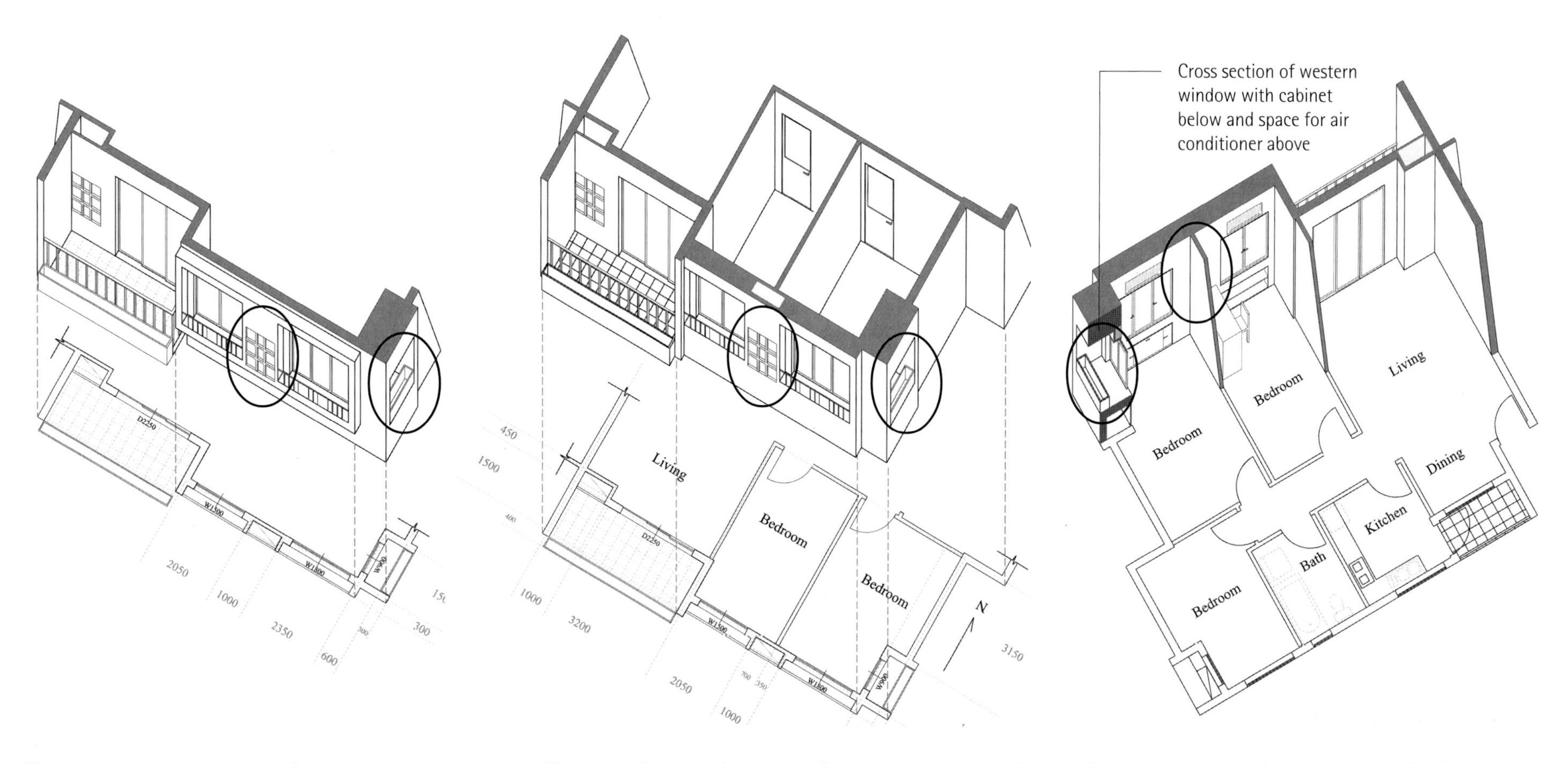

Figure 43 Axonometric façade study of sample unit from buildings J, K, and M - option 1. The possible locations of the air conditioner condensing unit are shown circled.

Figure 44 Axonometric façade studies of sample unit from buildings J, K, and M - option 2 (left) exterior view and (right) interior view. The possible locations of the air conditioner condensing unit are shown circled.

	Cooling Season	Mixed Season	Heating Season
Early morning	26°C (30 if away)	Off (fans on)	22°C (20 if away)
Mid morning	30°C (27 if home)	30°C (27 if home)	20°C (22 if home)
Mid day	30°C (27 if home)	30°C (27 if home)	20°C (22 if home)
Mid afternoon	30°C (27 if home)	30°C (27 if home)	20°C (22 if home)
Evening	26°C (30 if away)	27°C (30 if away)	20°C (22 if home)
Late evening	27°C (30 if away)	Off (fans on)	22°C (20 if away)

Table 10 Recommended temperature set points

Inside Temperature	Outside Temperature	User Action
Above the cooling set point	At least 2°C below the cooling set point	Turn off A/C, use natural or mechanical ventilation
Above the cooling set point	Above the cooling set point	Turn on A/C, turn off fan and close windows
Below the cooling set point, above the heating set point	Below the cooling set point	Turn off A/C, use natural or mechanical ventilation
Below the cooling set point, above the heating set point	Above the cooling set point	Turn off A/C and fan and close windows.
Below the heating set point	Below the heating set point	Turn off A/C and fan and close windows.
Below the heating set point	Above the heating set point	Turn off A/C, use natural or mechanical ventilation

Table 11 Chart showing user actions based on temperature set points

RESIDENTIAL USER MANUAL

As discussed in chapter 4, a great majority of a building's energy use is for operation during its lifetime. Therefore it is important that the end users be aware of operating procedures in order to maximize the benefits of energy-efficient housing. The residential user manual summarized in Table 11 and was written for future occupants as an informative tool to guide them in intelligently controlling their homes for both comfort and energy efficiency. The information serves as a guide to obtain optimum energy efficiency, which should be considered malleable according to personal preference and habits.

Use of Thermostat to Minimize Energy Use

The programmable thermostat allows the user to come home or wake up in a comfortable home without air-conditioning while they are sleeping or away from home. A thermostat, which monitors indoor and outdoor conditions, and controls the heating, cooling, and ventilation equipment, can be used to turn the equipment off even when the user is not at home. With the use of a programmable digital thermostat, the temperature settings recommended in Table 10 should be followed throughout the year. Users should alter these settings to suit personal comfort levels with the understanding that energy use and costs may increase. At times, high humidity may require the use of air-conditioning even when the temperature outside dictates natural ventilation.

Use of Temperature Readings to Control Air-Conditioning and Ventilation

Occupants can optimize air conditioner and fan use by setting thermostats according to Table 10 and running the air conditioner and mechanical ventilation according to Table 11. If the daytime temperature is uncomfortable, daytime ventilation should be reduced and instead, occupants should use night cooling to provide interior comfort.

Use of Shading Devices to Minimize Heat Gain from the Sun

To minimize the heat gain from the sun, the residence should have the blinds or other shading devices covering the windows while the sunlight is in direct contact with the window. If residents are able to shade the windows during the day when the home is unoccupied, the space will be much more comfortable toward the end of the day. Blinds or shades should be used even if the window is open, as sunlight can be blocked while still letting air in for natural ventilation.

Use of Natural Ventilation or Mechanical Ventilation to Remove Hot Air

Occupants should open windows and interior doors so that wind will move hot air out of the unit and make the home comfortable. During the spring and fall, one may be able to use natural ventilation all day without needing an air conditioner. During the warmer months, natural ventilation may only be possible during nighttime or morning hours (or not at all). The use of a thermostat with an outdoor sensor is invaluable in determining when ventilation will make the unit more comfortable. If it is uncomfortable inside, and the indoor/outdoor thermostat recommends the use of natural ventilation when the wind is not strong enough, then it is recommended to turn the exhaust fan on while simultaneously opening the windows and interior doors.

SUMMARY

The design for Shenzhen Wonderland phase IV involved close collaboration between Vanke Architecture Technology Research Center in Shenzhen and MIT faculty and students from several disciplines, including building technology, architecture, and urban design. The design was approached using a coordinated study of issues ranging from site design to detailed design, and included specifications such as insulation values for the roofing and recommendations for thermostat settings. Technical and design studies were carried out simultaneously, and the schemes evolved in a reciprocal fashion. The final scheme incorporated an extensive examination of urban design concepts such the creation of public spaces and semi-private courtyards as well as a diversity of building and unit types that reflected the rigor of area requirements requested by the developer. The primary goals were to optimize the use of wind and daylighting while still minimizing heat gain. The high-density low-rise solution, which was adopted from the beginning, has proven appropriate for the achievement of both of these goals while at the same time creating a contextually unique architectural environment.

ACKNOWLEDGMENTS

The *Technical Recommendations* section of this chapter is adapted from *Technical Recommendations for Shenzhen Wonderland*, authored by Sephir Hamilton, Brian Dean, Qingyan Chen, Leon Glicksman, Leslie Norford, and Andrew Scott in June of 2000 and Hamilton's *Passive Solar Building Design in Shenzhen Comparing Shading Techniques throughout the Year* in April of 2000 for the Sustainable Urban Housing in China Project. Results for the technical recommendations were summarized by Hamilton.

The sections entitled *Determining Sizes of Overhangs and Fins* and *Integrated Design of Window, Shading Device, and Air Conditioner* are adapted from Ozgur Basak Alkan's paper, *Determining Sizes of Overhangs and Fins*, written in August of 2000 for the Sustainable Urban Housing in China Project.

REFERENCES

AEC. 2006. *VisualDOE 4.0.* Architectural Energy Corporation. <http://www.archenergy.com/products/visualdoe>

U.S. Department of Energy. 2006. Sizing a Solar Water Heating System. Online publication available at <http://www.eere.energy.gov/consumer/your_home/water_heating/index.cfm/mytopic=12880>.

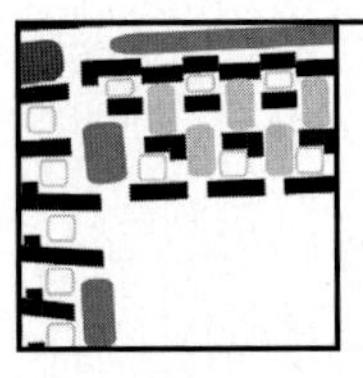

CHAPTER FOURTEEN

CASE STUDY FIVE - BEIJING HUI LONG GUAN

Juintow Lin and Leon Glicksman

PROJECT DESCRIPTION

The MIT Building Technology Group and Tsinghua University advised Tian Hong, a development firm, in the conceptual design of a residential development outside of Beijing named Hui Long Guan. MIT undertook the site planning and architectural design for two residential plots within the larger master plan for the Tian Hong Project. These included development parcels C02 and C06, consisting of 89,000 and 148,000 square meters respectively (Figure 1).

Tian Hong requested a design strategy with an emphasis on ecological, environmental, and economic sustainability. For the Beijing climate, with cold winters and hot summers, the primary design strategy was to reduce energy consumption in the buildings. The objective was to simultaneously decrease heat loss, maximize solar gain, and reduce the heating load in the winter months, while providing means to reduce energy use for cooling in the summer months.

In addition to adherence of environmental design principles, the design team was interested in demonstrating that designing for the community should be an integral aspect of the general scope of design. The intersection of Chinese cultural interests and appropriate sustainable principles prompted a design strategy that provided communal gathering and activity spaces wedded to both the built and natural environments.

L. Glicksman and J. Lin (eds), Sustainable Urban Housing in China, 212-237

DESIGN APPROACH

The C02 and C06 sites, located on the outskirts of Beijing, are part of a much larger, primarily residential development, Hui Long Guan, which is typical of modern urban development in Beijing and throughout China (Figures 2 and 3). A primary concern for an "environmental" approach was to provide a socially sustainable development in a high-density setting with a central design goal of achieving symbiotic and environmentally sensitive relationships between the individual dwelling and communal areas of the development. Many high-rise residential developments in the area neglect a sociological perspective characterized by community spaces with poorly articulated relationships between public and private space.

Beijing is an extremely active and rapidly changing city. Residential developments must address the sociological issues related to numbers of unrelated persons living in close proximity. The MIT group sought to accomplish this by providing public spaces of various scales and types for social interactions, as well as comfortable private living areas.

It is important to also describe several limitations that were imposed on the design by either local planning regulations or the developer. The first requirement was the winter sunlight code, discussed in earlier chapters. In addition, each unit must also have a south-facing entrance, a requisite with its basis in *feng shui*. And lastly, the palette of basic building materials was limited to KPI brick and concrete, with small amounts of steel.

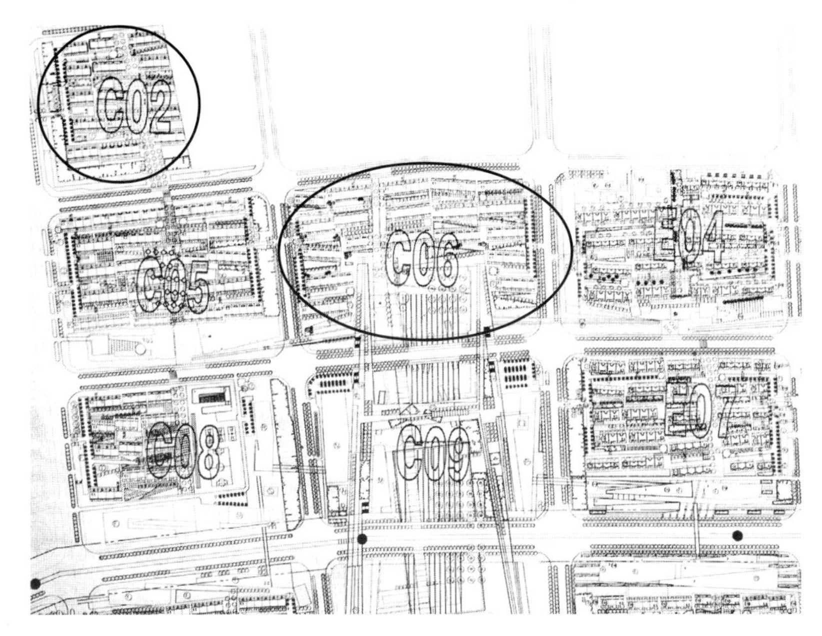

Figure 1 Tian Hong residential development site map showing C02 and C06 parcels

Figure 2 Typical six-story building construction at Hui Long Guan development

Figure 3 Row house construction typical of urban housing in China

Installation Quality	Heating Season ACH	Cooling Season ACH	Window U-Value $W/m^2 K$
High (airtight)	0.5	0.5	2.9
Medium	1.0	0.7	2.9
Low (leaky)	2.0	1.5	4.1

Table 1 Infiltration rates - assumed air leakage as a function of construction quality

Climatic Strategies

General climatic strategies for building in the Beijing area took into account several local environmental factors. Winter heat loss, mainly caused by infiltration, is a persistent problem in other, comparable residential developments in the area. Reduction of heat loss due to infiltration can be primarily addressed on the level of craft skills and construction techniques employed by the local builders. The MIT design team also responded to this problem by reducing the winter wind speed within the site through architectural planning. Any reduction in the winter heating load has substantial benefits to the end users, as winter heating bills for existing homeowners are very high, sometimes as much as 25 percent of one's salary.

In addition, high temperatures and high humidity in the summer create substantial demands on either passive or mechanical cooling systems. The demands of summer cooling should be addressed by providing appropriate shading and by taking advantage of prevailing wind conditions for evaporative cooling.

Additional climate concerns include water conservation and air quality. Water conservation was addressed by specifying landscaping practices that would encourage a slower release of storm water runoff and allow for a high level of aquifer recharge. Large grass-covered areas also added to the overall site surface permeability and helped to combat carbon dioxide emissions.

ENERGY STUDIES

MIT performed a series of energy simulations to determine the design improvements that should be recommended for the project. The energy analysis was performed using the Building Energy Calculator, available with this book. The MIT team analyzed three designs, based on a generic geometry of a 70 meter x 10 meter building oriented east-west. The building was six stories high, and the air temperature was the same throughout the building. All four sides of the building had 35 percent of the exterior wall occupied by windows. Details of the assumptions and conditions used in the program are given in the appendix. To simulate different levels of construction quality, the infiltration rates given in Table 1 were used for three levels of construction quality, poor, medium, and good.

For the low quality simulation, double-glazed low-e windows without a thermal break in the metal frame were used with a U-value of 4.1 W/m^2K. For medium and high quality, a U-value of 2.9 W/m^2K was used for double-glazed low-e windows with a thermal break. Two levels of insulation were considered, 10 millimeters of foam in the walls and the roof, and 50 millimeters of foam in the walls and roof.

It was determined that there was a substantial increase in energy efficiency when moving from the low quality, low insulated building to the high quality, highly insulated building. The seasonal heating energy was reduced from 161 kWh/m^2 to 35 kWh/m^2 (Figure 4). Note when single-glazed windows were used with low-quality construction and the same infiltration rate, the heating energy rose to 182 kWh/m^2. Thus, the use of double-glazed windows that do not reduce the infiltration rate for poor construction had only a modest influence on the energy requirement of the building. Changing the construction quality had a much more modest effect on the summer cooling energy requirement. No exterior shading was assumed for any of the cases.

Figure 5 shows the heating energy for six options, three levels of construction quality, and two levels of insulation. Starting with low construction quality, the largest improvement in energy efficiency was obtained by reducing the infiltration rate through better construction quality and better building products. Improving the insulation level on a poorly constructed building had a smaller influence.

Figure 6 shows the heating energy per unit of floor area as a function of winter infiltration rates. However, adding a modest amount of interior insulation to a 200 millimeter concrete wall had a substantial improvement over a wall without any insulation. As shown in Figure 7, thicker levels of insulation had a progressively smaller influence.

Infiltration

The infiltration rate for any given building is primarily a function of construction standards and the wind speed outside of the building. Without reductions in infiltration, the effect of the other specifications is not as significant. Placing taller buildings on the winter upwind side of the project could reduce the infiltration rate by reducing the local wind speed around adjacent buildings.

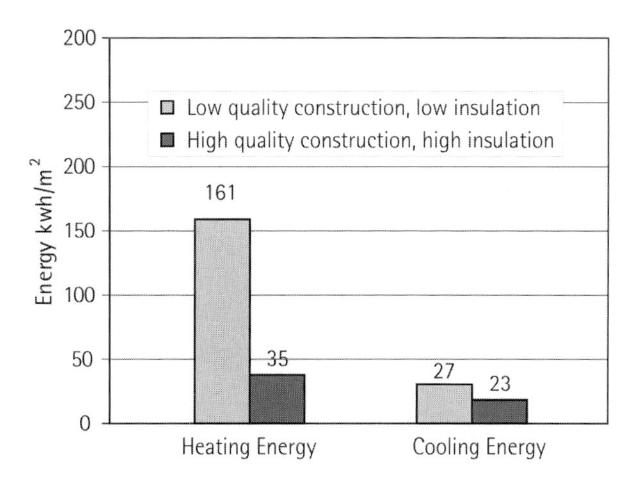

Figure 4 Comparison of yearly heating and cooling energy for high quality construction and high insulation versus low quality construction and low insulation for multi-family residential buildings in Beijing

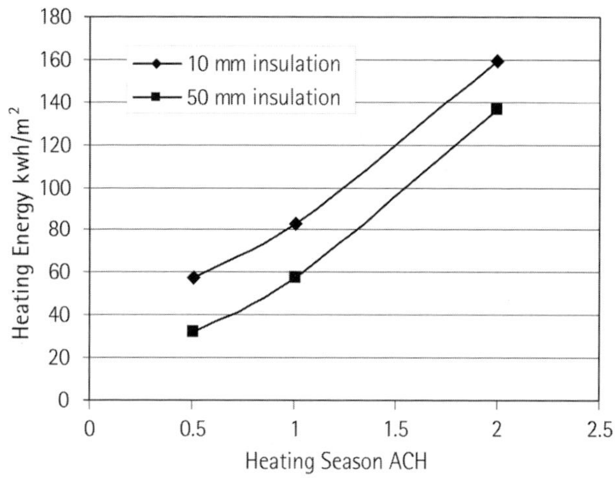

Figure 6 Heating energy versus winter infiltration rates. Increasing the interior insulation for a 200 mm concrete wall has a substantial improvement over an uninsulated wall. Thicker levels of insulation have a progressively smaller influence.

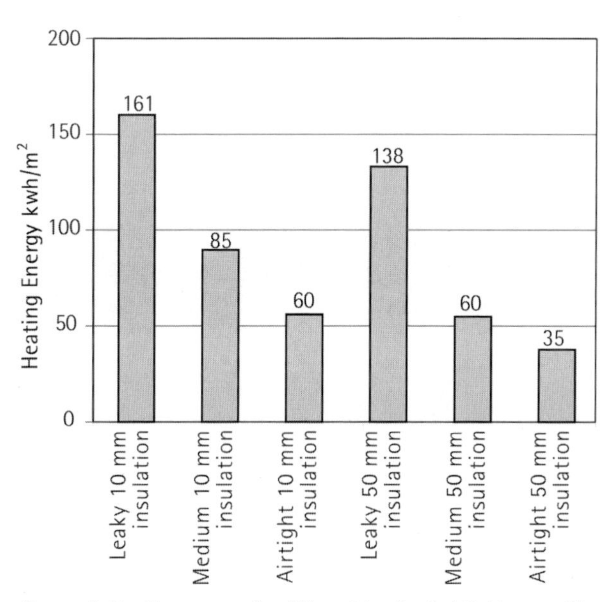

Figure 5 Heating energy for different levels of airtightness with levels of foam insulation

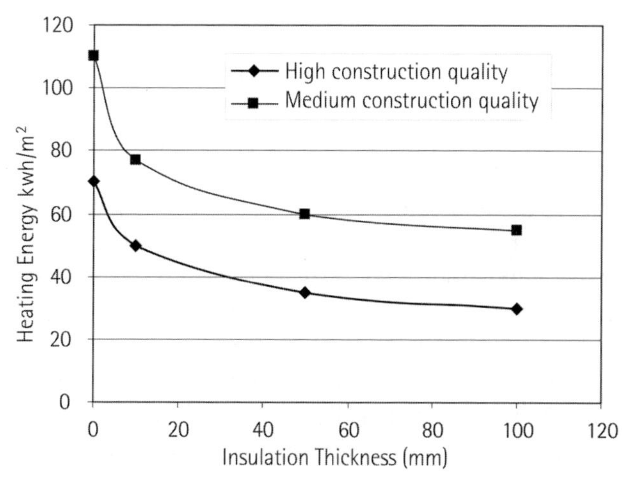

Figure 7 Heating per unit floor area versus thickness of foam insulation for medium- and high-quality construction

All of the buildings should have modest window area and the best quality construction for the windward side of the building. Figure 8 shows that reducing the outdoor airspeed by 50 percent could reduce the infiltration rate by about 50 percent or more. In addition to airspeed, infiltration can be reduced by means of following careful construction practices used to create an airtight building, in particular around openings such as doors, windows and vents.

Throughout most of the summer, the reverse is true; increased ventilation reduces the indoor temperatures (Figure 9), especially at night when it is cooler outside than inside. During the warmest and most humid days, the building should be closed and mechanical cooling is required if comfortable conditions cannot be obtained by other means.

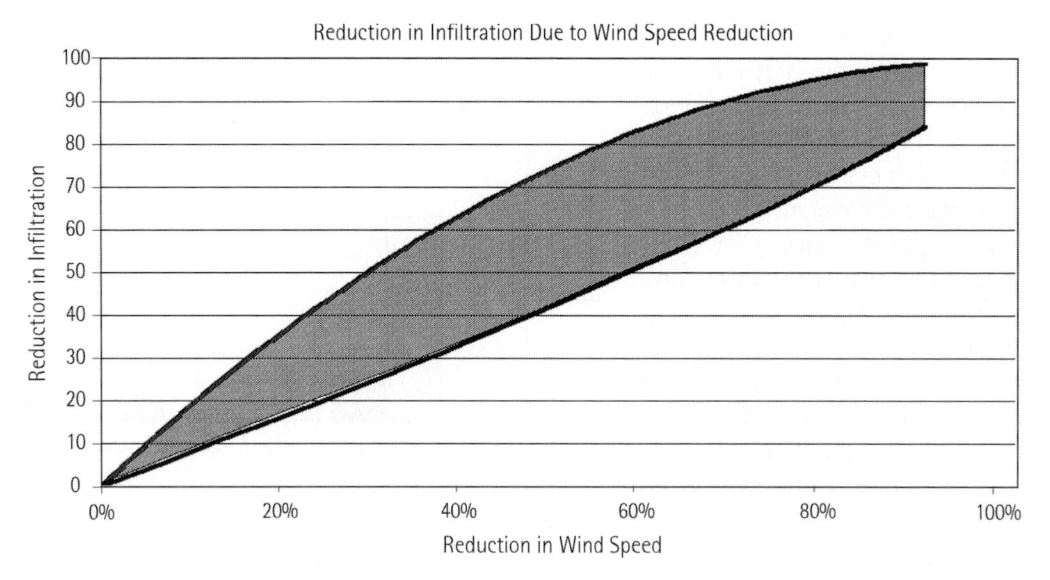

Figure 8 Effects of wind speed on infiltration

Insulation

The energy-reduction effects of insulation and window changes were comparable. Results from the study of interior insulation thickness showed that little energy savings were achieved beyond 50 millimeters of high-R foam insulation. Therefore, 50 millimeters of insulation was recommended for this project. Figure 10 compares the effective R-values of three types of insulation, and Figure 11 illustrates typical construction details for each type. Although exterior insulation is better, it is more expensive to install and requires more-highly trained craftsmen. Therefore, exterior insulation was not recommended as the small improvement in the insulating quality would more than likely not justify the additional expense or the risk of poor construction quality.

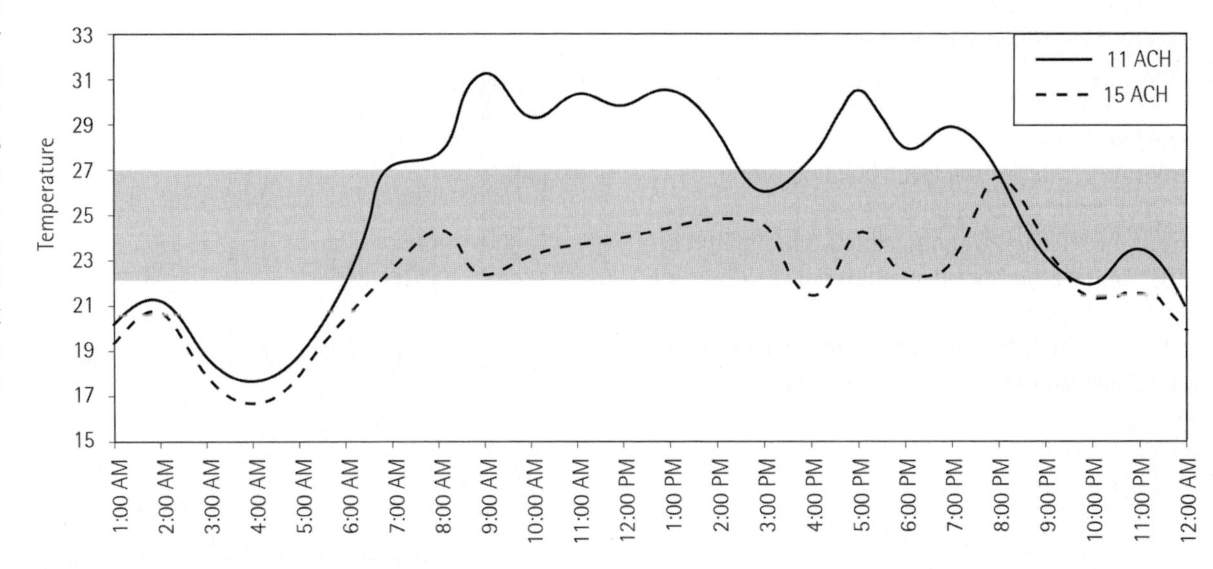

Figure 9 Effect of summer ventilation rates on indoor temperatures on 21 July

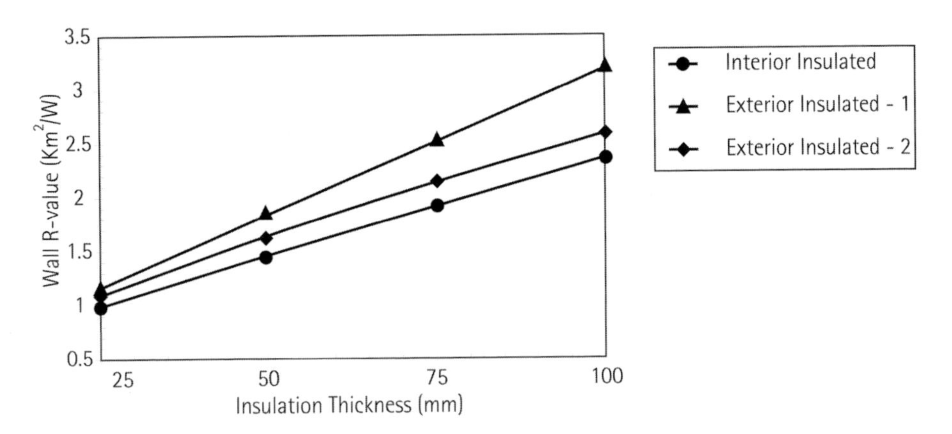

Figure 10 Wall R-value versus insulation thickness

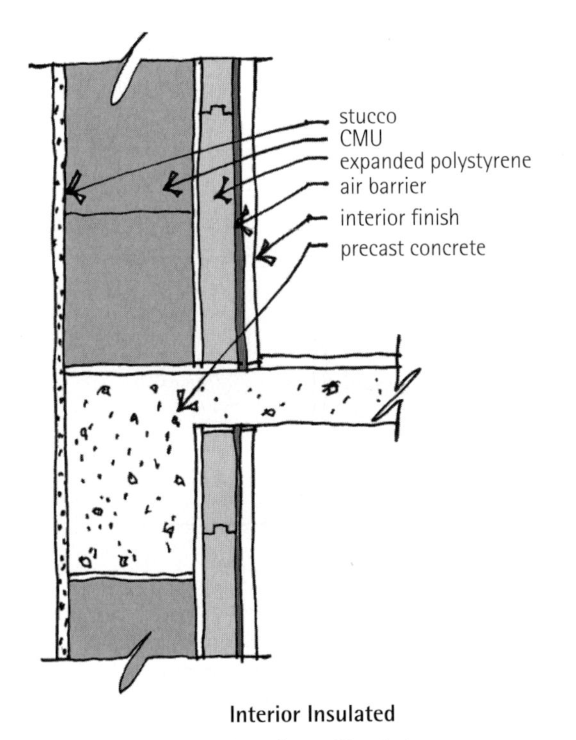

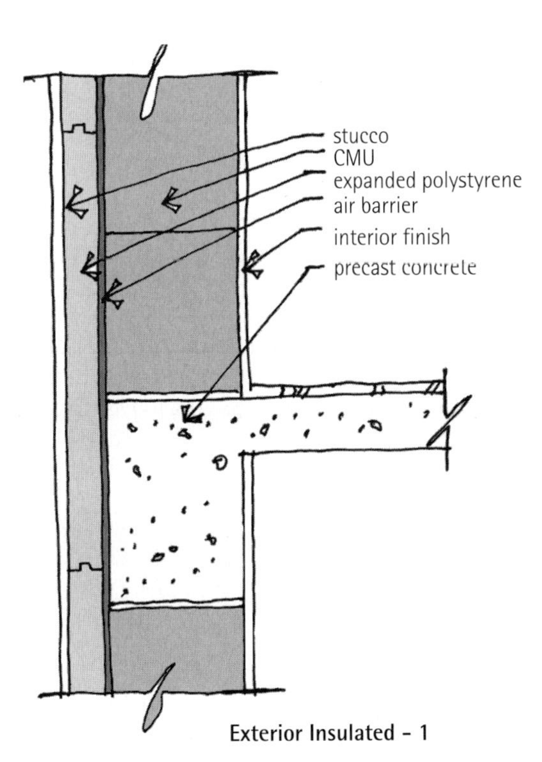

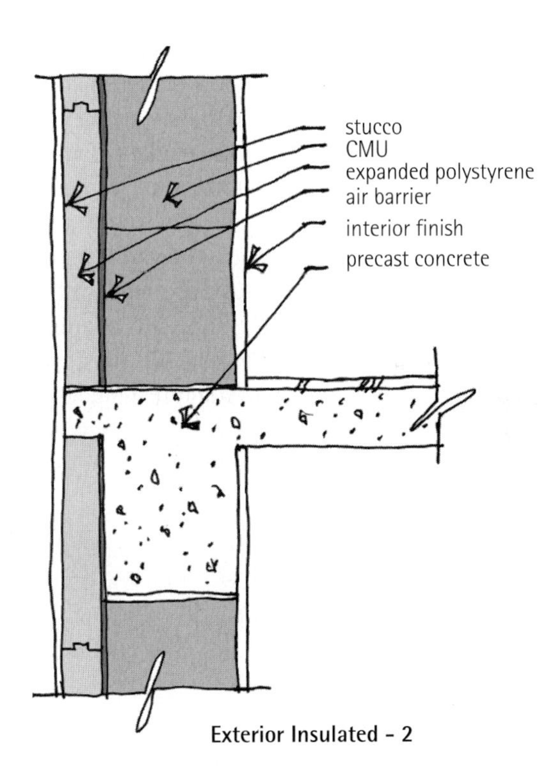

Figure 11 Typical wall sections of insulation types

Passive Solar Gain

It is recognized that in northern climates such as Beijing, adding well-insulated window area on the south façade can have a positive influence on heating energy requirements. The values for Figures 12 and 13 were calculated using the Building Energy Calculator included with this book. These studies assumed a well-insulated 6-story building, 70 meters east-west by 10 meters north-south with tight construction in Beijing, with a window area of 20 percent for all north, east, and west walls. Increasing the area of double-glazed windows on the south decreased heating requirements, while doing the same for single-glazed windows had the opposite effect because the windows conduct heat too readily (Figure 12). Figure 14 illustrates the effect on solar radiation of varying glass pane types described in Figures 12 and 13, including single-glazed clear, double-glazed clear, and double-glazed low-e. Note that an increase of window area caused an increase of the cooling load, as shown in Figure 13, unless the windows are carefully shaded to reduce direct solar energy in the summer.

Therefore, it was recommended that windows be carefully shaded. Figure 15 illustrates a summary of the recommended overhangs for Beijing. The shading worksheets shown in Figures 16a and 16b illustrate possible energy savings based on shading techniques.

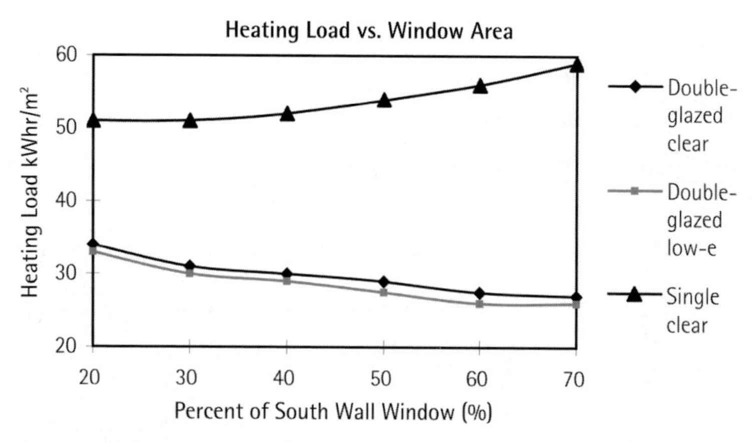

Figure 12 Effects on heating load with increasing southern window area for three window types

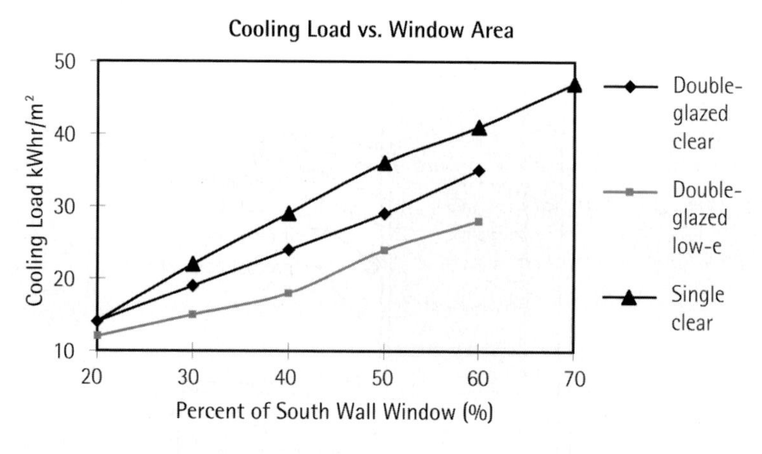

Figure 13 Cooling load versus southern window area for three different window types

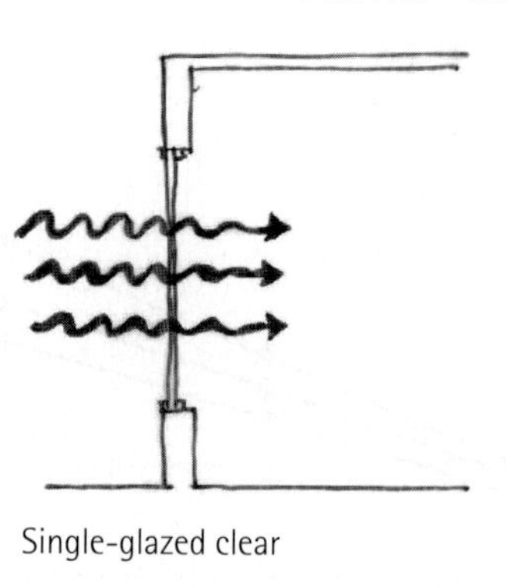

Single-glazed clear

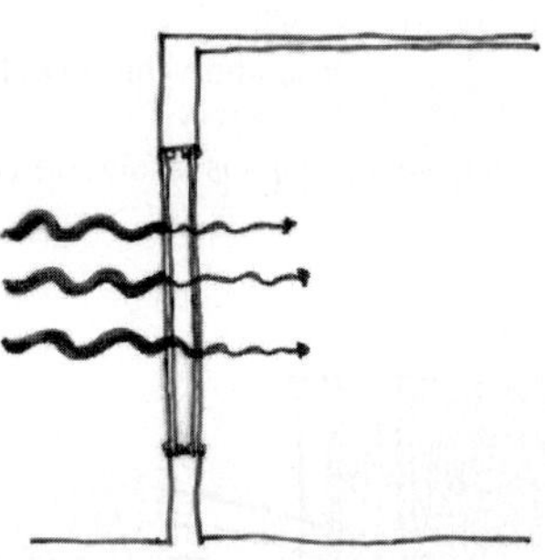

Double-glazed clear

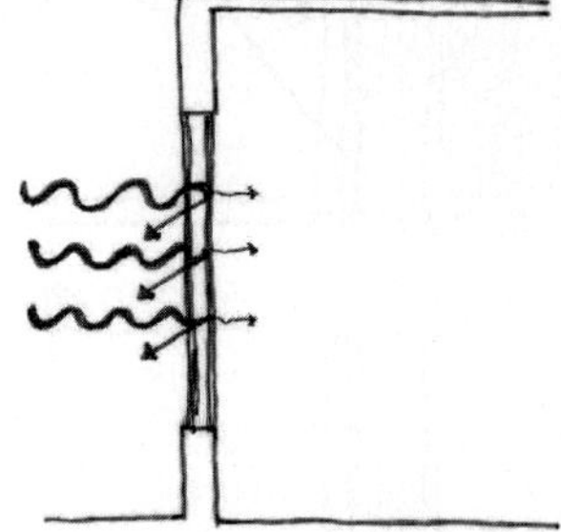

Double-glazed, low-e

Figure 14 Illustration showing effect of glass pane types on solar radiation within a building

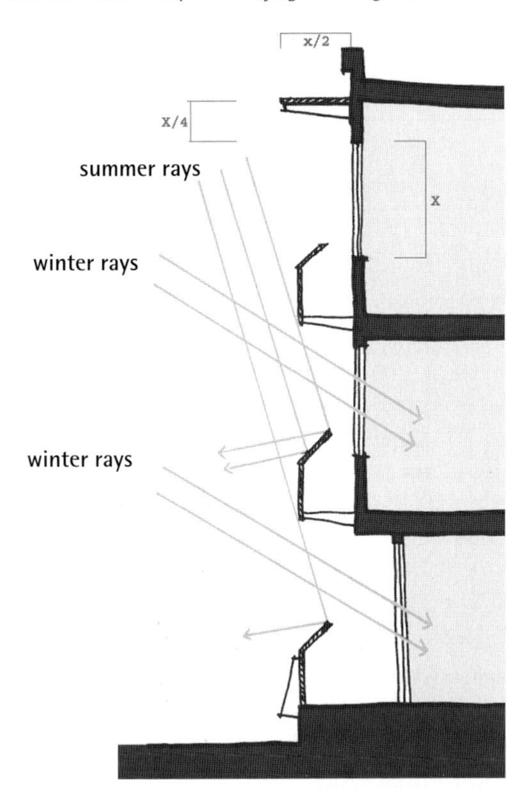

Figure 15 Section diagram of winter and summer sunlight and overhang and offset sizing relative to window height

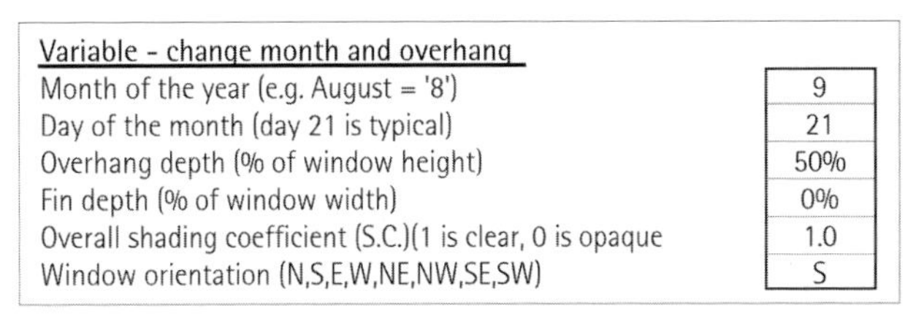

Variable - change month and overhang	
Month of the year (e.g. August = '8')	9
Day of the month (day 21 is typical)	21
Overhang depth (% of window height)	50%
Fin depth (% of window width)	0%
Overall shading coefficient (S.C.)(1 is clear, 0 is opaque	1.0
Window orientation (N,S,E,W,NE,NW,SE,SW)	S

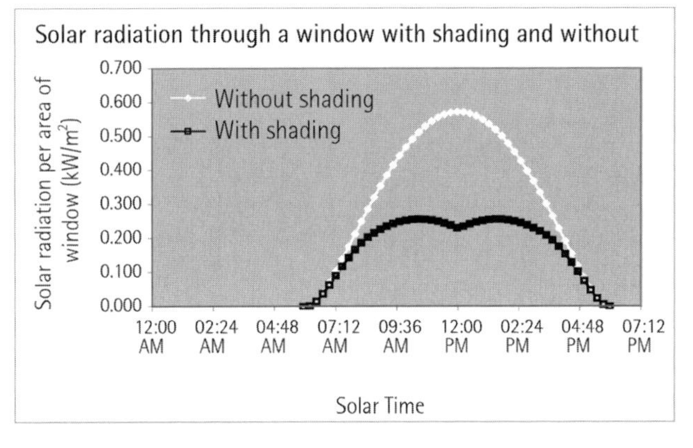

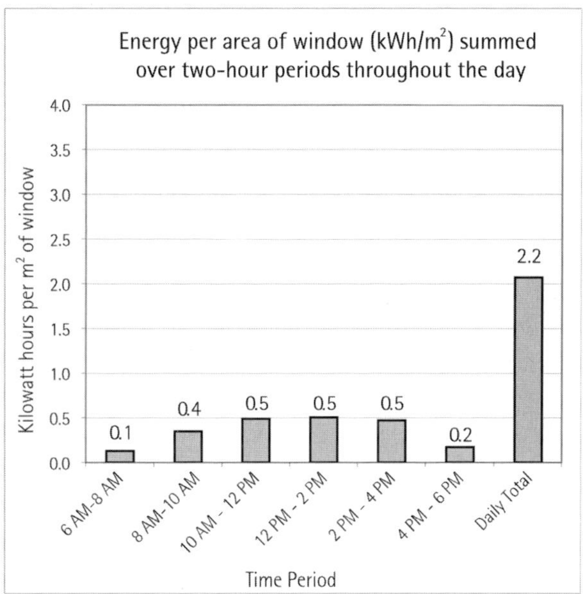

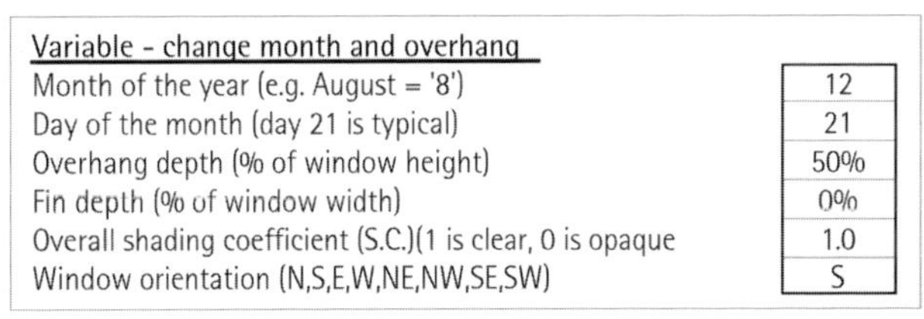

Variable - change month and overhang	
Month of the year (e.g. August = '8')	12
Day of the month (day 21 is typical)	21
Overhang depth (% of window height)	50%
Fin depth (% of window width)	0%
Overall shading coefficient (S.C.)(1 is clear, 0 is opaque	1.0
Window orientation (N,S,E,W,NE,NW,SE,SW)	S

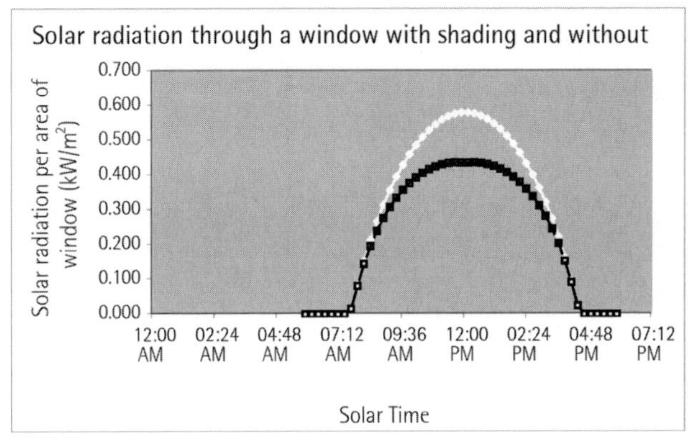

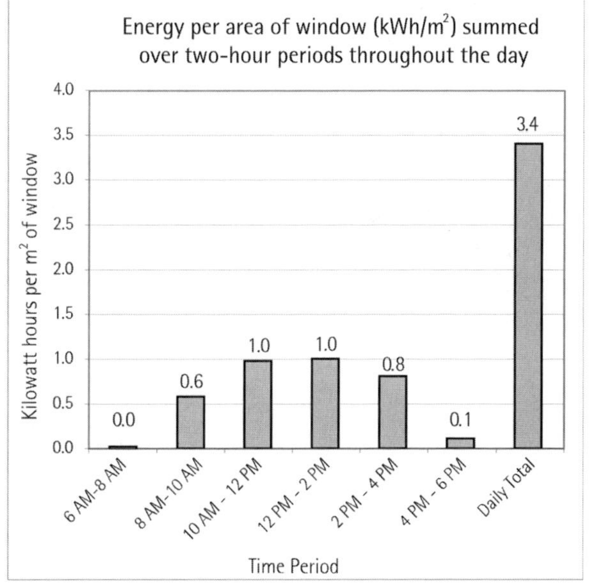

Figure 16a Beijing shading calculations worksheet showing solar radiation and energy per area of window for a south-facing window with an overhang 50% of the window height in September and December

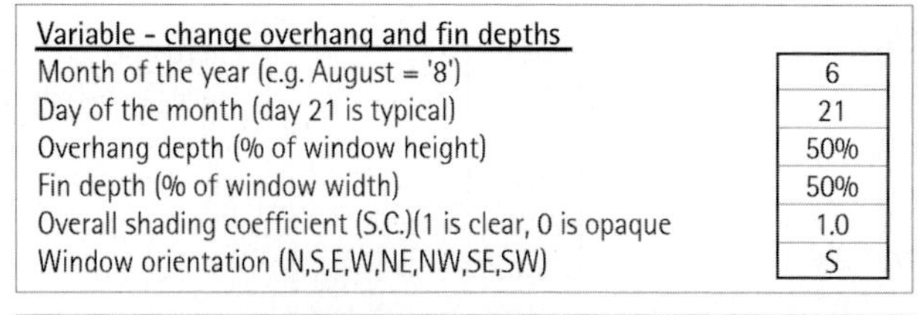

Variable - change overhang and fin depths	
Month of the year (e.g. August = '8')	6
Day of the month (day 21 is typical)	21
Overhang depth (% of window height)	50%
Fin depth (% of window width)	50%
Overall shading coefficient (S.C.)(1 is clear, 0 is opaque	1.0
Window orientation (N,S,E,W,NE,NW,SE,SW)	S

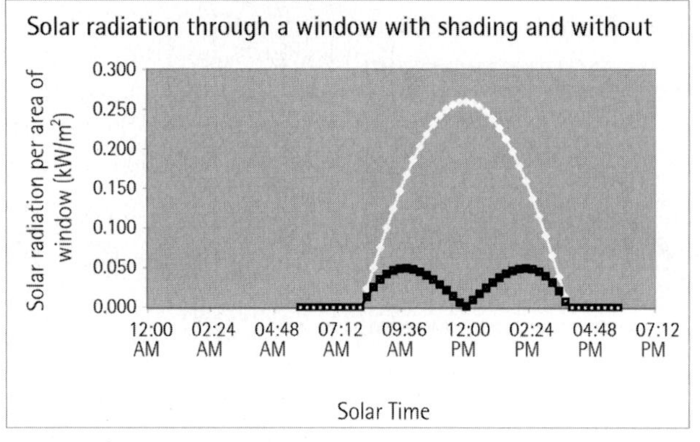

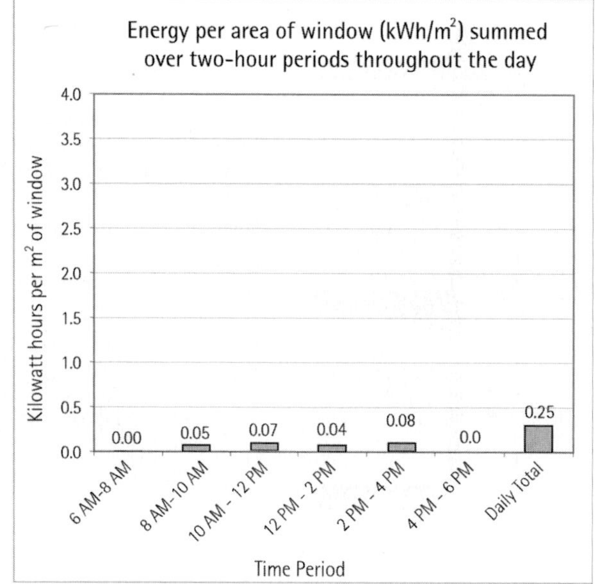

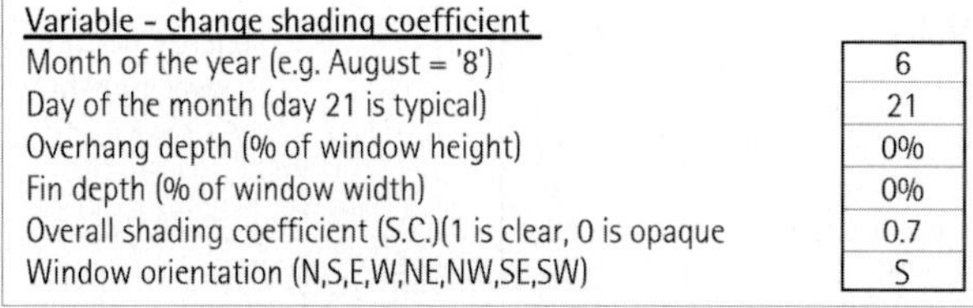

Variable - change shading coefficient	
Month of the year (e.g. August = '8')	6
Day of the month (day 21 is typical)	21
Overhang depth (% of window height)	0%
Fin depth (% of window width)	0%
Overall shading coefficient (S.C.)(1 is clear, 0 is opaque	0.7
Window orientation (N,S,E,W,NE,NW,SE,SW)	S

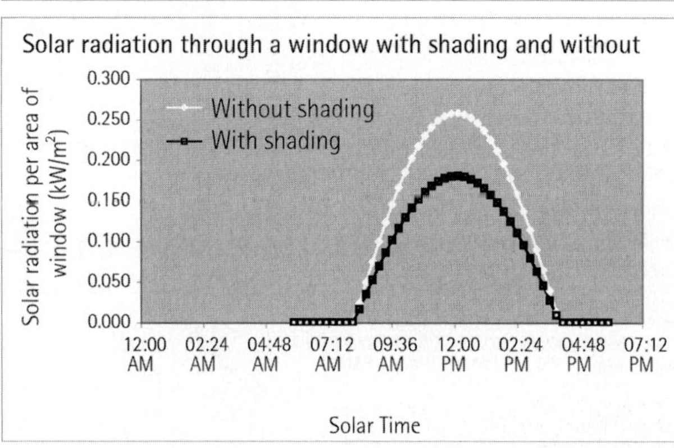

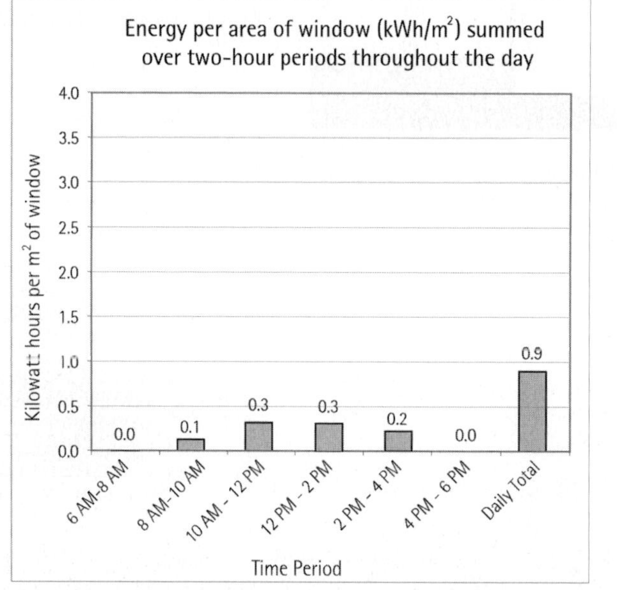

Figure 15b Beijing shading calculations worksheet showing results of overhang, vertical fins, and window shading coefficients on solar radiation and energy per area of window for a south-facing window in June

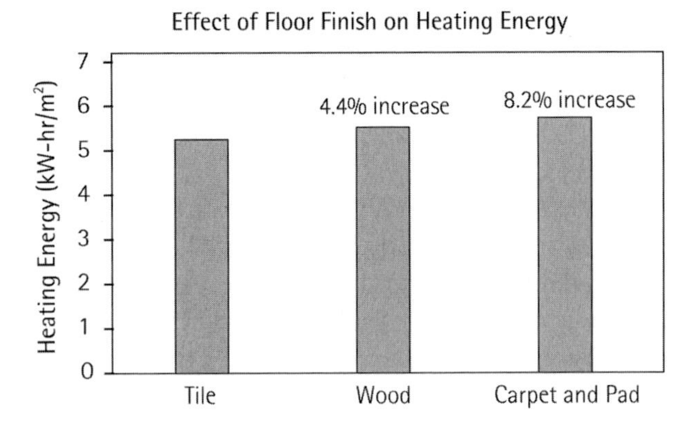

Figure 17 Effect of floor finish on heating energy

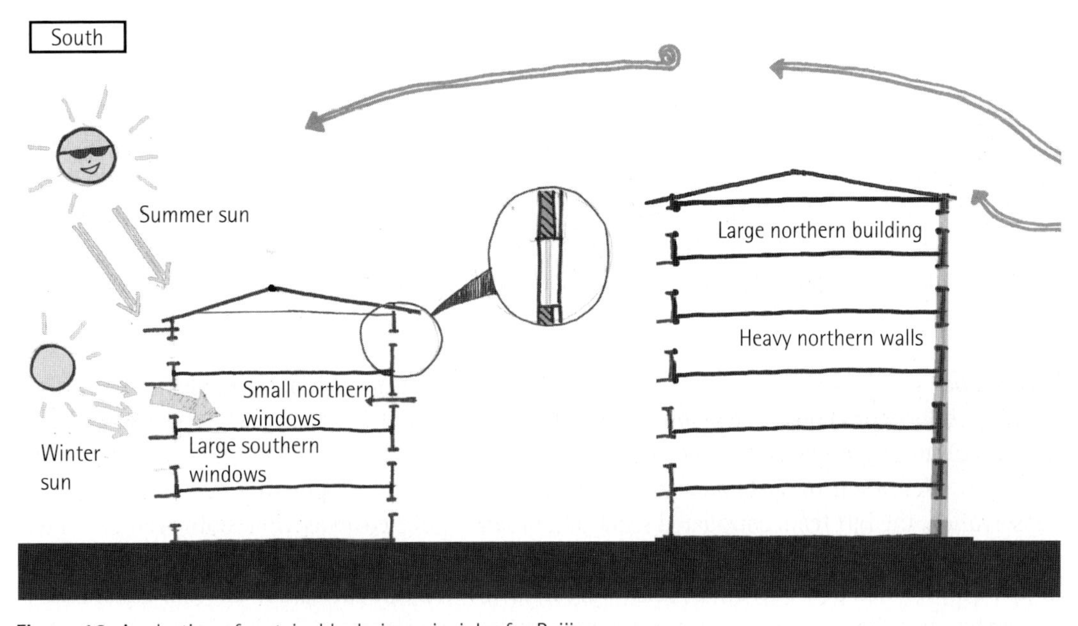

Figure 18 A selection of sustainable design principles for Beijing

Floor Finishes

The concrete floor slab, representing a large thermal mass, could store solar energy during the day for release later at night. This would utilize excess solar energy at night that otherwise would not be harnessed, and could help to reduce peak heating loads late at night. However, heat transfer to and from the slab could be hampered by the use of insulating finishing materials such as carpet or woods, as shown in Figure 17. Though the heating rate from the floor was substantially improved when the slab was very thick, the consequent increased construction costs and materials did not justify the benefits.

Summary

The recommendations based on these technical studies emphasize the advantages gained by better construction of the building envelope. High-quality construction and better building products will help to control infiltration, which will in turn reduce winter heating loads. Buildings constructed with 50 millimeter foam insulation or equivalent will limit heat loss in the winter. Double-glazed low-e windows will control heat losses, and overhangs or fins limit solar gains in the cooling season. Window area should occupy about 75 percent of the south façade. In addition, interior architectural features, such as finishes can also play a role in energy efficiency. The diagram in Figure 18 summarizes some of these findings.

Figure 19 Rather than replicate the six-story row houses with large uniform courtyards, the design team sought to create unique community spaces

C02 DEVELOPMENT PARCEL

Community

The existing architectural style of Beijing Hui Long Guan was that of a very large development consisting of numerous similar buildings with little relation to human scale. In Beijing, developers often build uniform six-story row-house-type buildings to maximize the building height without the requirement of elevators. The goal of the MIT design was to avoid the "row housing" typology in which the site contains no defined private or public spaces. Instead, the design should create a sense of a semi-private "neighborhoods," interspersed with spaces that encourage community interaction (Figure 19). To this end, a hierarchy of spaces was created, from public to private and in-between, in order to accommodate the various social needs and activities of the residents.

Public areas that were centered on junctions between major paths or in the central site area were designed to be places where all residents of the housing development would feel welcome. Community events, informal gatherings, and general social interaction could be facilitated through such public spaces, strategically located to maximize their accessibility and proximity to foot traffic. In addition, numerous "private spaces" were created by means of small green areas partially secluded from the roads and the pedestrian pathway traffic.

Landscape

Creating connections between community activities and the natural landscape was a major theme in the planning. Taking the definition of sustainability beyond energy conservation, the MIT team considered social and urban sustainability, defined as the preservation of the natural environment for ecological, psychological, and sociological benefits. Gardens, open green spaces, groves of trees, and a water canal were interspersed through the site to provide a connection to nature throughout.

The water canal was a centerpiece of the development's environmental theme (Figures 20 and 21). A pathway paralleled and weaved across the canal, facilitating pedestrian activity along the water and neighborly interactions. Because the canal looped and traversed the entire site, pedestrians could experience a variety of outdoor areas (Figures 22 and 23). Open fields along the western canal, playgrounds, and other outdoor recreation spaces provided an impetus for interacting with the larger environment. Diagrams showing a wide range of contextualized and integrated building and landscape conditions are included in Figures 21 through 28.

Neighborhoods

The buildings were arranged in groups by site area – north, south, east, west, and central – each with its own distinct community space. For instance, each of the buildings along the eastern portion of the site had an individual courtyard, whereby central building groups enclosed large semi-private plazas. Grass fields lined the northern edge, and a public plaza was located near the center of the site for all residents to use (Figures 24 and 25). In summary, a community space for the site as a whole was provided, as well as "neighborhood" areas geared towards use by residents of each individual building group.

Extensive landscaping included grass, trees, and water to provide both positive psychological and recreational amenities. The landscaping was also helpful in making the development more sustainable. The large grass-covered areas, while both aesthetic and recreational, also added to the overall site surface permeability and helped to combat carbon dioxide emissions. Coniferous trees were strategically placed on the north side of the

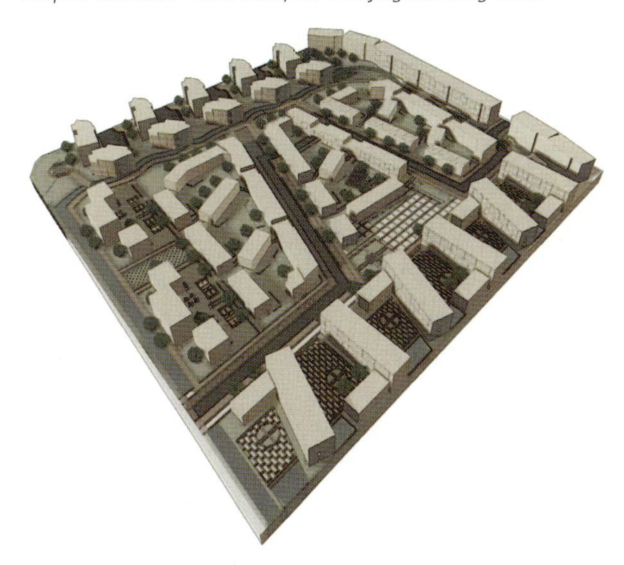

Figure 20 Aerial view of C02 scheme - a water canal flows through the site, bringing nature closer to the residences (north is to the upper right corner)

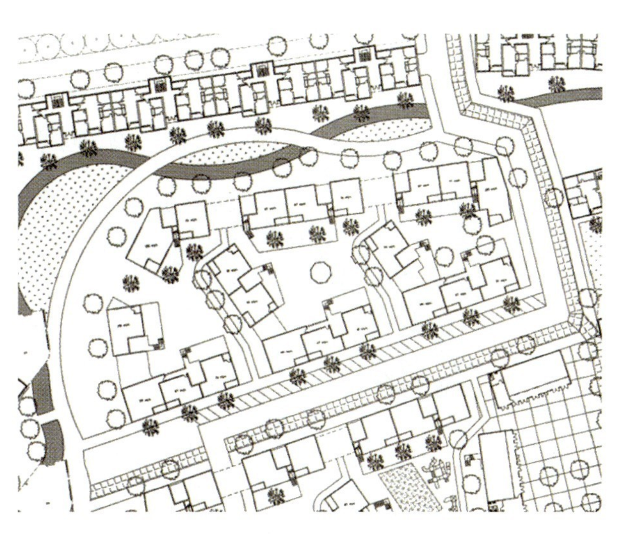

Figure 22 Building clusters are centered around a semi-private courtyard

Figure 24 C02 site diagram - green space

Figure 21 C02 site diagram - location of water features (north is to the top in Figures 21-25)

Figure 23 C02 site diagram - private courtyards

Figure 25 C02 site diagram - hierarchy of spaces

Figure 26 C02 site diagram - access and circulation (north is to the top in Figures 26-29)

Figure 27 C02 site diagram - paved and covered areas

Figure 28 C02 site diagram - overall diagram

buildings to block wind in the winter, while deciduous trees were placed on the south side of the buildings to block sun in the summer and admit warming sunlight in the winter.

Wind

One of the primary objectives of the site design was to create a scheme that blocked the winter wind while allowing summer winds from the south to penetrate throughout (Figures 29a and 29b). The design consisted of several high, relatively impermeable buildings along the northern edge of the site that blocked cold north winds in the winter. In addition, the landscape strategy recommended planting trees along the northern and western edges of buildings to help minimize the penetration of northerly and northwesterly winter winds into the site. Lower and smaller buildings formed a permeable southern edge to facilitate penetration of cooling winds during the summer.

CFD studies were performed to compare existing Tian Hong designs with the site designs by MIT. In the Tian Hong scheme, large building masses with long open spaces between them promoted dramatic wind velocities, with almost half the site averaging 4 m/s (Figure 30a).

The MIT scheme was also analyzed for summer and winter winds. Smaller building masses and shorter distances between the buildings, tended to alleviate the more extreme wind conditions, with the proposed configuration yielding an average wind speed of 1.5 m/s. In the summer months, the analysis showed air movement even within the central cluster spaces at an average 0.5 m/s (Figure 30b). The majority of the site saw some air movement only slightly lower than the original wind at 2 m/s from the south. We could therefore expect some benefit from natural ventilation in the summer time.

During the winter months, the CFD simulations showed greatly reduced air movement throughout the site, particularly within the central cluster spaces that have average wind speeds of 1.5 m/s (Figures 30c and 30d). The wind vectors in Figure 29 demonstrate that the northern building edge could be highly effective in redirecting the wind along the sides of the site. The average wind speed throughout the site was about 2 m/s, which was well below the original wind at 5 m/s from the north.

Shading

Shading studies were performed to assess daylighting and shading, as well as to show that the new scheme, with a larger number of shallower buildings, provided the desired density number while satisfying the winter sunlight code. The reduction in building width also increased daylight and natural ventilation penetration into the unit interiors. The shading studies are shown in Figure 31.

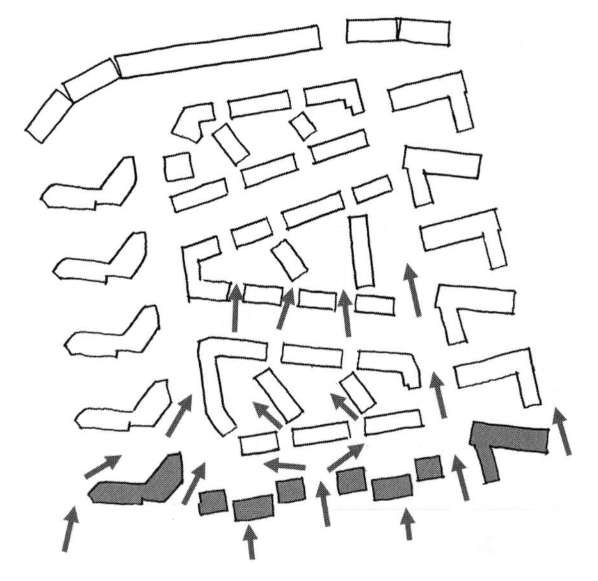

Figure 29a Site design intention - allow summer winds from the south to penetrate

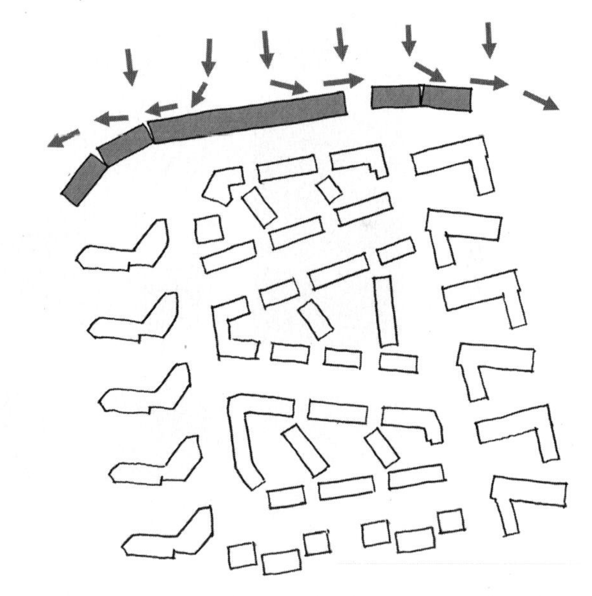

Figure 29b Site design intention - block winter winds from the north

Figure 30a Winter winds (5 m/s from north; velocity at 2 m above ground level) in Tian Hong's original scheme with large building masses (north is to the top in Figures 30a - 30d)

Figure 30c Winter winds (5 m/s from north; velocity at 2 m above ground level) in MIT's proposed scheme with smaller buildings

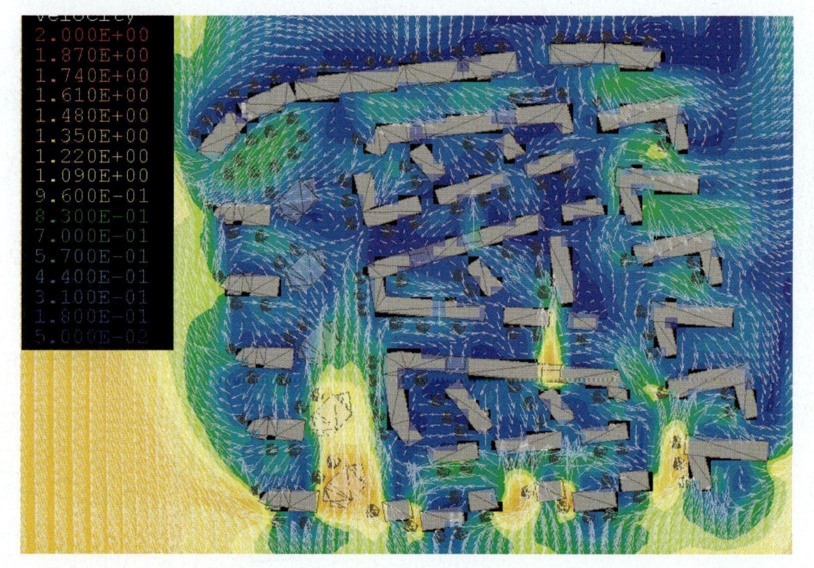

Figure 30b Summer winds (2 m/s from south; velocity at 2 m above ground level)

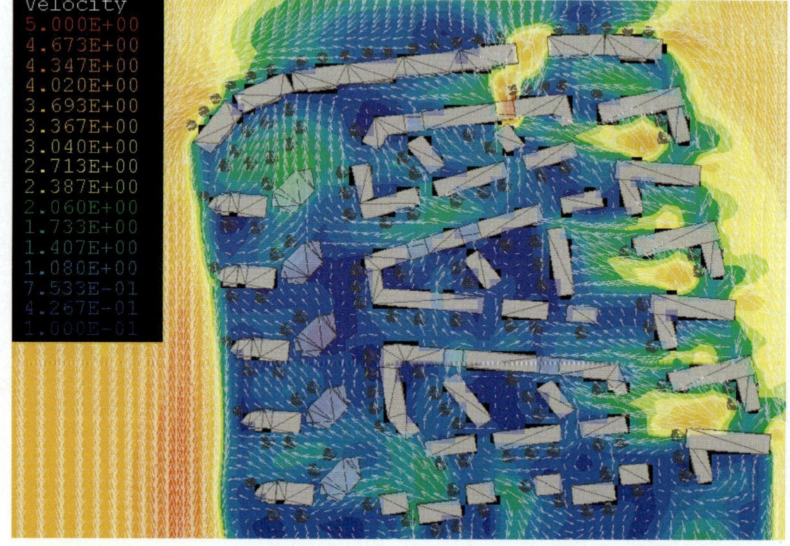

Figure 30d Summer winds (5 m/s from north; velocity at 2 m above ground level) in MIT's proposed scheme with smaller buildings

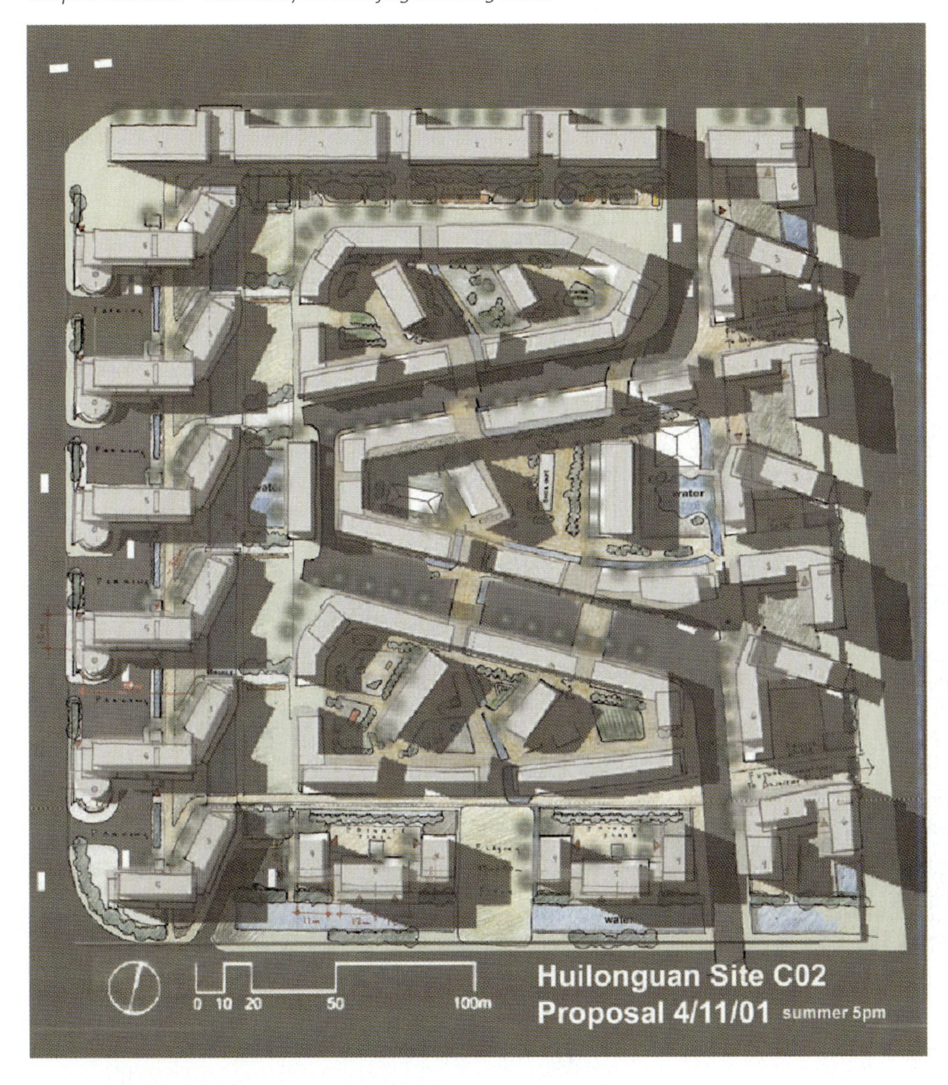

Figure 31a Shading study showing site shadows on 21 June at 5 PM

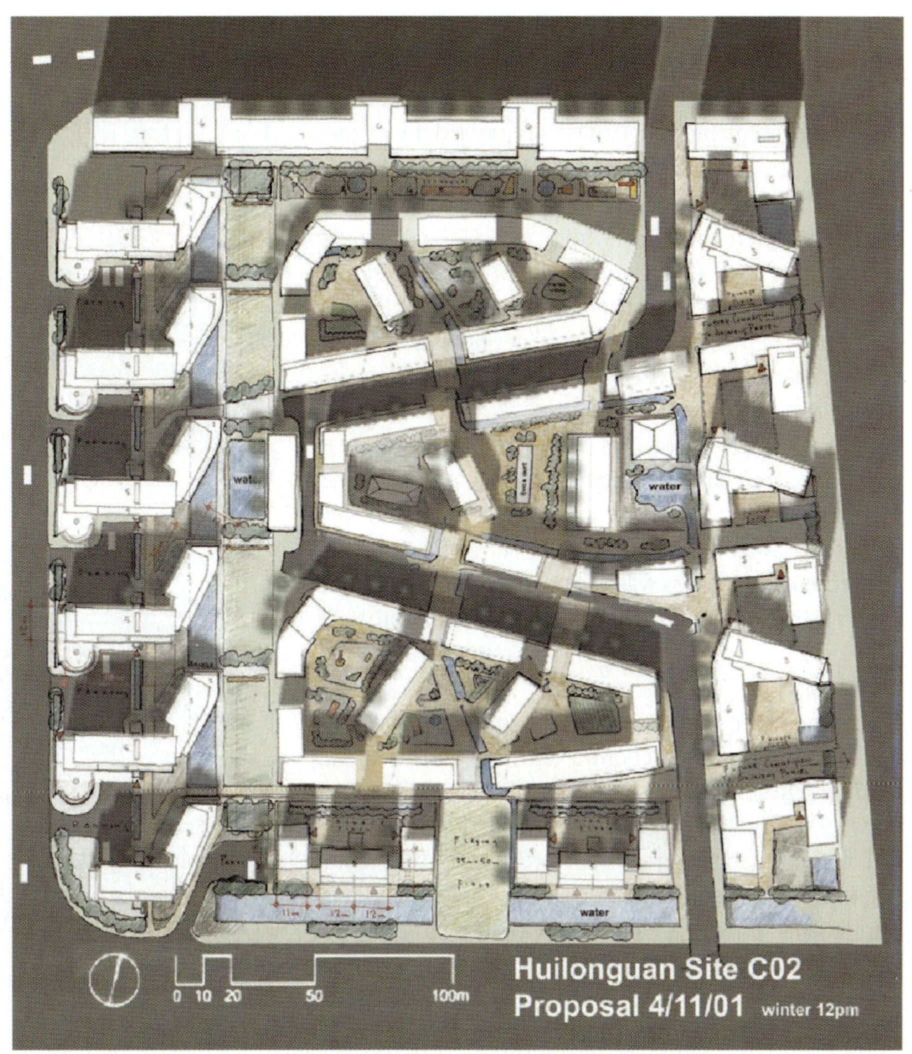

Figure 31b Shading study showing site shadows on 21 December at 12 PM

C06 DEVELOPMENT PARCEL

Community

During the design process, Tian Hong changed the project site from C02 to C06. Many of the goals and objectives discussed earlier for C02 were applied to the development of the site plan for C06. A total of 1,500 residences consisting of 148,000 square meters were designed. The new site was located to the southeast of C02 (Figure 32). Similar high-density residential developments occur to the east and west. The area to the north of the site was not yet planned and was a generally flat, undeveloped space. However, the area to the south was very articulated with a long green park space beginning more than half a mile to the south that terminated well within the larger neighborhood block of C06 (Figure 33). The proposed park area was considered a valuable public amenity that would greatly benefit the community.

Landscape

The design process addressed the energy needs of winter heating and summer cooling, as well as the social requirements of establishing meaningful connections between public and private space. The design scheme by MIT incorporated a long southern green space that vastly extended into the development, and was broken down into smaller private areas. The creation of a "serrated edge," which allowed the public green space to flow into the private courtyards, established a relationship between the central, public green space and the more private courtyards, located between individual buildings. Different outdoor spaces played varying roles in the community, with larger areas serving as community gathering spaces and gardens, and smaller areas being used for morning exercises and children's playgrounds. A secondary advantage of this strategy was an increased

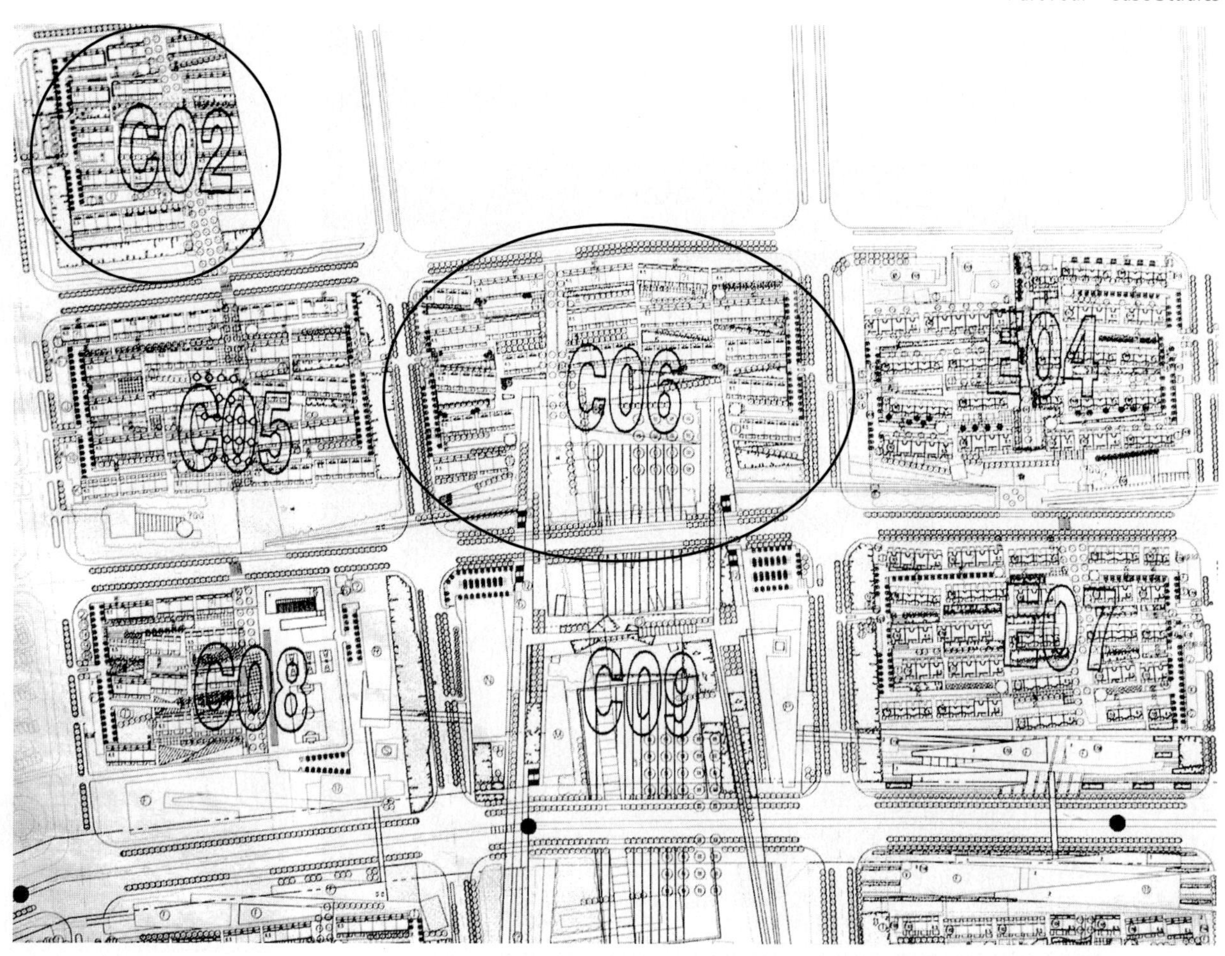

Figure 32 Hui Long Guan residential development site map of Tian Hong's master plan and original designs for the C02 and C06 parcels

Figure 33 Site plan of MIT's design for parcel C06. The site concept established a relationship between the central, public green space and the more private courtyards located between individual buildings. The creation of a "serrated edge" allowed the public green space to flow into the private courtyards, facilitating a transition between public spaces and more intimate spaces, and creating a strong connection to the park.

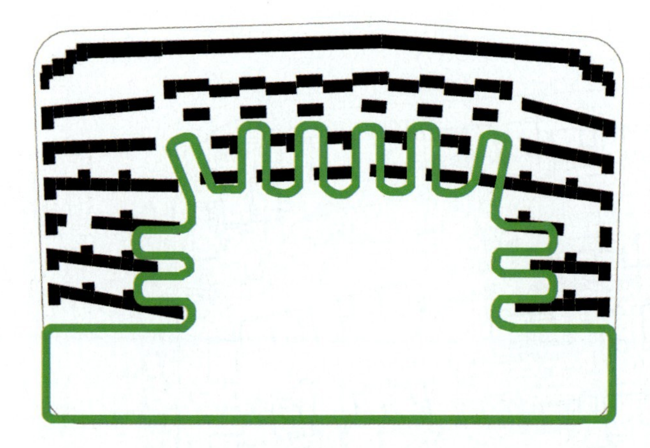

Figure 34 C06 site design concept: penetration of green space into the center of the site allowed an increased visual access to the park, higher air quality for the residents, and recreational access

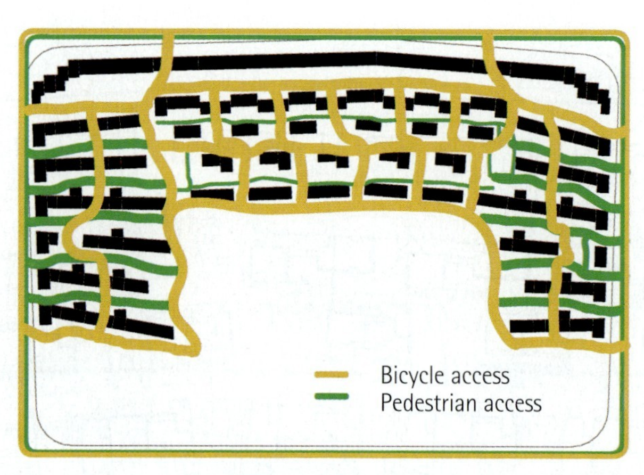

Figure 36 Bicycle and pedestrian access: in order to encourage use of more sustainable modes of transportation, bicycle and pedestrian paths were provided throughout the site

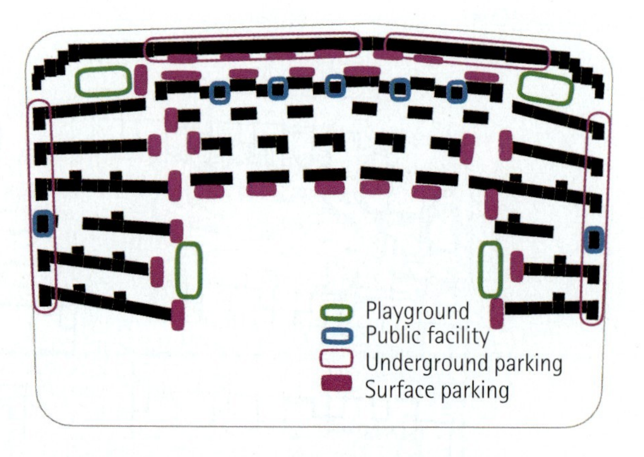

Figure 38 Amenities: surface parking followed the arrangement of car access, further defining a car-free center for every neighborhood

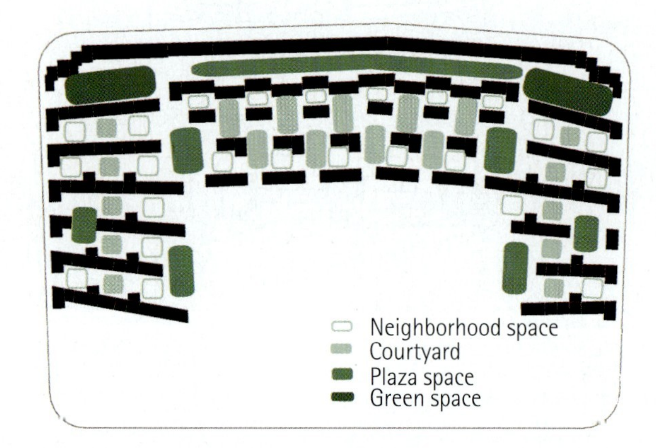

Figure 35 Hierarchy of public spaces: building form and spacing articulated various outdoor gathering areas

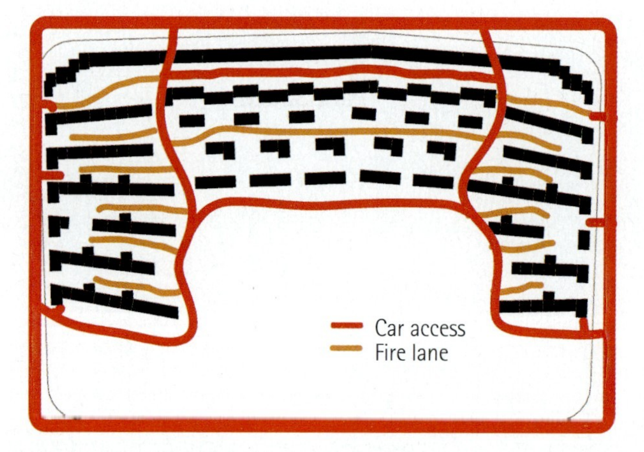

Figure 37 Vehicle access: while access for emergency vehicles was allowed, the site design kept most car traffic to the periphery of the neighborhoods

visual access to the park, which was achieved by allowing the spatial penetration of the southern green space into the residential areas. The entrance of predominantly southern winds was allowed into the heart of the development. By extending this green space into the development, the "heat island" common to man-made surfaces was also avoided. A further advantage was increased air quality due to positive effects of foliage on polluted air particulates and carbon dioxide.

Hierarchy of Public Spaces

The first step in the larger planning strategy was to establish the relationship between the public spaces outside of the development and the more private interface of the green spaces within the development. The design team further developed a hierarchy of public spaces and used building form and spacing to define various outdoor gathering areas (Figures 34 and 35). These spaces were articulated at a variety of scales for activities, ranging from large community gatherings to family picnics. Large areas were designed to be accessible to greater numbers of people, while the more frequented and smaller open spaces were dispersed throughout the neighborhoods for easy accessibility to smaller gatherings.

Circulation

In order to encourage the use of more sustainable modes of transportation, bicycle and pedestrian paths were integrated to allow as much movement as possible and were conceived of as the dominant forms of transportation within the housing development (Figure 36). While access for emergency vehicles was allowed, the site design kept the majority of vehicular traffic and parking to the periphery of neighborhoods (Figure 37). This arrangement should reduce noise and increase air quality within the neighborhoods. Surface parking followed a similar arrangement, further defining a car-free center to every neighborhood (Figure 38).

Wind

During the summer months, the prevailing winds in Beijing are from the south (Figure 39a). The standard design of other neighborhood developments in the area are of uniform height, rectangular-shaped buildings laid out in rows, with the long sides facing north to south. This configuration does not allow winds to penetrate deeply into the developments as individual buildings block airflow from the north or south. MIT's proposal staggered both the height and the plan of the buildings, allowing southern breezes to penetrate the site for passive cooling purposes.

The prevailing winter winds in Beijing are from the north (Figure 39b). The site design therefore reduced wind speeds throughout the site by placing the tallest buildings on the northern edge of the site in order to block and divert winter winds from the north. A CFD model was used to verify the effectiveness of this design, and results shown in Figures 40 through 42 demonstrate that the winter air speed in the interior of the site was generally reduced by more than 50 percent on the exterior of the site.

Shading

Shading studies were performed in order to show that the new scheme provided the desired density number while still satisfying the winter sunlight code. Shading studies shown in Figures 43 through 45 demonstrate winter shadows at 9 AM, 12 PM, and 3 PM. Note that the proposed study had additional rows of thinner buildings. The buildings were arranged in this manner as a means to obtain better daylight access while also providing improved summer wind circulation.

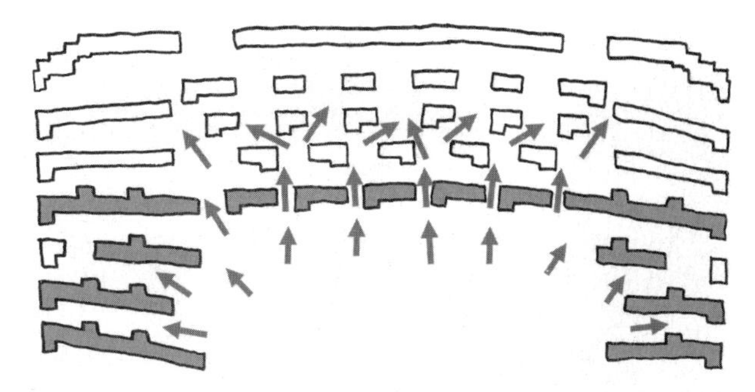

Figure 39a Summer winds: the residences were arranged to allow the cooling summer breezes to penetrate into the site

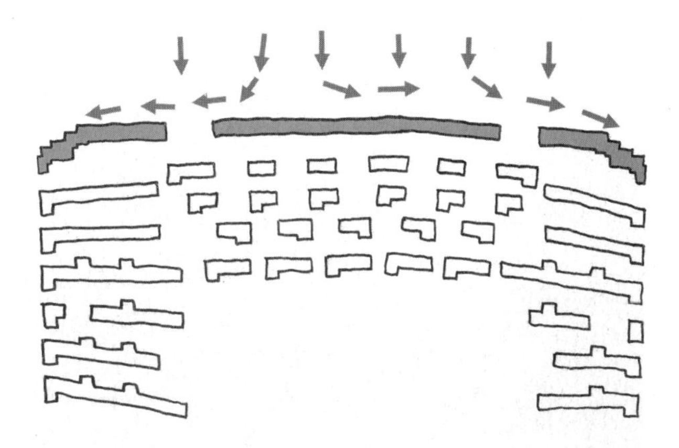

Figure 39b Winter winds: a heavy northern perimeter reduced harsh winter winds that penetrated into the site

Figure 40a MIT design CFD results showing summer winds (1.8 m/s from south; velocity at 6 m above ground level) - during the summer months, the prevailing southern breezes easily penetrate into the site to cool the neighborhoods

Figure 40b MIT design CFD results showing winter winds (6.8 m/s from north; velocity at 6.6 m above ground level): the majority of the site is protected from cold winter winds from the north. High-velocity winds, indicated by orange and yellow areas, are greatly reduced within the interior of the site, as indicated by blue areas.

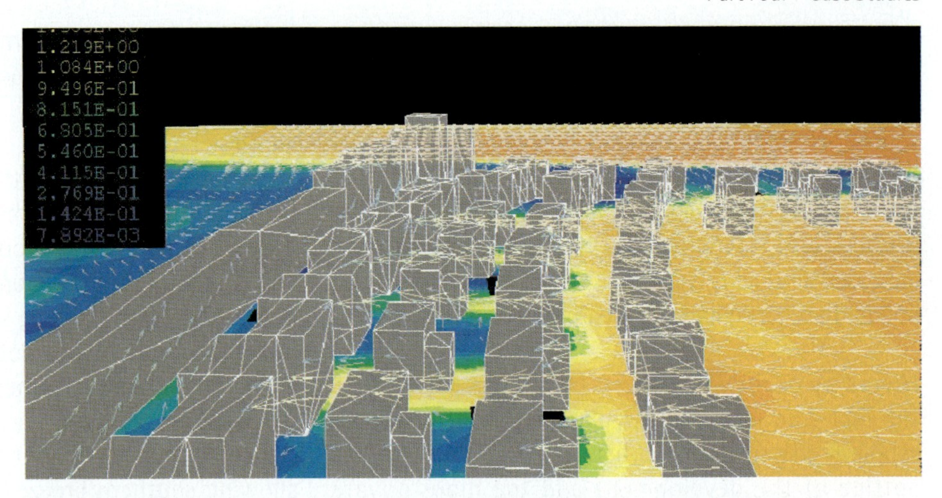

Figure 41 MIT design: lower buildings with staggered heights allow summer breezes to penetrate deep into the site

Figure 42 Tian Hong's original design: beneficial summer breezes from the south are stopped by uniform six-story buildings

Solar Orientation

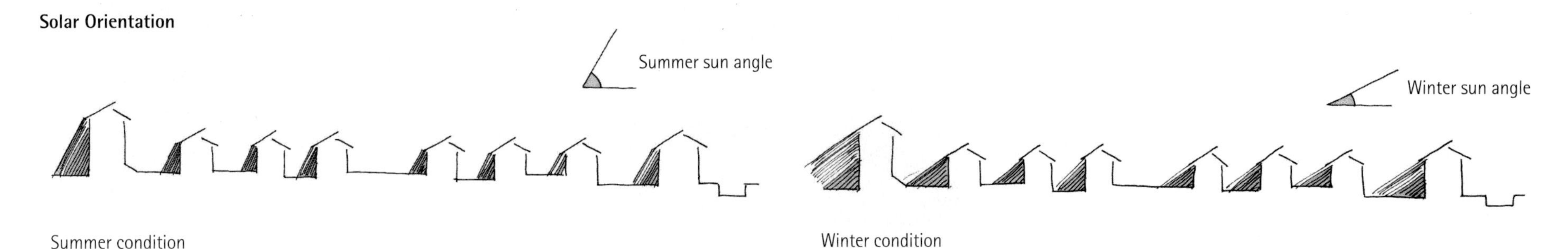

Figure 43 Building spacing and roof heights were determined with sun angle studies, utilized to verify that the winter sunlight code was adhered to for the site plan of C06; building heights correlate to building spacing (section looking east; north is to the left)

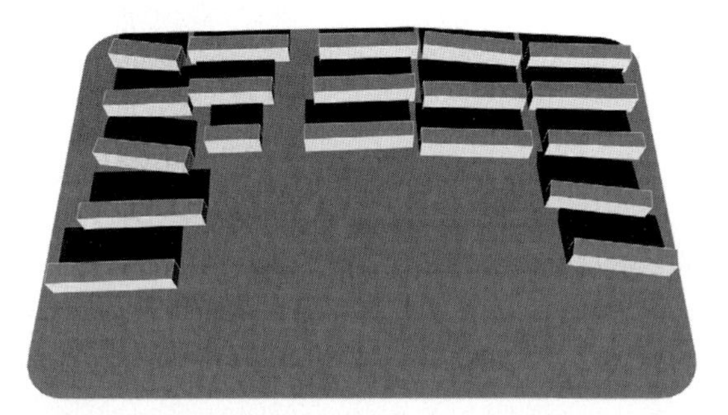

Figure 44 Shading study showing the Tian Hong's original design for C06 on 21 December at 12 PM (north is to the top in Figures 44-45)

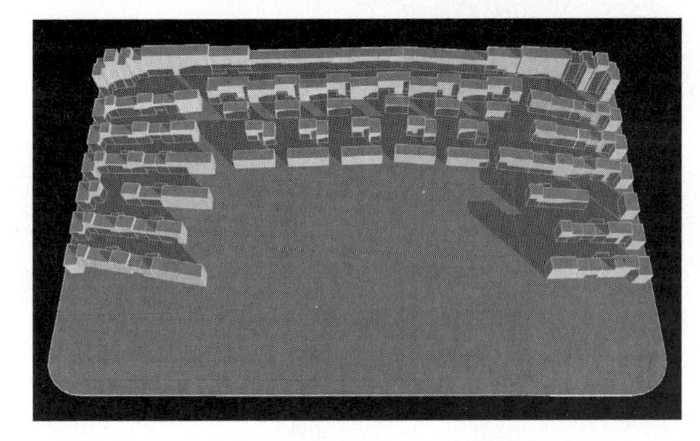

Figure 45b Shading studies of MIT's proposed scheme showing winter sun conditions on 21 December at 9 AM

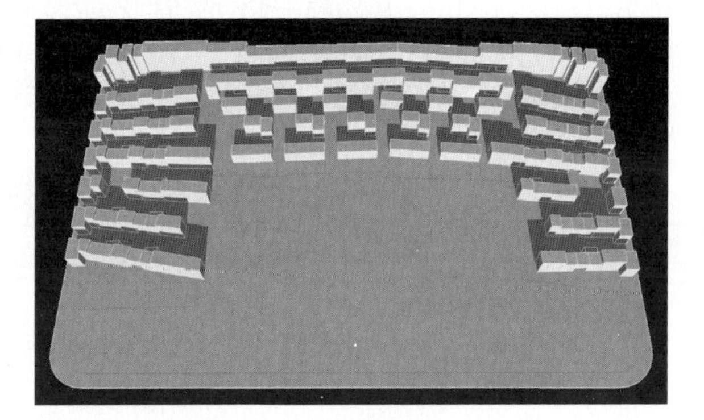

Figure 45a Shading studies of MIT's proposed scheme showing winter sun conditions on 21 December at 12 PM

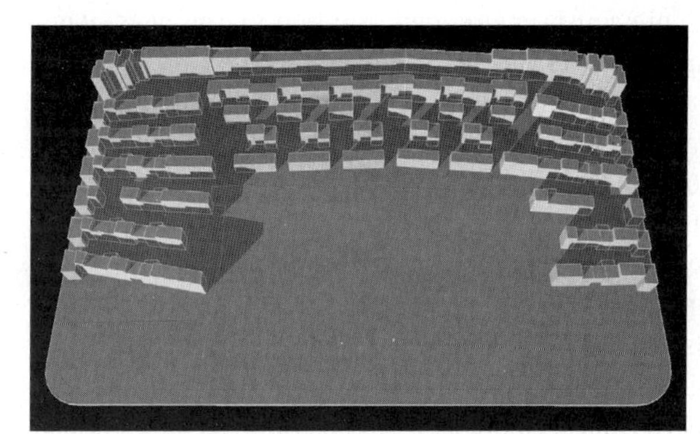

Figure 45c Shading studies of MIT's proposed scheme showing winter sun conditions on 21 December at 3 PM

Building Typologies

The building design balanced the needs of economy achieved through simplicity of design with a spatial richness articulated in a variety of neighborhood characteristics. The spatial diversity of the site was created through the combination of four basic building modules, each of which was a combination of the five unit types. To make the construction process more efficient, the units themselves were derived from a ten-meter-deep unit that was seven, ten, or thirteen meters wide, and a variety in architectural form was created from a single structural model (Figures 46-50).

The building depth was less than that of conventional Beijing designs. The narrower buildings allowed more daylight and natural ventilation. The staggered building arrangement also allowed the narrower building to achieve the same floor area ratio (FAR) as the wider conventional designs of the same building height. The various architectural forms were in turn composed in a variety of ways to create both small and large neighborhoods, as well as various wind and shading conditions.

Façade Treatment

As previously mentioned, the one of the developer's requirements was that all buildings have south-facing entrances. This, combined with the required density, encouraged the design of buildings oriented towards the south. The sun was shaded using deciduous trees and shading devices that protected interior spaces from the high-angle summer sunlight, while allowing the low-angle winter sun in. To optimize the performance of passive winter heating energy, overhangs were sized for all windows; 75 percent of all south-facing walls were glazed; and a combination of fixed and operable double-glazed low-e windows were used to achieve the lowest possible U-value (Figures 51 and 52). Note, if the windows

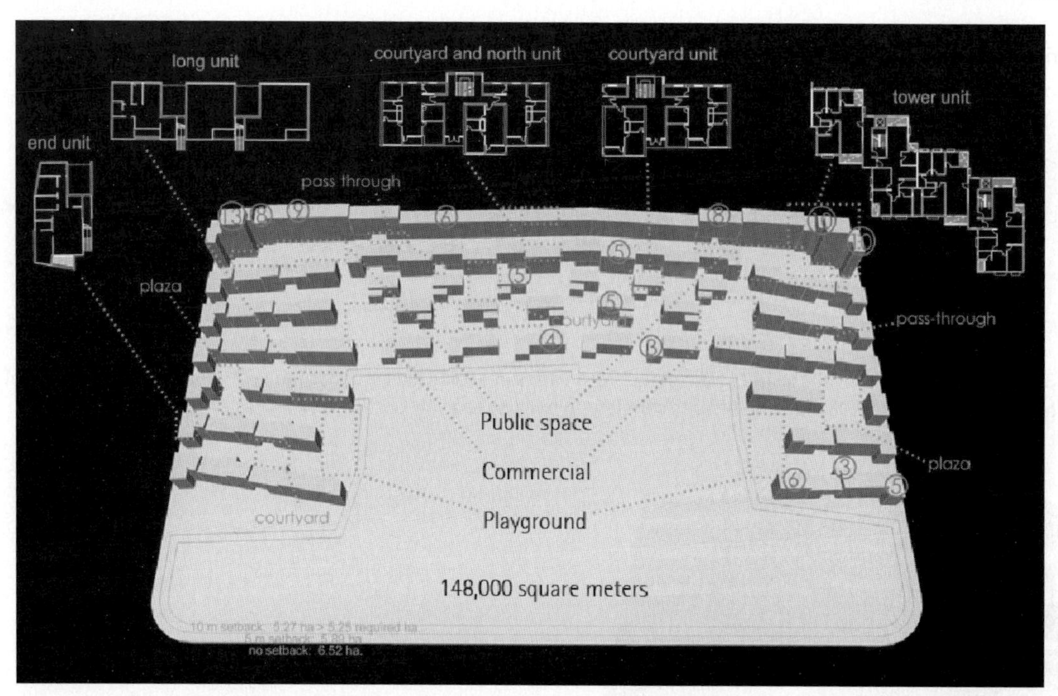

Figure 46 Preliminary diagram of building modules and unit plans

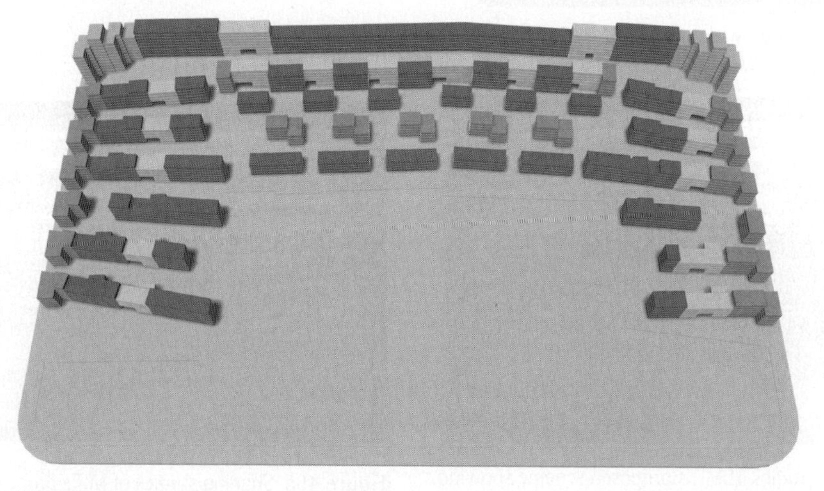

Figure 47 Building typologies and buildings - varying shades of grey indicate the four building modules

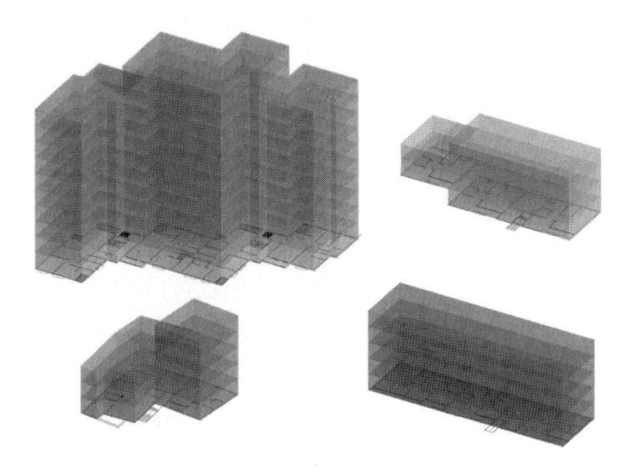

Figure 48 Axonometric sketches of the four building modules

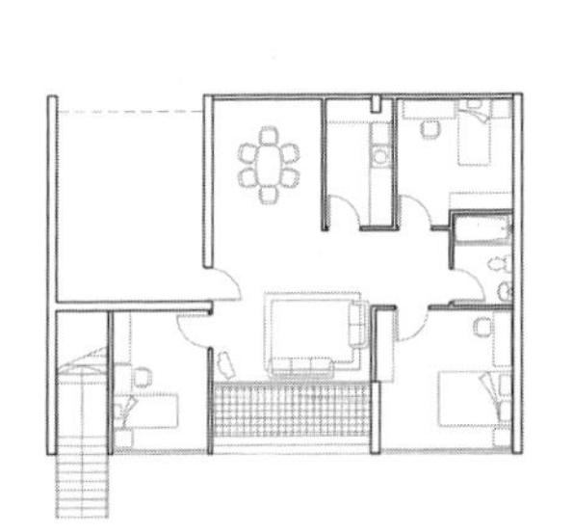

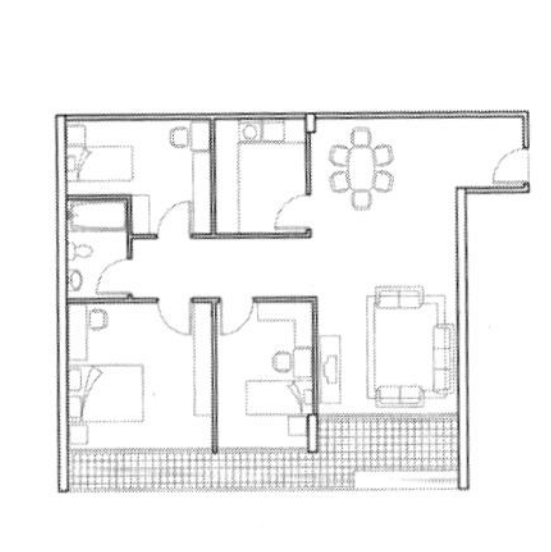

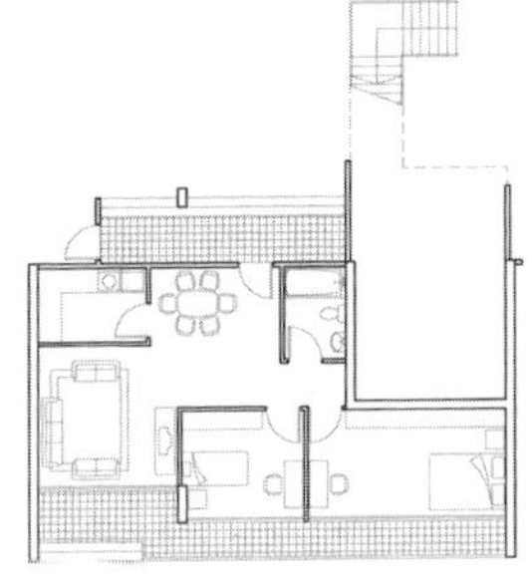

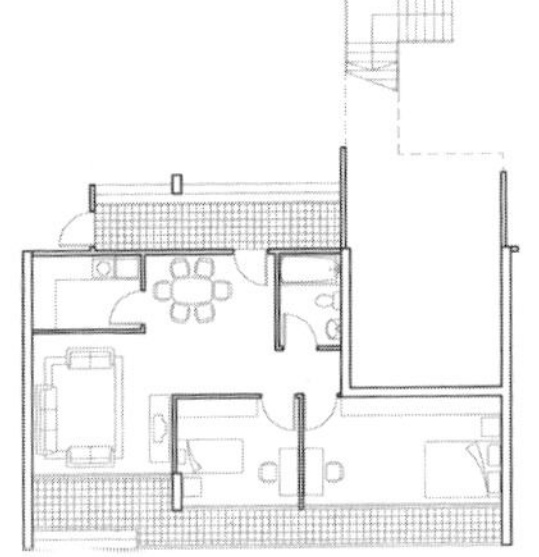

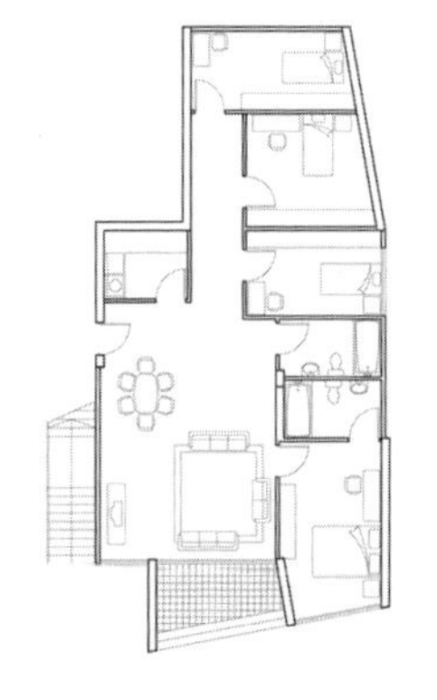

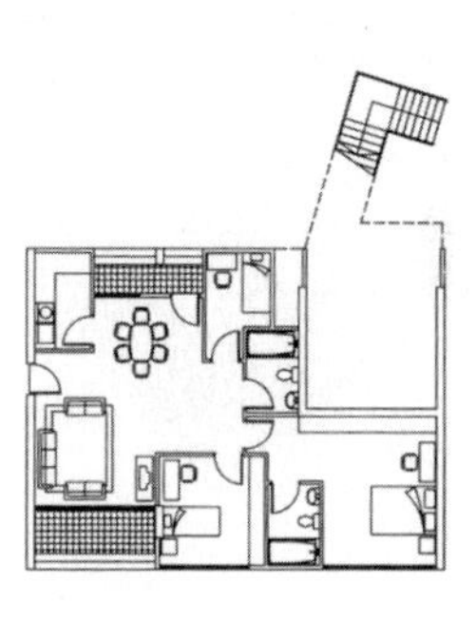

Figure 49 Five typical unit plans

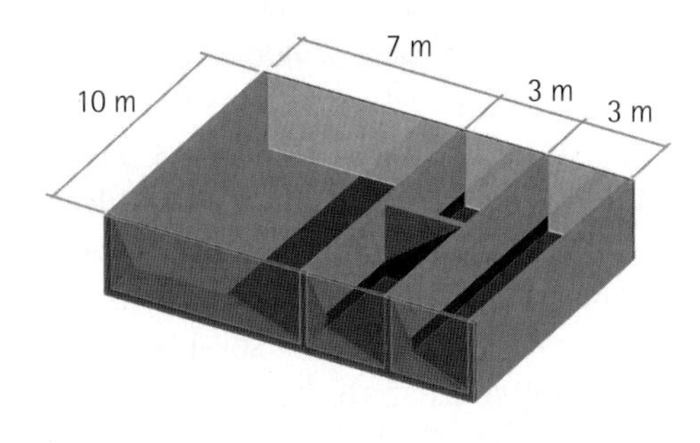

Figure 50 The individual living units are all derived from a 10 m deep module that is either 7 m, 10 m, or 13 m wide

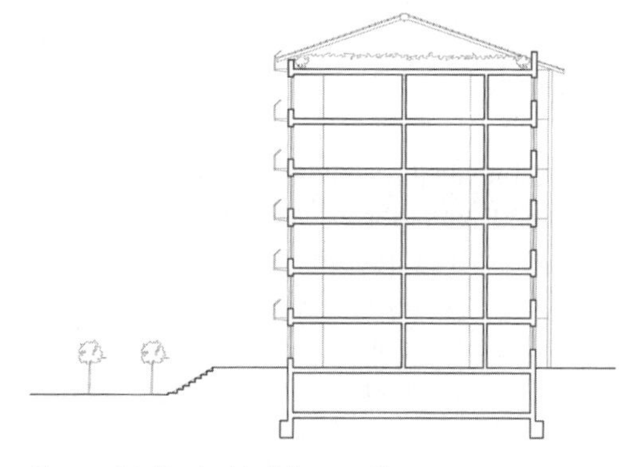

Figure 51 Typical building section

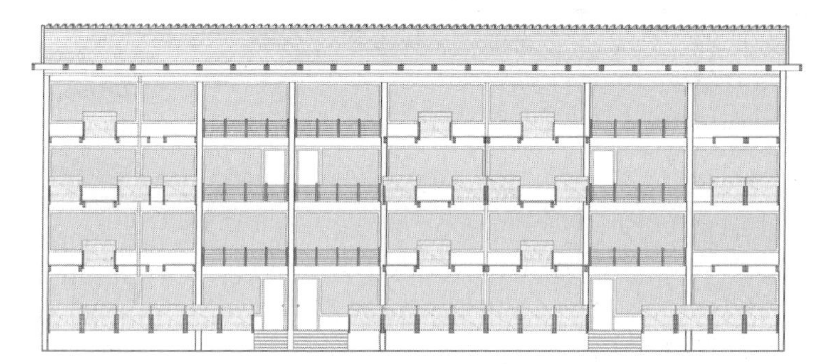

Figure 52 Typical southern façade: this example demonstrates 75% glazing to optimize winter solar gain while overhangs reduce summer solar gain

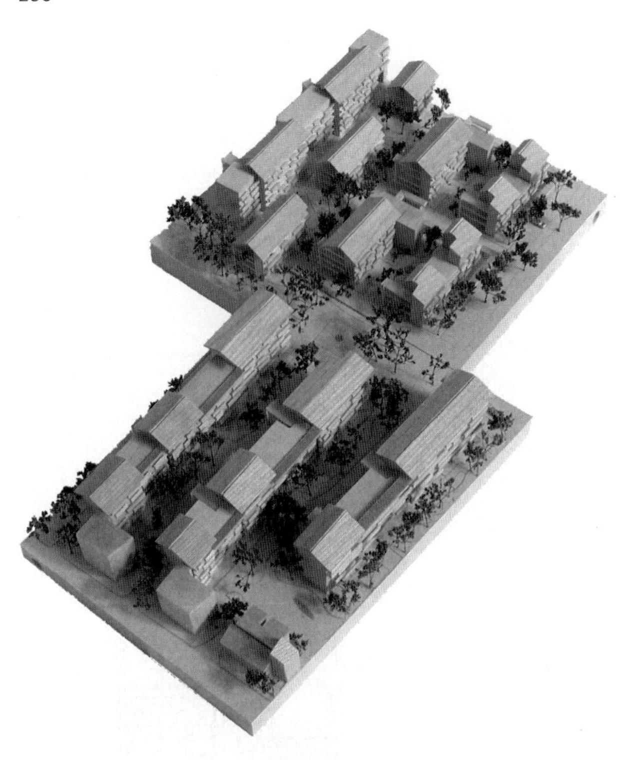

Figure 53 Partial site model showing interior courtyards with a mixture of soft- and hardscape

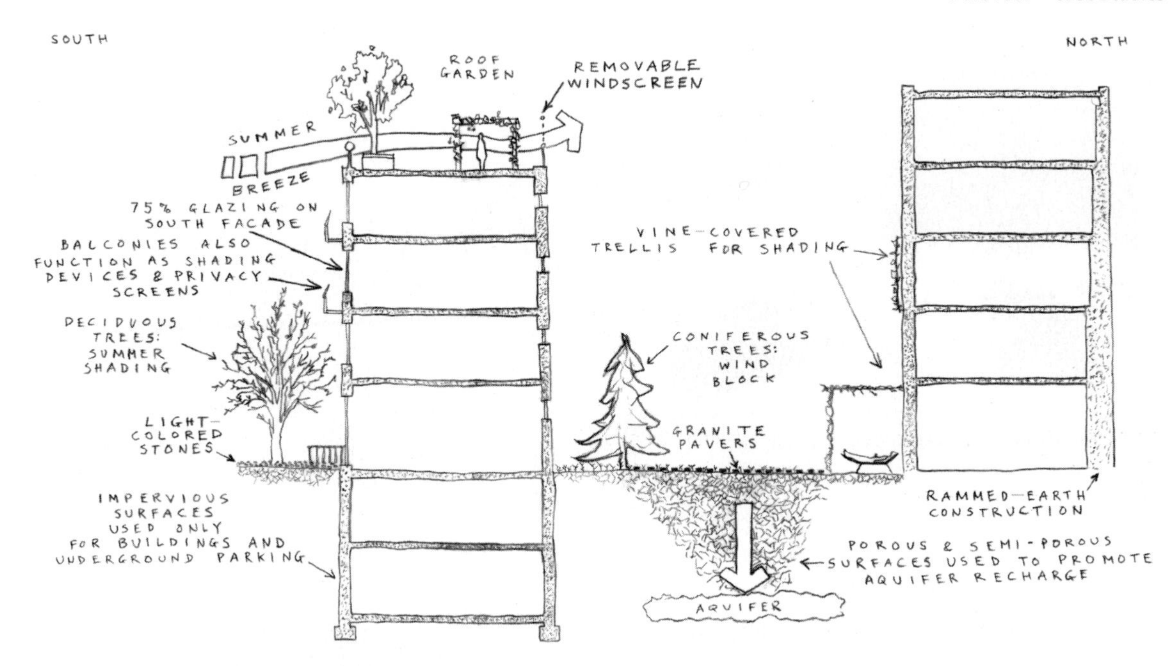

Figure 54 Diagram of preliinary material selection for Hui Long Guan

are extensively shaded by neighboring buildings this recommendation needs to be revised.

Porous Paving

The surface treatment of any large development is an essential aspect of its sustainability. While similar projects often feature vast areas of non-porous surfacing, such as concrete and blacktop, they require an enormous expense involved in building substantial storm water runoff systems. These systems also fail to recover rainwater that can be used to recharge local aquifers.

Beijing experiences both periods of heavy rain whereby the volume of water overwhelms local sewer systems, as well as periods of drought. Porous paving materials allow more storm water to be absorbed into the earth at the point of contact. Savings are achieved in the installation of infrastructure systems because of the reduction in storm water runoff. Porous paving materials also allow for more water to reach deep, underground aquifers that are a major source of much of Beijing's drinking water. Planted surfaces, or areas with foliage distributed within a matrix of soil-controlling materials, also do not concentrate heat, and serve to promote outdoor activities. The MIT design for Hui Long Guan provided impervious materials in semi-private courtyards and along circulation corridors (Figures 53 and 54).

SUMMARY

This design for Hui Long Guan parcels C02 and C06 provided equal attention to the development of both social and environmentally sustainable objectives. A truly sustainable example of design must succeed in both reducing energy loads on the site and in demonstrating responsiveness to the social needs of the occupants.

While developing a modular and systematic method by which the residences could be constructed, the MIT design team presented opportunities for variety and gradations of space within the site plan. The hierarchy of public spaces also worked hand in hand with the strategy for acknowledging the seasonal wind flows within the site. The circulation paths were designed to encourage the existing tradition of bicycle and foot traffic, while limiting the growing trends of polluting and energy-intensive car travel.

Social and environmental sustainability are important drivers for design in any context. In a city such as Beijing that is undergoing rapid changes, there is an even more urgent mandate to develop new solutions to housing needs. This project explored many aspects of Chinese culture that intersect with sustainable principles in the form of communal gathering and activity spaces, and the overlap of the built and natural environment. The design sought to create a balance between nature and urbanity through providing both an ecologically and sociologically beneficial development.

ACKNOWLEDGMENTS

The authors would like to thank Zachary Kron for his work in developing earlier versions of this chapter.

REFERENCES

Smith, J. 2004. *Building Energy Calculator: A Design Tool for Energy Analysis of Residential Buildings in Developing Countries.* Master's thesis. MIT, Cambridge, MA.

Part Five
Future Steps

CHAPTER FIFTEEN

FUTURE STEPS

Leon Glicksman and Lara Greden

INTRODUCTION

The rapid growth of the Chinese economy has improved the lives of many of its people. With increased income, people desire improved living conditions. The demand for larger and more comfortable housing is increasing. Each year, more than 10 million new residential units are built each year in China, almost ten times more than in the United States. The demand for durable household goods, such as air conditioners, is also increasing dramatically. As a result, demand for energy and materials is accelerating each year. As is common throughout many parts of the developed and developing world, current growth cannot be sustained. The carrying capacity of China's environment is at risk.

Numerous visits by the authors to new affordable and middle-class housing developments in Beijing, Shanghai, and Shenzhen reveal that in many instances, the buildings fall far short of the requirements of the established Energy Conservation Design (ECD) Standard. Even new buildings still often have little or no insulation in the exterior envelopes, loose single-glazed windows, short material lifetimes, and a disregard for siting to promote wind-driven ventilation. A significant problem suffered by recently constructed residential buildings lies in the shortcomings of construction quality. In many ways, traditional buildings in China were more environmentally friendly than new contemporary

L. Glicksman and J. Lin (eds), Sustainable Urban Housing in China, 240-245

buildings. They used building mass and wide roof overhangs to shade windows and walls to limit summer overheating. Proper site planning of traditional buildings limited exposure to severe winter winds and promoted more community interaction.

Current substandard design and construction trends have resulted in an increased demand for heating, cooling, and electricity. Without careful attention to sustainable design and construction, China's residential development path is likely to follow the trajectory of the Western world, and the building sector's share of total national energy consumption will rise dramatically. The buildings will also suffer from rapid deterioration, limiting their lifespan and requiring substantial investment in manpower and materials to replace the building stock. The poor building performance will also create a burden for Chinese homeowners, imposing high costs for energy and maintenance. For example, the authors visited a new development where heating is metered, and heating costs represented a substantial portion of household income. The residents had little opportunity to reduce heating energy consumption because the features that largely determine heating (i.e., insulation, window quality, and low infiltration) are outside of their reach once the building has been constructed. This example points to the need to address the disconnect between developers and homeowners when it comes to providing and benefiting from sustainable design.

In the move to accommodate a growing population, to make use of valuable urban land, and to satisfy the "Western-like" typology trend, designers have overlooked many traditional values. Copying Western solutions may not be the most rational choice for Chinese builders. Rather, they should make use of available local materials, and the limited labor skills in the housing construction industry (Figure 1). Designs should substantially change for different areas of China that have distinct climatic patterns and population densities. Variations of advanced building styles are illustrated in previous chapters.

For sustainable design to succeed, a close cooperation between developers, architects, engineers, and government officials must take place during the design and construction phases of a project. Building design must be integrated and occur simultaneously with consideration of sustainable technologies. Sustainable features properly integrated during initial design and construction can be shown to not only reduce the overall cost of owning and operating a residence; they will also provide more comfortable and healthy living conditions. Thus, education for homeowners is also a critical part of the sustainable housing equation. Their understanding of the benefits will reinforce demand for sustainable construction.

The pace of change in China is accelerating, and developers and consumers must be fast learners. There are many technical opportunities to reduce the energy consumption in new residential units while maintaining comfortable and spacious features. Improvements that reduce energy for heating by at least one-third over present standards are straightforward. They require the use of insulation materials and improved windows that reduce air leakage and capture more solar energy in winter. Window systems require efficient window units as well as proper framing design to ensure that air leakage is minimized. With proper building design, including orientation to encourage natural ventilation, landscaping, and shading the interior from solar energy in the summer, a substantial portion of air-conditioning energy can be saved. If the materials are available and construction methods learned, these technological changes could be made for little or no additional cost, especially when energy savings are considered.

Figure 1 Use of renewable resources in construction, such as bamboo scaffolding on this Shanghai construction project

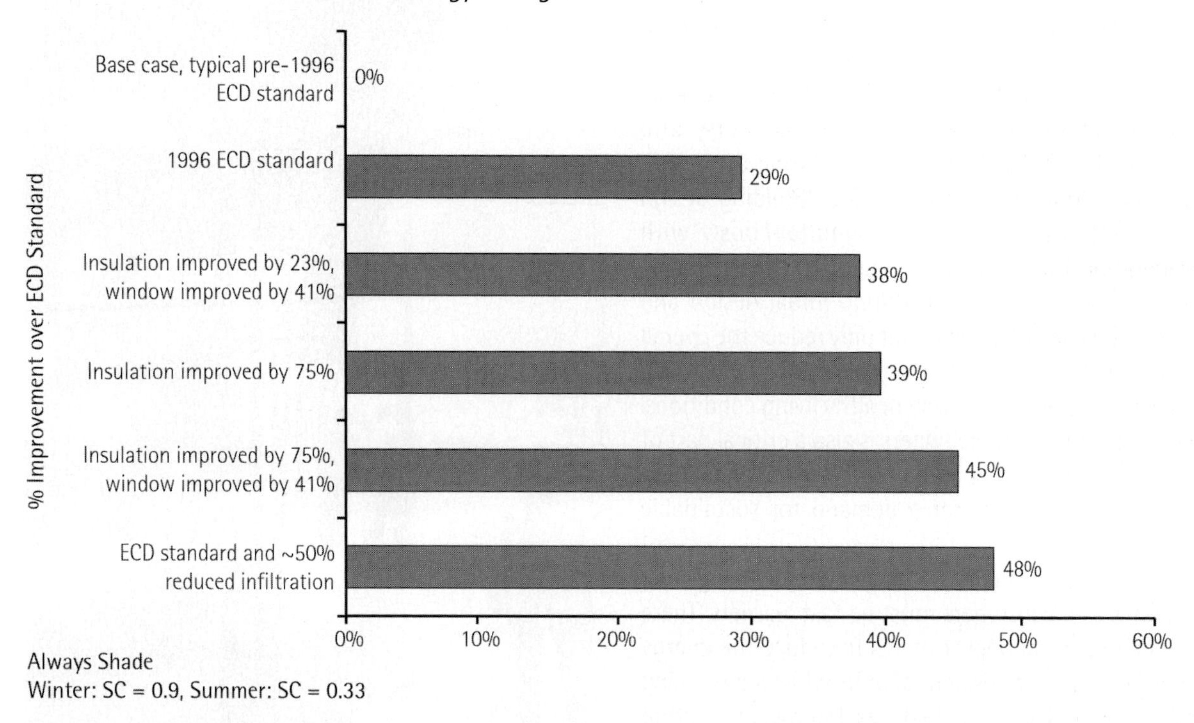

Figure 2 Energy savings from different conservation measures for a typical residential unit in Beijing

During the current stage of development, emphasis should be placed on the application of straightforward technologies. In some cases, such as ventilation and daylighting, these require good technical planning during the conceptual design. Once that is accomplished, the construction is straightforward.

Designs that increase the potential for natural ventilation, daylighting, and solar heat gain in the winter require rethinking building and community design. Buildings of shallow depth that can allow ventilation and daylighting to penetrate through most of the interior may be most appropriate. The spacing between adjoining buildings and staggering of individual buildings should also be planned in order to facilitate good air circulation between buildings as well as solar access in winter. Sustainable designs may emulate some more traditional Chinese architecture with low-rise units close enough to foster good community interaction. These designs are in contrast to many conventional Chinese residential buildings that are rather deep in plan and are formed into a series of parallel rows. Alternatively, some current developments have relatively widely spaced high-rise towers where individual units are rather deep.

Given the potential advantages, there is also the temptation to apply more advanced technologies to new Chinese residential buildings. Technologies such as ground source heat pumps, solar-powered dehumidification, and advanced windows may be projected to provide substantial gains in efficiency; however, they require expertise in design and more highly trained personnel in construction, operation, and maintenance. It must be emphasized that substantial gains can be made by properly using more conventional technology as demonstrated by Figure 2. Beyond these, the addition of more advanced technologies may only offer a more limited return.

IMPLEMENTATION

The rapid pace of building makes it imperative that sustainable design is adopted and correctly applied on a widespread basis throughout the country. The Chinese consumers as well as the entire country will have to live with the repercussions of current construction for a good portion of the century. Concepts for good design and technology are not enough; there must be compelling reasons for the government to mandate such steps, for consumers to demand it, and for developers to produce truly sustainable buildings. For this to happen, however, there are a number of steps required, including both policy and education.

POLICY

Policy must create demand for energy conservation and sustainability in the housing market from all angles: by mandating and enforcing codes and standards, by creating real market prices for energy use, and by educating consumers. Removing subsidies from energy costs and charging each residence for individual electricity and heating use would also make individuals more conscious of energy efficiency. This must be instituted with care to reduce dislocation for those in the lowest income level.

If current construction met the requirements of recent Chinese building codes, the country would make considerable progress. Stricter building codes will only be effective when they are coupled with stronger inspection and enforcement. In the current system, design institutes staffed by architects are responsible for producing designs that conform to the building codes in areas such as energy efficiency. Unfortunately, there is little if any government inspection of projects during construction or commissioning. Thus construction flaws and substitution of inferior products often goes unnoticed. This could be rectified with a voluntary post-construction inspection program that awards a sustainability rating to a developer that could be used in advertising. The Ministry of Construction (MOC) is proposing such a program in conjunction with academic institutions in China and those associated with the Alliance for Global Sustainability.

The Role of Policy

On a national level, historical efforts to improve energy efficiency in the residential arena pale beside reform of industrial initiatives. The growth of the residential sector is very significant, however. For example, while the residential sector consumed only 12 percent of the total supply of electricity in 2002, this grew from 7 percent in 1990. In 2002, the residential sector comprised 11 percent of the total Chinese energy consumption (China Statistical Yearbook 2004). A small but significant group is leading work to improve the energy efficiency of buildings, appliances, and the pricing and provision of fuel to the residential sector.

The MOC's Office of Building Energy Efficiency, responsible for implementing the national Energy Conservation Law of 1997, is developing prescriptive codes for energy conservation in new residential buildings. These design standards impose maximum heat transfer coefficients for building envelopes based on region, and guidelines for heating system design. Additionally, the MOC is organizing an educational plan that includes a textbook for architects and engineers designed to help them pass a certification test. They are also working to advertise energy conservation as a solution to quality. The MOC is also working to reform the billing and pricing of heating and developing building standards for new development in China's transient and southern climate zones. MOC engineers who are responsible for helping localities implement the codes noted that enforcement remains a huge problem. Beijing, although apparently the leader among cities in adopting new guidelines for energy efficiency, still lacks the personnel necessary to carry out post-construction enforcement at a meaningful level.

Clearly, incentives must exist for architects and builders to adopt building standards. Similarly, enforcement will help by encouraging architects to attempt to comply with standards; however, enforcement will only be effective if qualified inspectors are available to carry out inspections. Approved building designs are supposed to receive a stamp from an inspector; however, post-construction inspection is necessary to ensure that inferior construction methods do not negate the design's intentions. Economic incentives may play a role in encouraging the implementation of new standards, if properly designed. The MOC implemented a five percent tax incentive for buildings complying with energy efficiency guidelines in the last decade, but rescinded it after its apparent failure. At that time, energy efficiency guidelines were not well defined, and nearly all building managers found some way to qualify for the incentives.

Perhaps the largest potential gains for energy conservation exist in making institutional changes to heating provision. In the past, heating, like housing, was a welfare good provided by the state. As with most publicly provided goods, residents have little incentive to conserve energy for heating if they do not have to pay for what they use. Because more than 80 percent of building energy consumption in China's colder regions is attributable to heating, improvements in conventional building practices could have a large impact. This necessary transformation of the housing market requires a multidisciplinary solution incorporating social, economic, and technical elements. Tsinghua University and the World Bank are currently working on a three-phase program to promote the pricing and billing reform of centralized space heating in China.

One new affordable housing development in Beijing, Tian Hong, is solving the problem of metering by placing a high-efficiency gas furnace in each unit. Although well-designed district heating systems combined with co-generation of electricity is ideally the most energy-efficient heat source, a tradeoff had to be made to make the market work so that heating consumption could be directly charged to the user. Occupants pay for heat by swiping the meter located in their unit with a prepaid card. Astoundingly, heating gas bills in the winter are costing residents anywhere from 400 to 1,000 RMB per month. If the building were to be designed with energy conservation in mind, occupants could effortlessly conserve 30 to 50 percent of that, saving a significant 5 to 10 percent of their already far-stretched incomes.

Another significant area of policy activity designed to improve energy efficiency in China's housing sector has been in the development of energy standards for appliances. Programs such as China Green Lights and the CFC-Free Energy-Efficient Refrigerator Project have extended the traditional technology-based efforts by reaching consumers, dealers and other stakeholders with the product information needed to make responsible choices. Furthering the success of these projects is a new, unified energy informational label developed by the China Sustainable Energy Project (CSEP). The project encapsulates three key aspects of China's policy initiatives: mandatory energy efficiency standards, mandatory (and unified) energy information labeling, and voluntary energy efficiency labeling. In August of 1999, China formally launched the label, awarding the first one to refrigerators. These ideas must be transferred to the building material industry to give the necessary instruction and incentive to supply certified, quality insulation, windows, and other products.

We recommend that the MOC go beyond the standard approach of building standards and seize the opportunity to use new market mechanisms to drive demand for energy conservation. Targeting inexperienced consumers now with education and incentives about the clear advantages of energy-efficient homes will help form a market that inherently values energy efficiency and sustainability. With many markets in the United States trying to undertake market transformation, it seems that China can benefit from near-term investments in market *formation*. We recommend that the government consider a labeling system for homes that would communicate a sustainability-indexed checklist for consumers, based on the unified energy information label for appliances. Such a labeling system will require a solid base of performance-based building standards. It also offers a solution to decision-making problems faced by consumers and serves as a catalyst for all other stakeholders: for government implementation of building standards, as a foundation for architects to assess the sustainability of their designs, and as a way for developers to credibly promote sustainable features of their properties. The ability of developers to legitimize higher selling prices and market differentiation for sustainable buildings should provide a market pull and spur growth of sustainable buildings.

EDUCATION

Almost all of the developers, architects, and potential homebuyers encountered by the authors in China indicated that sustainable design was an important and desirable attribute of new homes. Developers have also included statements to that effect in their advertisements for new projects. In most cases however, the understanding of proper designs and technologies was severely limited. In some instances, developers and potential buyers limited their attention to visible attributes such as exterior landscaping. Designers need to understand what strategies will best meet a particular project's need. A program is needed to show them how new designs using relatively straightforward techniques can achieve substantial improvements. They must also understand how to structure the design process.

It appears that the Chinese consumer is learning fast. In a survey of potential homebuyers, location was recorded as the highest priority. They have learned to vote with their feet in evaluating new projects. At this time, they have no way to assess the validity of claims for sustainable performance. Consumers must be informed so they can press for these improvements. Instructional literature about attributes of potential features would be helpful. An impartial sustainability rating program, possibly with projected yearly costs for heating, cooling, and lighting of residential units, would aid potential homebuyers. This will also provide incentives for designers and developers to produce more competitive housing.

SUMMARY

Opportunities should be seized now while many new buildings are going up, as such buildings will remain in the housing stock for a half-century or more. Policy must create demand for energy conservation and sustainability in the housing market from all angles: by mandating codes and standards, by creating real market prices for energy use, and by educating consumers. Removing subsidies from energy costs and charging each residence for individual energy and heat use should raise awareness of energy efficiency. This should be implemented carefully to avoid dislocating those in the lowest income level.

China's consumers, developers, and government officials appear to be responsive to the need for sustainable development although it can sometimes be frustrating to institute new technological opportunities that appear to be win-win scenarios. Education, incentives, stricter codes, and enforcement will help to translate current opportunities into future benefits for building in China.

REFERENCES

China Statistical Yearbook 2004. 2004. China Statistical Publication House, Beijing. <http://www.stats.gov.cn/english/statisticaldata/yearlydata/yb2004-e/indexeh.htm>, accessed 21 March 2005.

APPENDIX
BUILDING ENERGY CALCULATOR

OVERVIEW

The primary goal of the software design tool called Building Energy Calculator is to guide consumers, developers and architects toward more energy-efficient buildings. The software is included with the CD at the back of the book, and Figure 1 shows the program interface. The program was written by Jonathan Smith at MIT. Mr. Smith and MIT hold the copyright. Users are free to use the software but may not charge others for copies or charge a fee to others for its use. For a quick start of the program, turn to the section entitled *Installing the Program.*

The program allows the user to evaluation the impact of major design and material choices as well as construction standards. The program performs hourly simulations and approximates the annual heating and cooling loads for different buildings input by the user. Chinese residential buildings have been targeted specifically because of the current growth in the sector, which is one important factor driving the recent increases in both energy production and carbon dioxide emissions in China. This software can help architects see how some of their important design decisions impact building energy performance. The hope of this work is that by providing such a tool, energy performance will be considered as an important part of a project from its inception. Early consideration of these issues is one of the keys to achieving more efficient buildings because there is still a great deal of flexibility to incorporate sustainable features into the design.

The program is designed to be used in the conceptual design phase of a project. Taking this fact into account, the program was created using several guiding principles: (1) ease of use, (2) minimal required input, and (3) output that is simple, graphical, and quantitative. Ease of use is especially important for this tool since it is intended for use early in the design process. Input to the program is restricted to variables that can be considered at the early design stages, and a limited number of choices for each variable are presented. The program produces graphical results that allow for side-by-side comparison of different scenarios. The relative strength of one design compared to another in term of energy efficiency should be readily apparent from the output of the program. Creating a program following these guiding principles requires making simplifying assumptions and using typical values for many parameters. Simplifications of this sort are appropriate since the goal of the program is to provide rough estimates that guide the user by indicating which decisions have the greatest impact on building energy use. The approach is fundamentally different from that taken by most other programs that work in a similar way. Typically, programs that do energy simulations are extremely detailed in nature and are useful late in the design process or after a design is complete. This software makes more assumptions and limits the input of the user to a several basic choices. By tailoring the options for materials, cities, etc. and appropriately setting model parameters, it targets Chinese residential buildings.

Figure 2 illustrates the forms of energy transport and boundary conditions used in the program. The heat transfer through the envelope includes conduction, convection and infrared radiation through a wall, roof and foundation made up of insulation and concrete. Solar radiation absorbed by the wall and roof surface is included as well as solar gains through the window system. All of the solar gains through the windows are assumed to be absorbed by the floors. Blinds are placed on the interior of the windows.

The blinds are kept open during the heating season and partially closed during the cooling season reducing the interior solar gains by 25 percent. The transmission through particular windows is found by using ASHRAE tables for solar heat gain coefficient (SHGC).

The roof, floor, and foundation are all assumed to be 100-millimeter-thick concrete slabs; the transient response is calculated including thermal gradients through the slab thickness with the temperature calculated each hour of the year. The roof is a horizontal flat roof. The foundation is slab on grade and losses to the ground are estimated by use of a perimeter loss coefficient.

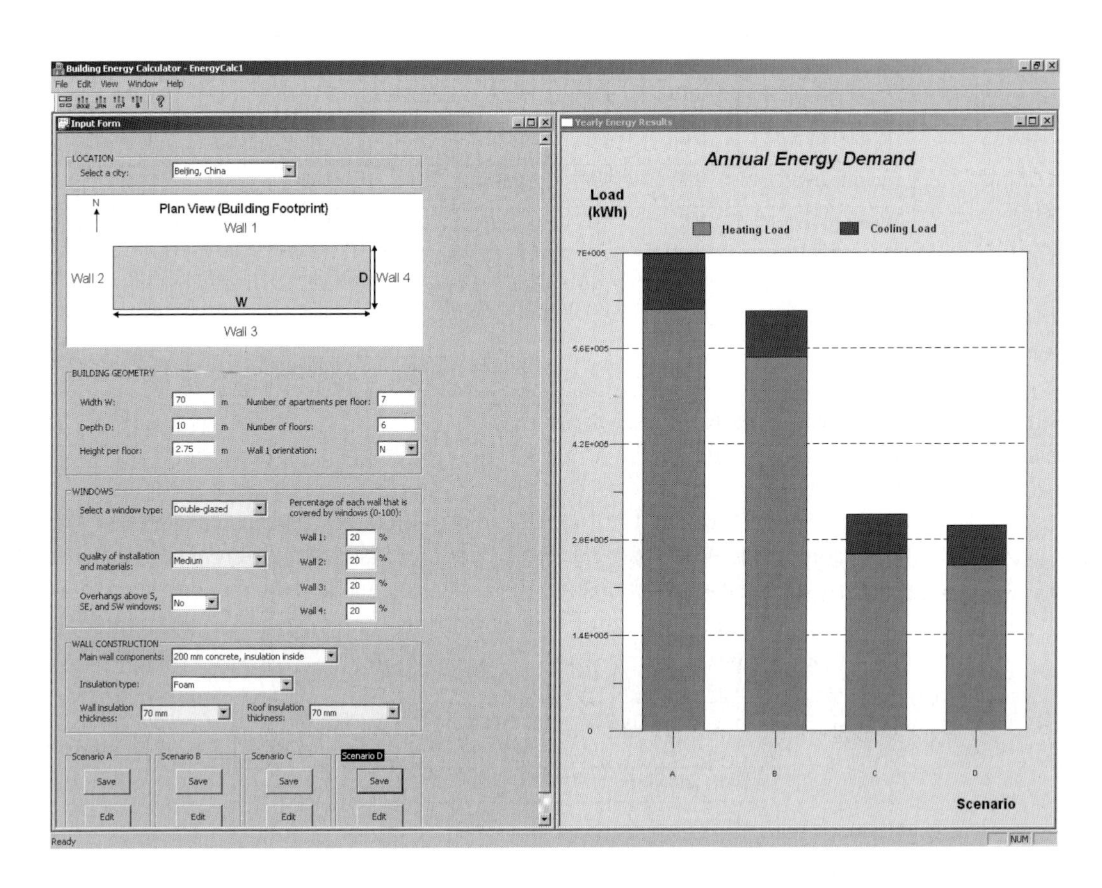

Figure 1 Screen shot of the Building Energy Calculator program interface

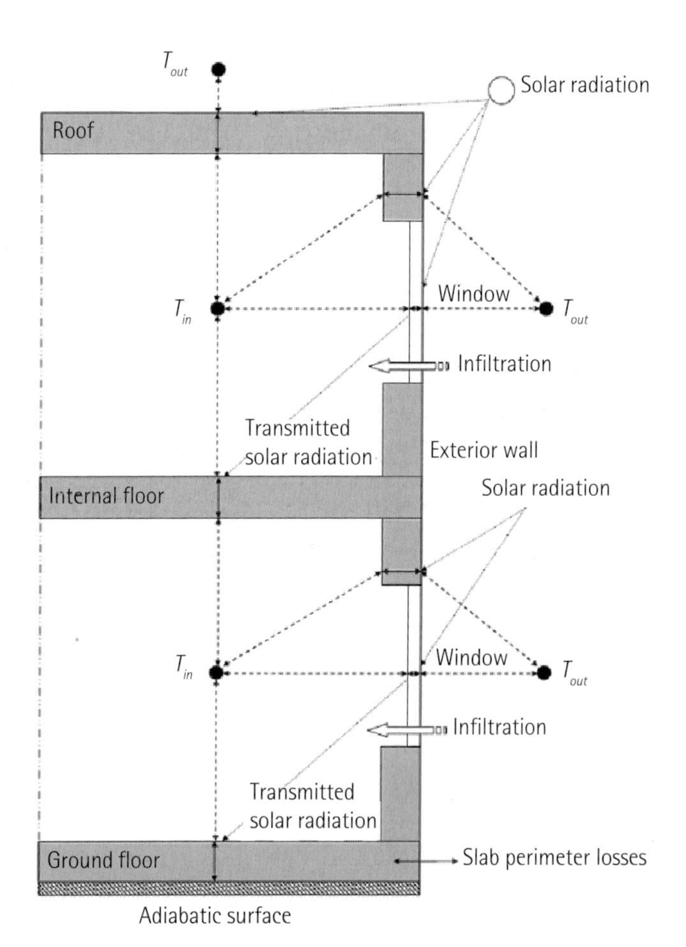

Figure 2 Partial building cross section showing forms of energy transport and boundary conditions

Interior Air Temperature

Energy associated with airflow into the building by controlled ventilation and leaks are considered as a function of construction quality. The air inside is assumed to be one well-mixed temperature zone for the entire building. The air temperature is allowed to float between 18°C and 26°C each hour, thus allowing for the influence of the thermal mass. Heating and/or cooling are only supplied when the air temperature exceeds one of these temperature limits within each hour. The amount of heating or cooling is supplied for the next hour to maintain the interior air temperature at 18°C and 26°C, respectively.

The windows are open, allowing five air changes per hour (ACH) in the cooling season when the outdoor air temperature is between 18°C and 26°C. Otherwise, the windows are closed and the air changes are given by the leakage rates discussed in the next section. The heating season is defined as any month when the average outdoor temperature is below 18°C. The hourly weather data is taken from typical meteorological year data for each city in the program. The data contains hourly temperature, direct, and diffuse solar radiation.

Several different window types can be modeled for the building as given by Table 1. With different quality of construction, the windows have differing heat coefficients due to the frame construction with or without thermal breaks.

Description	U1*	U2**	SHGC (As a Function of Incident Solar Angle)				
	W/m^2K	W/m^2K	0°	40°	60°	80°	Diffuse
Single-glazed	6.1	7.2	0.86	0.84	0.78	0.42	0.78
Double-glazed	3.4	4.6	0.76	0.74	0.64	0.26	0.66
Double, low-e	2.9	4.1	0.65	0.64	0.56	0.23	0.57
Triple-glazed	2.6	3.8	0.68	0.65	0.54	0.18	0.57

Table 1 Window overall heat transfer coefficient and solar heat gain coefficient (SHGC)
U1* Overall heat transfer coefficient for windows with thermal break: high- or medium-quality construction.
U2** Overall heat transfer coefficient for windows without a thermal break: low-quality construction.
(Source: ASHRAE Handbook of Fundamentals 2001)

Air Leakage

More importantly, the quality of construction influences the air leakage into the building expressed as ACH. Table 2 gives the assumed values for different construction qualities. This is a key assumption, and the air leakage values can only be regarded as approximate. The infiltration rates will increase as the construction and materials quality are degraded.

Installation Quality	Heating Season ACH	Cooling Season ACH
High (airtight)	0.5	0.5
Medium	1.0	0.7
Low (leaky)	2.0	1.5

Table 2 Assumed air leakage, or ACH, as a function of construction quality

The convective heat transfer coefficients are assumed to remain constant for each surface. The values are given in Table 3.

Surface	Heat Transfer Coefficient (W/m^2K)
Inside wall	8.3
Ceiling	6.1
Floor	9.3
Outside	34

Table 3 Surface heat transfer coefficients, constant throughout the year

Internal Gains

The number of units per floor multiplied by the number of floors gives the total number of residential units in the building, which is used to estimate the internal gains that add heat to the indoor air. Recall that solar gains are modeled separately. The internal gains are highly uncertain at the preliminary design stages of a building, but they are an important factor in determining cooling loads and must be estimated. The assumption made in the program is that, on average, each residential unit has two people giving off 67 W each and an appliance load of 250 W. The total internal gains are therefore estimated to be 384 W per unit multiplied by the total number of units, and this value is held constant throughout the year. In reality, the gains would vary throughout the course of a day as people come to and go from a unit, but attempting to account for occupancy schedules is beyond the scope of this program.

Insulation Type

There are many different types of insulation that are used in buildings, and they have varying levels of performance. Foam insulation is the most common type in buildings that have concrete frames. The insulation type input allows the user to choose between four different types of insulation, and the selection sets the thermal conductivity, k, for the primary insulation layer. The available options are foam (0.025 W/mK), glass fiber (0.036 W/mK), loose fill cellulose (0.042 W/mK), and cement fiber slabs (0.074 W/mK).

Installing the Program

The program can be used directly from the CD or it can be installed on the hard drive. Copy the files and folders exactly as they are arranged on the CD to your hard drive. Open the folder "EnergyCalc" and then open the program Energy Calculator (EnergyCalculator.exe).

Using the Interface

The various parameters that must be input by the user to describe a building are explained in the Input Description section, and the output graphs that are available are explained in the Output Description section.

At the bottom of the input form, there are four groups of buttons (Scenario A, Scenario B, Scenario C, and Scenario D). Clicking on a "Save" button performs a simulation using the current information displayed on the input form. After the simulation completes (which should take only a few seconds), the results will be immediately displayed in a bar graph, provided the graph view has not been closed. The labels at the bottom of the bar graph (A, B, C, and D) correspond to the scenarios on the input form. Enter a "base case" in the input form and save it as Scenario A. When other cases are saved as Scenario B, C, or D, anything input that is different from Scenario A will be highlighted in yellow to remind the user what was changed. The input for a saved scenario can be loaded onto the form again by clicking on the corresponding "Edit" button. The user can tell which scenario was last saved or edited because the label above the buttons has white text and a black background.

An entire set of scenarios can be saved for future use by using the "Save" or "Save As" options from the "File" menu. This functionality allows the user to review previously saved building scenarios after exiting the program.

Input Description

A. Location Group

   i. Select a city

   A variety of cities in China and the United States are available in the drop-down list. The selection made in this box determines which weather data file is used in the simulations. If the city in which the building is located is not available from the list, a city with a similar climate can be selected for very rough estimates of building energy use.

B. Building Geometry Group

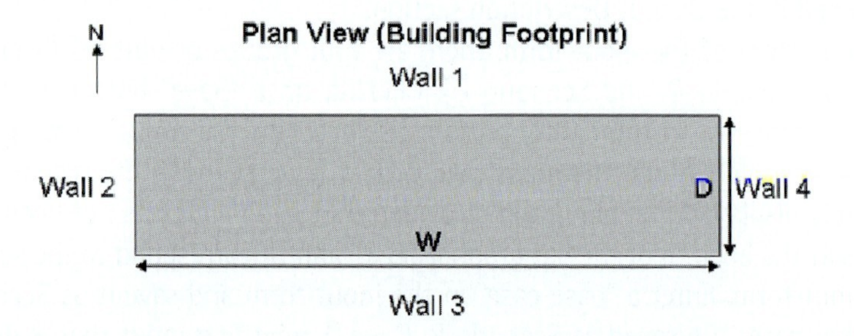

Figure 3 Plan view of building footprint

   i. Width W

   The width of the rectangular building footprint, in meters, is entered here. This dimension is labeled W in Figure 3, and it corresponds to the outside building dimension rather than the inside. Note that only rectangular building footprints are supported at this time.

   ii. Depth D

   The depth of the rectangular building footprint, in meters, is entered here. This dimension is labeled D in the figure above, and it corresponds to the outside building dimension rather than the inside. Note that only rectangular building footprints are supported at this time.

   iii. Height per Floor

   The height of each of the building's floors, in meters, is entered here. This quantity multiplied by the number of floors should equal the total outside height of the building.

   iv. Number of Floors

   The number of floors in the building is entered here. This quantity multiplied by the height per floor should equal the total outside height of the building.

   v. Number of Apartments per Floor

   The number of apartment units on each floor of the building is entered here. This input is used to approximate the number of people and appliances in the building. It is assumed that, on average, two people are present in each unit at all times. There is an assumed appliance load of 250 watts in each unit.

   vi. Wall 1 Orientation

   This input determines the orientation of the building. The compass arrow in the figure adjusts each time the input is varied so that it always points north. "Wall 1 orientation" simply means the direction that the wall labeled "Wall 1" in the figure faces.

C. Windows Group

i. Window Type

Select from the basic window types listed, which include single-glazed, double-glazed, double-glazed with a low-emissivity (low-e) coating, and triple-glazed. The same type of window is assumed to be used throughout the building. Multiple glazings (panes of glass) can substantially reduce the amount of heat transmitted through the windows without much impact on the transmission of sunlight.

ii. Quality of Installation and Materials

The amount of care taken during the installation of windows, the skill of the construction workers, and the quality of the materials used all have an impact on building energy performance. Although the selection made for this input is somewhat subjective, it demonstrates the importance of good construction practice and materials. High quality of installation and materials means that windows are virtually airtight with no leakage. Low quality means that windows are quite leaky and allow a significant amount of outdoor air to enter the building even when they are closed. Medium quality is somewhere in between these two extremes. The selection made affects two parameters in the model: the overall window heat transfer coefficient (U-value) and the air change rate in the building.

When low quality of installation and materials is selected, the windows are assumed to have aluminum frames with no thermal break. Aluminum frames with thermal breaks are used in the model when medium or high quality is selected. Frames with no thermal break result in a slightly higher overall heat transfer coefficient (U-value) for the windows.

The air change rate in the building is by far the most important influence that this input has in the model. The air change rates used in the model for each of the qualities are listed in Table 4 below. Particularly in harsh climates, allowing excess infiltration of outdoor air is extremely detrimental to building energy performance.

Installation Quality	Heating Season ACH	Cooling Season ACH
High (airtight)	0.5	0.5
Medium	1.0	0.7
Low (leaky)	2.0	1.5

Table 4 Air changes per hour (ACH) in the building for different qualities in the heating and cooling seasons

iii. Overhangs above S, SE, and SW Windows
Set this input to "yes" to include overhangs (i.e., horizontal projections) above all windows on walls facing south, southeast, and southwest. The overhang geometry assumed in the model is shown in Figure 4 below. Note that none of the dimensions in the figure are used to determine the overall window area – the window percentage inputs (described next) fully determine the total window area on each wall. The geometry shown below is only used to determine the fraction of the total window area that is shaded at any given time.

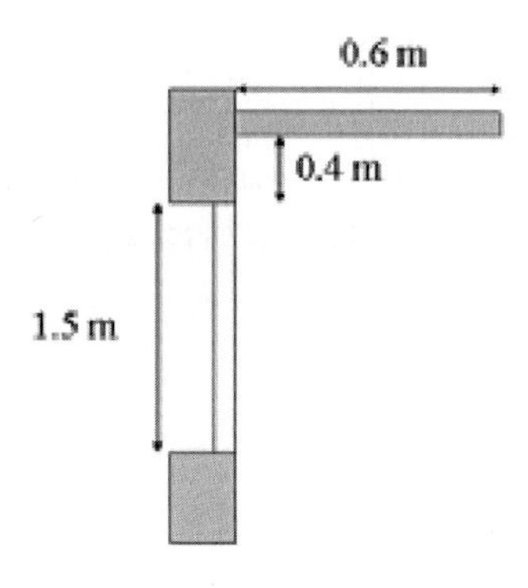

Figure 4 The height of the window is 1.5 m; the width of the window is 0.9 m; the overhang length is 0.6 m; the spacing between the top of the window and the overhang is 0.4 m (note: drawing not to scale)

iv. Wall 1 Window Percentage
Enter the percentage of the wall labeled "Wall 1" in the figure that is covered by windows. For example, if the total area of a wall surface is 100 m^2, entering 20 percent (entered as 20, NOT 0.20) means that there are 20 m^2 of windows (and thus 80 m^2 of the exterior wall material).

v. Wall 2 Window Percentage
Enter the percentage of the wall labeled "Wall 2" in the figure that is covered by windows. For example, if the total area of a wall surface is 100 m^2, entering 20 percent (entered as 20, NOT 0.20) means that there are 20 m^2 of windows (and thus 80 m^2 of the exterior wall material).

vi. Wall 3 Window Percentage
Enter the percentage of the wall labeled "Wall 3" in the figure that is covered by windows. For example, if the total area of a wall surface is 100 m^2, entering 20 percent (entered as 20, NOT 0.20) means that there are 20 m^2 of windows (and thus 80 m^2 of the exterior wall material).

vii. Wall 4 Window Percentage
Enter the percentage of the wall labeled "Wall 4" in the figure that is covered by windows. For example, if the total area of a wall surface is 100 m^2, entering 20 percent (entered as 20, NOT 0.20) means that there are 20 m^2 of windows (and thus 80 m^2 of the exterior wall material).

D. Wall Construction Group

i. Main Wall Components
Select from the available wall constructions. The choices include concrete slab walls with thicknesses of both 200 mm and 300 mm. The primary insulation layer can be either on the outside or inside surface of the concrete layer. Other materials assumed to be present in the wall constructions are (listed from inside the building to outside):

- insulation inside constructions: gypsum wallboard, plywood, insulation layer, concrete; and
- insulation outside constructions: gypsum wallboard, plywood, concrete, insulation layer, sheathing.

ii. Insulation Type
Select a type of insulation to be used in the building from the available choices. The same type of insulation is assumed to be used in the exterior walls and the roof.

iii. Wall Insulation Thickness
Select the thickness of the primary insulation layer in the exterior walls. The same thickness is used in all four walls of the building, and the material used is the one selected in the Insulation Type input.

iv. Roof Insulation Thickness
Select the thickness of the primary insulation layer in the roof. The layer is assumed to be inside of the concrete layer in the roof regardless of the selection made in the Main Wall Components input. The material used is the one selected in the Insulation Type input.

Output Description

All of the output graphics in the program are stacked bar graphs. The red portion of the bar represents heating energy or cost, and the blue portion represents cooling energy or cost. All of the energy results are presented in kilowatt-hours (kWh). Results are shown for up to four scenarios (A, B, C, and D) at the same time, which correspond to the four sets of buttons on the input form.

A. Yearly Energy Results
The results shown in this graph represent total loads for an entire year. These quantities are equal to the total energy that must be added to or removed from the indoor air in order to maintain a comfortable indoor temperature (18°C-26°C is the acceptable range currently set in the program). They represent the energy that must be delivered by the heating and cooling systems. Heating and cooling equipment efficiencies are not incorporated into these figures.

This output graph is the one that is displayed by default when the program starts. When it is not on the screen, it can be generated again by selecting "Yearly Energy Graph" from the "View" menu or by clicking on the toolbar button that says "2002." Only the input form and one graph can be displayed at a time.

B. Monthly Energy Results
The results shown in this graph represent the total loads for each month of the year. They are simply those of the Yearly Energy Results broken down by month.

This output graph can be generated by selecting "Monthly Energy Graph" from the "View" menu or by clicking on the toolbar button that says "JAN". Only the input form and one graph can be displayed at a time.

C. Yearly Energy Results per Unit Floor Area
The results shown in this graph represent the total annual loads divided by the total floor area of the building. They are simply those of the Yearly Energy Results divided by the total floor area of the building. Such a figure is often useful in comparing buildings of different sizes.

This output graph can be generated by selecting "Yearly Energy Graph per Unit Floor Area" from the "View" menu or by clicking on the toolbar button that says "m^2." Only the input form and one graph can be displayed at a time.

D. Yearly Energy Costs
This output graph displays the total energy costs for an entire year. In order to display the costs, some additional input is required of the user. When this option is selected, a dialog box is presented to the user that asks for four quantities:

i. Heating Energy Cost
Enter the cost of heating energy per kilowatt-hour (kWh). It may be the cost of electricity or the cost of natural gas, depending on the type of heating system. Any monetary unit can be used; the output graph will be in the same units as those entered here. The same monetary unit must be used for heating and cooling energy costs, however.

ii. Cooling Energy Cost
Enter the cost of cooling energy per kilowatt-hour (kWh). Typically, it will be the cost of electricity. Any monetary unit can be used; the output graph will be in the same units as those entered here. The same monetary unit must be used for heating and cooling energy costs, however.

iii. Heating Equipment Efficiency
Enter the efficiency of the heating equipment as a percentage. This figure should be available for any type of heating system. The efficiency is simply the amount of heat output per unit energy input the system produces. Efficiencies as high as 90% are common.

iv. Cooling Equipment Coefficient of Performance (COP)
Enter the coefficient of performance (COP) for the cooling system. This figure should be available for any type of air-conditioning system. The COP is equal to the heat energy removed from the space per unit energy input. COPs of 3.0 or higher are common.

Once these values are entered, the output graph is displayed showing the energy costs for an entire year. These figures account for the fact that the actual consumed energy is different from the loads (shown by the other output graphs) since equipment efficiencies vary. The units of the costs are the same as the currency units input by the user for the cost of energy.

This output graph can be generated by selecting "Yearly Energy Costs" from the "View" menu or by clicking on the toolbar button that says "$". Only the input form and one graph can be displayed at a time.

ACKNOWLEDGMENTS

The appendix and is excerpted from the thesis by Jonathan Smith: Smith, J. 2004. *Building Energy Calculator: A Design Tool for Energy Analysis of Residential Buildings in Developing Countries*. Master's thesis. MIT, Cambridge, MA.

REFERENCES

ASHRAE. 2001. Handbook of Fundamentals. American Society of Heating, Refrigerating and Air-Conditioning Engineers. Atlanta, GA.

Smith, J. 2004. *Building Energy Calculator: A Design Tool for Energy Analysis of Residential Buildings in Developing Countries*. Master's thesis. MIT, Cambridge, MA.

INDEX

Symbols

A

B

M

N

O

P

V

W

Z

IMAGE CREDITS

The authors would like to express gratitude to all the students who contributed to the production of images for this publication, in addition to the photographers and institutions that have granted us permission to use their photographs. Most figures, including graphs and tables, have been altered from their original state, and have been redrawn by Juintow Lin, Michael Fox, and Anshuman Khanna to be suitable for this publication.

While every reasonable effort has been made to trace copyright owners and credit the original source of images, it has not always been possible to do so. Certain images exist in the public domain, the use of which the authors are most grateful. Where known, the original place of publication for an image is also credited.

COVER DESIGN

Front and back cover design by Juintow Lin

Introduction

Figure 3 – China map; image by MIT China Workshop

PART ONE – BACKGROUND

Section header – Shanghai 2000; image courtesy of Mario Cipresso

Chapter One – Sustainability and the Building Sector

Chapter logo – Diagram of sustainable features; image by MIT China Workshop

Chapter Two – China – Environment and Culture

Chapter logo – Solar-powered water kettle; photograph by Juintow Lin
Figure 5 – Air conditioners installed by occupants; photograph by Leon Glicksman
Figure 6 – Permanently closed-in balconies; photograph by Leon Glicksman
Figure 7 – Beijing housing survey results; image by MIT China Workshop

PART TWO – DESIGN PRINCIPLES

Section header – Bamboo scaffolding in Shanghai; photograph by Juintow Lin

Chapter Three – Design Principles

Chapter logo – Ventilation diagram of Beijing Hui Long Guan; image by MIT China Workshop
Figure 1 – Typical low-rise housing block in Beijing; photograph by Andrew Scott
Figure 2 – Shanghai urban landscape; photograph courtesy of Mario Cipresso
Figure 3 – Circular housing in Fujian; photograph by Juintow Lin
Figure 4 – Beijing courtyard housing; photograph by Juintow Lin
Figure 6 – Wind tower in Kerman; photograph courtesy of Conway Library, Courtauld Institute of Art, photograph by Robert Byron
Figure 7 – Double-layered façade; photograph by Andrew Scott
Figure 8 – Mud house, Datong, China; photograph from the MIT Rotch Visual Collection
Figure 9 – Cave dwellings; image courtesy of Graduate Center of Architecture, Peking University
Figure 10 – Plaza de la Corredera, Cordoba, Spain; photograph courtesy of Oliver Radford
Figure 11 – Piazza Maggiore, Bologna, Italy; photograph courtesy of William Rawn
Figure 12 – New Orleans porch, Vieux Carre; photograph courtesy of MIT Rotch Visual Collection, from the Kidder Smith slide archives
Figure 13 – Circular housing in Fujian; photograph by Juintow Lin

Figure 14 – Old Swan House; photograph courtesy of Oliver Radford
Figure 15 – Oxford University building; photograph by Andrew Scott
Figure 16 – Murcott's Boyd Education Centre; photograph courtesy of Li Lian Tan
Figure 17 – Double-skin glass louvered façade; photograph by Andrew Scott
Figures 18-22 – Thompson Island diagrams, model, thermal imaging, and photographs; images by Andrew Scott
Figures 23-27 – Shanghai low-rise diagram, CFD studies, drawings, and model; images by MIT China Workshop
Figures 28-31 – Beijing Star Garden site drawings, CFD studies, sketch, and rendering; images by MIT China Workshop
Figures 32-36 – Hui Long Guan renderings and diagrams; images by MIT China Workshop
Figures 37-38 – Shenzhen Wonderland drawings and model; images by MIT China Workshop
Figures 39-40 – Beijing high-rise sketches, drawings, and model; images by MIT China Workshop
Figures 41-42 – Beijing Star Garden diagram and drawings; images by MIT China Workshop
Figures 43-46 – Beijing 12 x 12 house diagrams and models; images by MIT China Workshop

Chapter Four – Materials and Construction for Low-Energy Buildings in China

Chapter logo – Moisture infiltration problem areas; diagram by John Fernandez
Figures 5-12 – Images of concrete, steel, engineered lumber, and exterior envelope; photographs by John Fernandez
Figure 13 – Diagrams of air infiltration problem areas; images by John Fernandez
Figure 14 – Image of unsealed piping for air-conditioning units; image by Leon Glicksman
Figures 15-19, 22-24 – Diagrams of infiltration problem areas, wall assembly, seals, and windows; images by John Fernandez

PART THREE – TECHNICAL FINDINGS

Section header – Shanghai Taidong Residential Quarter CFD studies; image by MIT China Workshop

Chapter Five - Energy in Building Design

Chapter logo – Beijing wind rose; image by MIT China Workshop
Figure 2, Table 1 – Climate data; images by MIT China Workshop
Figures 4-9, Tables 2-3 – Study 1: Beijing; images by MIT China Workshop
Figures 10-11 – Study 1: Shanghai; images by MIT China Workshop
Figures 12 – Study 2: Beijing; images by MIT China Workshop
Figures 13-16 – Series I simulations; images by MIT China Workshop
Figures 19-28, Tables 4-5 - Study 3: Shenzhen; images by MIT China Workshop

Chapter Six – Wind in Building Environment Design

Chapter logo – Northern winter winds buffered by trees; image by MIT China Workshop
Figure 4 – Wind tunnel model; photograph courtesy of National University of Singapore
Figure 5a – Stata Center model; photograph courtesy of Frank O. Gehry and Associates
Figure 5b – Stata Center surroundings; image courtesy of MIT
Figures 6-7 – CFD images of MIT's Stata Center; images courtesy of MIT Building Technology Group

Figure 8 – Beijing Star Garden residential development; rendering courtesy of Beijing Vanke Co. Ltd.
Figure 9-10 – Beijing Star Garden sketch diagram and CFD simulations; images by MIT China Workshop
Figure 10 – Beijing Star Garden sketch diagram; image by MIT China Workshop
Figures 11-13 – Shanghai Taidong Residential Development drawings and CFD studies; images by MIT China Workshop
Figure 14 – Beijing unit interior CFD studies; images by MIT China Workshop
Figures 15-18 – MIT student dormitory CFD studies and diagrams; images courtesy of MIT Building Technology Group

Chapter Seven - Design of Natural Ventilation with CFD

Chapter logo – Interior CFD study; image by MIT China Workshop
Figure 2 – Beijing City Garden diagram; image by MIT China Workshop
Figure 3 – Beijing wind rose; image by MIT China Workshop
Figures 5-7 – Beijing City Garden CFD studies; images by MIT China Workshop
Figures 8-12 – Beijing Star Garden diagrams and CFD studies; images by MIT China Workshop

Chapter Eight - Light and Shading

Chapter logo – Photo-realistic rendering of unit interior; image by MIT China Workshop
Figures 1-2 – Daylighting in Shanghai school and residential building; photographs by Leslie Norford
Figure 3 – Imperial Palace overhanging roofs; photograph by Leslie Norford
Figure 4 – Houses on Beijing streets; photograph by Leslie Norford
Figure 5 – House with vegetation; photograph by Leslie Norford
Figure 6 – Beijing high-rise building; photograph by Leslie Norford
Figure 7 – Shading at a school in Ahmedabad; photograph by Leslie Norford
Figure 8 – Harare office building; photograph courtesy of Melissa Edmands
Figures 9-10 – Shanghai Taidong Residential Quarter lighting studies; images by MIT China Workshop
Figure 11 – Lighting studies of a school in Montana and a Palladian Villa; images by Eric Walter, Juintow Lin, and Eunice Lin for MIT's Department of Architecture
Figure 18 – Material costs vs. operating energy for a residential unit in Beijing; image by MIT China Workshop

PART FOUR – CASE STUDIES

Section header – Shenzhen Wonderland plan; image by MIT China Workshop

Chapter Nine - Case Studies

Chapter logo – Beijing Hui Long Guan site plan; image by MIT China Workshop

Chapter Ten - Case Study One – Beijing Prototype Housing

Chapter logo – Natural ventilation diagram; image by MIT China Workshop
Figures 1-17 – Beijing 12 x 12 house drawings, diagrams, sketches, models, and CFD studies; images by MIT China Workshop

Chapter Eleven - Case Study Two - Beijing Star Garden

Chapter logo – Axonometric drawing; image by MIT China Workshop
Figures 1-2 – Beijing Star Garden site plan and rendering; images courtesy of Beijing Vanke Co. Ltd.
Figures 3-29, Tables 1-4 – Beijing Star Garden drawings, diagrams, sketches, models, and CFD studies; images by MIT China Workshop

Chapter Twelve - Case Study Three - Shanghai Taidong Residential Quarter

Chapter logo – Adjustable shading device diagram; image by MIT China Workshop

Figures 2-47, Tables 1-4 – Shanghai Taidong Residential Quarter drawings, diagrams, sketches, models, and CFD studies; images by MIT China Workshop

Chapter Thirteen - Case Study Four - Shenzhen Wonderland Phase IV

Chapter logo – Sketch showing balconies and shading; image by MIT China Workshop

Figure 1 – Shenzhen Wonderland site plan; image by MIT China Workshop

Figures 2-4 – Shenzhen Wonderland existing development; photographs by MIT China Workshop

Figures 5-9, Tables 1-2 – Shenzhen Wonderland Phase IV photographs, sketches, diagrams, and graphs; images by MIT China Workshop

Figurc 10 – MIT building overhangs and fins; photograph by Sephir Hamilton

Figures 11-16, Table 3 – Shenzhen Wonderland Phase IV energy studies; images by MIT China Workshop

Figure 17 – Solar water heaters in Xi'An; photograph by Juintow Lin

Figures 18-39, Tables 4-9 – Shenzhen Wonderland Phase IV technical recommendations, drawings, diagrams, sketches, and CFD studies; images by MIT China Workshop

Figure 40 – Shenzhen Wonderland overhangs and air conditioners; photograph by Leon Glicksman

Figures 41-44, Tables 10-11 – Shenzhen Wonderland Phase IV drawings and residential user manual settings; images by MIT China Workshop

Chapter Fourteen - Case Study Five - Beijing Hui Long Guan

Chapter logo – Diagram of public spaces; image by MIT China Workshop

Figure 1 – Hui Long Guan site plan showing C02 and C06 parcels; image courtesy of Tian Hong

Figure 2 – Typical building in Hui Long Guan; photograph by MIT China Workshop

Figure 3 – Row housing, typical of China; photograph by Leon Glicksman

Figures 4-18, Table 1 – Beijing Hui Long Guan energy studies; images by MIT China Workshop

Figure 19 – Images from China; photographs by MIT China Workshop

Figures 20-31 – Beijing Hui Long Guan diagrams, drawings, renderings, and CFD studies; images by MIT China Workshop

Figure 32 – Hui Long Guan site plan showing C02 and C06 parcels; image courtesy of Tian Hong

Figures 33-54 – Beijing Hui Long Guan diagrams, drawings, sketches, models, and CFD studies; images by MIT China Workshop

PART FIVE – FUTURE STEPS

Section header – Dancing in the park; photograph by Zachary Kron

Chapter Fifteen – Future Steps

Chapter logo – Hui Long Guan rendering; image by MIT China Workshop

Figure 1 – Bamboo scaffolding in Shanghai; photograph by Juintow Lin

Figure 2 – Possible energy savings for a typical residential unit in Beijing; image by MIT China Workshop

Appendix - Building Energy Calculator

Figures 1-4, Tables 1-3 – *Building Energy Calculator* screen shots; images from *Building Energy Calculator software*

Alliance for Global Sustainability Series

1. F. Moavenzadeh, K. Hanaki and P. Baccini (eds.): *Future Cities: Dynamics and Sustainability*. 2002 ISBN 1-4020-0540-7
2. L. Molina (ed.): *Air Quality in the Mexico Megacity: An Integrated Assessment*. 2002 ISBN 1-4020-0452-4
3. W. Wimmer and R. Züst: *ECODESIGN Pilot. Product-Investigation-, Learning- and Optimization-Tool for Sustainable Product Development with CD-ROM*. 2003 ISBN 1-4020-0965-8
4. B. Eliasson and Y. Lee (eds.): *Integrated Assessment of Sustainable Energy Systems in China. The China Technology Program. A Framework for Decision Support in the Electric Sector of Shandong Province.* 2003 ISBN 1-4020-1198-9
5. M. Keiner, C. Zegras, W.A. Schmid and D. Salmerón (eds.): *From Understanding to Action. Sustainable Urban Development in Medium-Sized Cities in Africa and Latin America*. 2004 ISBN 1-4020-2879-2
6. W. Wimmer, R. Züst and K.M. Lee: *ECODESIGN Implementation. A Systematic Guidance on Integrating Environmental Considerations into Product Development*. 2004 ISBN 1-4020-3070-3
7. D.L. Goldblatt: *Sustainable Energy Consumption and Society*. Personal, Technological, or Social Change? 2005 ISBN 1-4020-3086-X
8. K.R. Polenske (ed.): *The Technology-Energy-Environment-Health (TEEH) Chain in China*. A Case Study of Cokemaking. 2006 ISBN 1-4020-3433-4
9. L. Glicksman and J. Lin (eds.): *Sustainable Urban Housing in China*. Principles and Case Studies for Low-Energy Design. 2006 ISBN 1-4020-4785-1

springeronline.com